Partial Dynamical Systems, Fell Bundles and Applications

Ruy Exel

Mathematical Surveys and Monographs

Volume 224

Partial Dynamical Systems, Fell Bundles and Applications

Ruy Exel

American Mathematical Society
Providence, Rhode Island

2010 *Mathematics Subject Classification.* Primary 46L55, 46L45, 37A55, 16S35, 16S40.

For additional information and updates on this book, visit
www.ams.org/bookpages/surv-224

Library of Congress Cataloging-in-Publication Data

Names: Exel, Ruy, 1956–
Title: Partial dynamical systems, Fell bundles and applications / Ruy Exel.
Description: Providence, Rhode Island: American Mathematical Society, [2017] | Series: Mathematical surveys and monographs; volume 224 | Includes bibliographical references and subject index.
Identifiers: LCCN 2017013411 | ISBN 9781470437855 (alk. paper)
Subjects: LCSH: C*-algebras. | Banach spaces. | Isometrics (Mathematics) | AMS: Functional analysis – Selfadjoint operator algebras (C^*-algebras, von Neumann (W^*-) algebras, etc.) – Noncommutative dynamical systems. msc | Functional analysis – Selfadjoint operator algebras (C^*-algebras, von Neumann (W^*-) algebras, etc.) – Decomposition theory for C^*-algebras. msc | Associative rings and algebras – Rings and algebras arising under various constructions – Twisted and skew group rings, crossed products. msc | Associative rings and algebras – Rings and algebras arising under various constructions – Smash products of general Hopf actions. msc
Classification: LCC QA326 .E94 2017 | DDC 515/.7–dc23
LC record available at https://lccn.loc.gov/2017013411

∞ The paper used in this book is acid-free and falls within the guidelines established to ensure permanence and durability.
Visit the AMS home page at http://www.ams.org/

10 9 8 7 6 5 4 3 2 1　　22 21 20 19 18 17

Contents

CHAPTER 1

Introduction

The concept of a partial dynamical system has been an essential part of Mathematics since at least the late 1800s when, thanks to the work of Cauchy, Lindelöf, Lipschitz and Picard, we know that given a Lipschitz vector field X on an open subset $U \subseteq \mathbb{R}^n$, the initial value problem

$$f'(t) = X\big(f(t)\big), \quad f(0) = x_0,$$

admits a unique solution for every x_0 in U, defined on some open interval about zero.

Assuming that we extend the above solution f to the maximal possible interval, and if we write $\phi_t(x_0)$ for $f(t)$, then each ϕ_t is a diffeomorphism between open subsets of U. Moreover, if x is in the domain of ϕ_t, and if $\phi_t(x)$ is in the domain of ϕ_s, it is easy to see that x lies in the domain of ϕ_{s+t}, and that

$$\phi_{s+t}(x) = \phi_s\big(\phi_t(x)\big).$$

This is to say that, defining the composition $\phi_s \circ \phi_t$ on the largest possible domain where it makes sense, one has that

$$\phi_s \circ \phi_t \subseteq \phi_{s+t},$$

meaning that ϕ_{s+t} extends $\phi_s \circ \phi_t$. This extension property is the central piece in defining the notion of a *partial dynamical system*, the main object of study in the present book.

Since dynamical systems permeate virtually all of mankind's most important scientific advances, a wide variety of methods have been used in their study. Here we shall adopt an algebraic point of view to study partial dynamical systems, occasionally veering towards a functional analytic perspective.

According to this approach we will extend our reach in order to encompass partial actions on several categories, notably sets, topological spaces, algebras and C*-algebras.

One of our main goals is to study graded C*-algebras from the point of view of partial actions. The fundamental connection between these concepts is established via the notion of *crossed product* (known to algebraists as *skew-group algebra*) in the sense that, given a partial action of a group G on an algebra A, the crossed product of A by G is a graded algebra. While there are many graded algebras which cannot be built out of a partial action, as above, the number of those that can is surprisingly large. Firstly, there is a vast quantity of graded algebras which, when looked at with the appropriate bias, simply happen to be a partial crossed product. Secondly, and more importantly, we will see that any given graded algebra satisfying quite general hypotheses is necessarily a partial crossed product!

Once a graded algebra is described as a partial crossed product, we offer various tools to study it, but we also dedicate a large part of our attention to the study of graded algebras per se, mainly through a very clever device introduced by J. M. G. Fell under the name of *C*-algebraic bundles,* but which is now more commonly known as *Fell bundles.* A Fell bundle may be seen essentially as a graded algebra which has been disassembled in such a way that we are left only with the scattered resulting parts.

Our study of Fell bundles consists of two essentially disjoint disciplines. On the one hand we study its internal structure and, on the other, we discuss the various ways in which a Fell bundle may be reassembled to form a C*-algebra. The main structural result we present is that every separable Fell bundle with stable unit fiber algebra must necessarily arise as the semi-direct product bundle for a partial action of the base group on its unit fiber algebra. The study of reassembly, on the other hand, is done via the notions of cross-sectional algebras and amenability.

A number of applications are presented to the study of C*-algebras, notably C*-algebras generated by semigroups of isometries, and the now standard class of graph C*-algebras.

Although not discussed here, the reader may find several other situations where well known C*-algebras may be described as partial crossed products. Among these we mention:

- Bunce-Deddens algebras [**46**],
- AF-algebras [**47**],
- the Bost-Connes algebra [**15**],
- Exel-Laca algebras [**57**],
- C*-algebras associated to right-angled Artin groups [**28**],
- Hecke algebras [**54**],
- algebras associated with integral domains [**14**], and
- algebras associated to dynamical systems of type (m, n) [**7**].

Besides, there are numerous other developments involving partial actions whose absence in this book should be acknowledged. First and foremost we should mention that we have chosen to restrict ourselves to discrete groups (i.e., groups without any topology), completely avoiding partial actions of topological groups, even though the latter is a well studied theory. See, for example, [**48**] and [**2**].

Although the computation of the K-theory groups of partial crossed product algebras is one of the main focus of the first two papers on the subject, namely [**45**] and [**81**], there is no mention here of these developments.

Twisted partial actions and projective partial representations are also absent, even though they form an important part of the theory. See, e.g. [**48**], [**102**], [**35**], [**11**], [**36**], [**87**], [**37**], [**22**], [**23**] and [**38**].

Even though we briefly discuss the relationship between partial actions and inverse semigroups, many interesting developments have been left out, such as [**101**], [**50**], [**102**], [**55**], [**61**], [**22**], [**23**] and [**24**]. The absence of any mention of the close relationship between partial actions and groupoids [**3**], must also be pointed out.

The study of KMS states for gauge actions on partial crossed products studied in [**58**] is also missing here.

The author gratefully acknowledges financial support from CNPq (Conselho Nacional de Desenvolvimento Científico e Tecnológico – Brazil).

Part 1

Partial Actions

CHAPTER 2

Partial Actions

The notion of a group action applies to virtually every category in Mathematics, the most basic being the category of sets. Correspondingly we shall start our development by focusing on partial actions of groups on sets.

▶ Throughout this chapter we will therefore fix a group G, with unit denoted 1, and a set X.

Definition 2.1. A *partial action* of G on X is a pair

$$\theta = \big(\{D_g\}_{g\in G}, \{\theta_g\}_{g\in G}\big)$$

consisting of a collection $\{D_g\}_{g\in G}$ of subsets of X, and a collection $\{\theta_g\}_{g\in G}$ of maps,

$$\theta_g : D_{g^{-1}} \to D_g,$$

such that

(i) $D_1 = X$, and θ_1 is the identity map,
(ii) $\theta_g \circ \theta_h \subseteq \theta_{gh}$, for all g and h in G.

By a *partial dynamical system* we shall mean a quadruple

$$\big(X,\ G,\ \{D_g\}_{g\in G},\ \{\theta_g\}_{g\in G}\big)$$

where X is a set, G is a group, and $\big(\{D_g\}_{g\in G}, \{\theta_g\}_{g\in G}\big)$ is a partial action of G on X. In case every $D_g = X$, we will say that θ is a *global action*, or that we have a *global dynamical system*.

We should observe that the composition "$\theta_g \circ \theta_h$" appearing in (2.1.ii) is not defined in the traditional way since the image of θ_h is not necessarily contained in the domain of θ_g. Instead this composition is meant to refer to the map whose domain is the set of all elements x in X for which the expression

$$\theta_g\big(\theta_h(x)\big)$$

makes sense. For this, x must be in $D_{h^{-1}}$ (the domain of θ_h), and $\theta_h(x)$ must be in $D_{g^{-1}}$ (the domain of θ_g). In other words, the domain of $\theta_g \circ \theta_h$ is the set

$$\{x \in D_{h^{-1}} : \theta_h(x) \in D_{g^{-1}}\} = \theta_h^{-1}(D_{g^{-1}}) = \theta_h^{-1}(D_h \cap D_{g^{-1}}). \tag{2.2}$$

For each such x, we of course define $(\theta_g \circ \theta_h)(x) = \theta_g\big(\theta_h(x)\big)$.

Still referring to (2.1.ii), the symbol "$\subseteq$" appearing there is meant to express the fact that the function on the right-hand-side is an extension[1] of the one on the left-hand-side. In other words (2.1.ii) requires that θ_{gh} be an extension of $\theta_g \circ \theta_h$.

[1] If one defines a function as a *set of ordered pairs*, in the usual technical way, then the symbol "$\subseteq$" should indeed be interpreted simply as set inclusion.

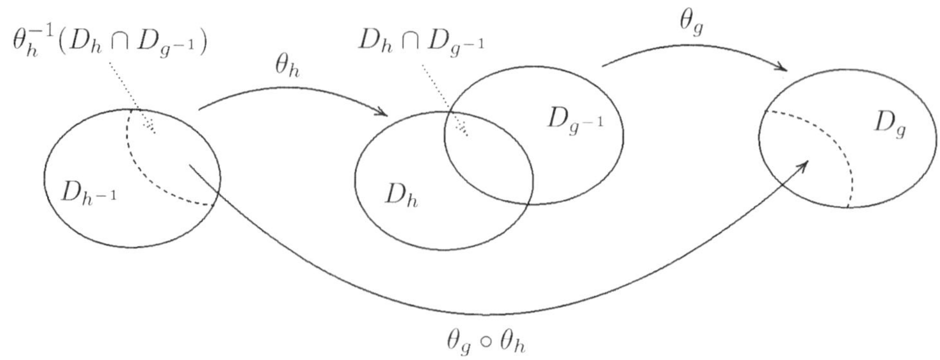

FIGURE 2.3. Composing partially defined functions.

It is possible to rephrase the definition of partial actions just mentioning the collection $\{\theta_g\}_{g\in G}$ of partially defined[2] maps on X, without emphasizing the collection of sets $\{D_g\}_{g\in G}$. In this case we could denote a posteriori the domain of $\theta_{g^{-1}}$ by D_g, and an axiom should be added to require that the range of θ_g be contained in D_g.

Proposition 2.4. *Given a partial action θ of G on X, as above, one has that each θ_g is a bijection from $D_{g^{-1}}$ onto D_g and, moreover, $\theta_{g^{-1}} = \theta_g^{-1}$.*

PROOF. By (2.1.ii), we have that $\theta_{g^{-1}} \circ \theta_g$ is a restriction of θ_1, which is the identity map by (2.1.i). Thus $\theta_{g^{-1}} \circ \theta_g$ is the identity on its domain, which is clearly $D_{g^{-1}}$. Similarly $\theta_g \circ \theta_{g^{-1}}$ is the identity on D_g. This concludes the proof. □

The following provides an equivalent definition of partial actions.

Proposition 2.5. *Let $\{D_g\}_{g\in G}$ be a collection of subsets of X, and let $\{\theta_g\}_{g\in G}$ be a collection of maps*

$$\theta_g : D_{g^{-1}} \to D_g.$$

Then $\big(\{D_g\}_{g\in G}, \{\theta_g\}_{g\in G}\big)$ is a partial action of G on X if and only if, in addition to (2.1.i), for all g and h in G, one has that:

(*i*) $\theta_g\big(D_{g^{-1}} \cap D_h\big) \subseteq D_{gh}$,

(*ii*) $\theta_g\big(\theta_h(x)\big) = \theta_{gh}(x)$, *for all* $x \in D_{h^{-1}} \cap D_{(gh)^{-1}}$.

PROOF. Before we begin, we should notice that (i) justifies (ii) in the following sense: for all x as in (ii), the fact that x is in $D_{h^{-1}}$ tells us that $\theta_h(x)$ is well defined, while x being in $D_{(gh)^{-1}}$ implies that $\theta_{gh}(x)$ is also well defined. In addition notice that

$$\theta_h(x) \in \theta_h\big(D_{h^{-1}} \cap D_{(gh)^{-1}}\big) \overset{\text{(i)}}{\subseteq} D_{g^{-1}},$$

whence $\theta_h(x)$ indeed lies in the domain of θ_g.

Assuming we have a partial action, we have already seen that the domain of $\theta_g \circ \theta_h$ is precisely $\theta_h^{-1}(D_h \cap D_{g^{-1}})$. Since this map is extended by θ_{gh}, whose domain is $D_{(gh)^{-1}}$, we deduce that

$$\theta_h^{-1}(D_h \cap D_{g^{-1}}) \subseteq D_{(gh)^{-1}}.$$

[2]By a partially defined map on X we mean any map between two subsets of X.

With the change of variables $h := g^{-1}$, and $g := h^{-1}$, and using (2.4), we obtain (i).

Given $x \in D_{h^{-1}} \cap D_{(gh)^{-1}}$, notice that by (i), we have $\theta_h(x) \in D_h \cap D_{g^{-1}}$, whence x lies in the domain of $\theta_g \circ \theta_h$, and then (ii) follows from (2.1.ii).

Conversely, assuming (2.1.i), (i) and (ii), let us first prove that

$$\theta_{g^{-1}} = \theta_g^{-1}, \quad \forall g \in G. \tag{2.5.1}$$

In fact, with $h = g^{-1}$, point (ii) states that

$$\theta_g\big(\theta_{g^{-1}}(x)\big) = x, \quad \forall x \in D_g,$$

which says that $\theta_{g^{-1}}$ is a right inverse for θ_g. Replacing g by g^{-1}, we see that $\theta_{g^{-1}}$ is also a left inverse for θ_g, thus proving (2.5.1).

Let us now prove (2.1.ii). For this, let x be in the domain of $\theta_g \circ \theta_h$, which, as already seen, means that

$$x \in \theta_h^{-1}(D_h \cap D_{g^{-1}}) \overset{(2.5.1)}{=} \theta_{h^{-1}}(D_h \cap D_{g^{-1}}) \overset{(i)}{\subseteq} D_{h^{-1}g^{-1}},$$

thus showing that the domain of $\theta_g \circ \theta_h$ is contained in the domain of θ_{gh}. Since x is evidently also in the domain of θ_h, we conclude that

$$x \in D_{h^{-1}} \cap D_{h^{-1}g^{-1}},$$

and then (ii) implies that $\theta_g\big(\theta_h(x)\big) = \theta_{gh}(x)$, which means that θ_{gh} indeed extends $\theta_g \circ \theta_h$, as desired. □

Condition (2.5.i) may be slightly improved, as follows:

Proposition 2.6. *Let $\theta = \big(\{D_g\}_{g\in G}, \{\theta_g\}_{g\in G}\big)$ be a partial action of G on X. Then*

$$\theta_g\big(D_{g^{-1}} \cap D_h\big) = D_g \cap D_{gh}, \quad \forall g, h \in G.$$

PROOF. Combining (2.5.i) with the fact that the range of θ_g is contained in D_g, we have that

$$\theta_g\big(D_{g^{-1}} \cap D_h\big) \subseteq D_g \cap D_{gh}. \tag{2.6.1}$$

Applying $\theta_{g^{-1}}$ to both sides of the above inclusion, and using (2.4), we then deduce that

$$D_{g^{-1}} \cap D_h \subseteq \theta_{g^{-1}}(D_{gh} \cap D_g).$$

Replacing g by g^{-1}, and h by gh, we then deduce the reverse inclusion in (2.6.1), hence equality. □

The natural notion of equivalence for partial actions is part of our next:

Definition 2.7. Let G be a group, and suppose that, for each $i = 1, 2$, we are given a partial action $\theta^i = \big(\{D^i_g\}_{g\in G}, \{\theta^i_g\}_{g\in G}\big)$ of G on a set X^i. A map

$$\phi : X^1 \to X^2$$

will be said to be *G-equivariant* when, for all g in G, one has that

(i) $\phi(D^1_g) \subseteq D^2_g$, and
(ii) $\phi\big(\theta^1_g(x)\big) = \theta^2_g\big(\phi(x)\big)$, for all x in $D^1_{g^{-1}}$.

If moreover ϕ is bijective and ϕ^{-1} is also G-equivariant, we will say that ϕ is an *equivalence of partial actions*. If such an equivalence exists, we will say that θ^1 is equivalent to θ^2.

Observe that a G-equivariant bijective map ϕ from X^1 to X^2 may fail to be an equivalence since property (2.7.i) may fail for ϕ^{-1}.

On the other hand, it is easy to see that a necessary and sufficient condition for a G-equivariant bijective map to be an equivalence is that equality hold in (2.7.i).

Definition 2.8. Let $\theta = \big(\{D_g\}_{g\in G}, \{\theta_g\}_{g\in G}\big)$ be a partial action of G on X. The *graph* of θ is defined to be the set

$$\mathrm{Graph}(\theta) = \{(y, g, x) \in X \times G \times X : x \in D_{g^{-1}},\ \theta_g(x) = y\}.$$

The graph of a partial action evidently encodes all of the information contained in the partial action itself. It is therefore a very useful device in the study of partial actions and we will re-encounter it often in the sequel. Meanwhile let us remark that it has a natural groupoid [**3**] structure, where two elements (w, h, z) and (y, g, x) in $\mathrm{Graph}(\theta)$ may be multiplied if and only if $z = y$, in which case we put

$$(w, h, z)(y, g, x) = (w, hg, x),$$

while

$$(y, g, x)^{-1} = (x, g^{-1}, y).$$

We leave it for the reader to check that $\mathrm{Graph}(\theta)$ is indeed a groupoid with these operations.

The usual notion of invariance for group actions has a counterpart in partial actions, as follows.

Definition 2.9. Given a partial action $\theta = \big(\{D_g\}_{g\in G}, \{\theta_g\}_{g\in G}\big)$ of G on X, we will say that a given subset $Y \subseteq X$ is *invariant* under θ, if

$$\theta_g(Y \cap D_{g^{-1}}) \subseteq Y, \quad \forall\, g \in G.$$

As in the global case we have:

Proposition 2.10. *Given a partial action $\theta = \big(\{D_g\}_{g\in G}, \{\theta_g\}_{g\in G}\big)$ of G on X, and an invariant subset $X' \subseteq X$, let*

$$D'_g = X' \cap D_g, \quad \forall\, g \in G,$$

and let θ'_g be the restriction of θ_g to $D'_{g^{-1}}$. Then

(*i*) *$\theta' = \big(\{D'_g\}_{g\in G}, \{\theta'_g\}_{g\in G}\big)$ is a partial action of G on X', and*
(*ii*) *the inclusion $X' \hookrightarrow X$ is a G-equivariant map.*

PROOF. By hypothesis it is clear that each θ'_g maps $D'_{g^{-1}}$ into D'_g. Observing that (2.1.i) is evident for θ', it suffices to check conditions (2.5.i)–(2.5.ii).

For all g and h in G, we have that

$$\begin{aligned} &\theta'_g(D'_{g^{-1}} \cap D'_h) \\ &\qquad = \theta_g(X' \cap D_{g^{-1}} \cap D_h) \\ &\qquad \subseteq \theta_g(X' \cap D_{g^{-1}}) \ \cap\ \theta_g(D_{g^{-1}} \cap D_h) \\ &\qquad \subseteq X' \cap D_{gh} = D'_{gh}, \end{aligned}$$

thus verifying (2.5.i). We leave the easy proof of (2.5.ii) for the reader. Point (ii) is also easy to check. □

Notes and remarks. Partial actions of the group $\mathbb{Z}$ on C*-algebras were first considered by the author in [**45**]. Soon afterward McClanahan [**81**] generalized this notion for arbitrary groups. Partial actions of groups on arbitrary sets were introduced in [**50**].

CHAPTER 3

Restriction and Globalization

One of the easiest ways to produce nontrivial examples of partial actions is via the process of *restriction* to not necessarily invariant subsets, which we would now like to describe.

Suppose we are given a global action η of a group G on a set Y. Suppose further that X is a given subset of Y and we wish to restrict η to an action of G on X. Evidently this requires that X be an η-invariant subset.

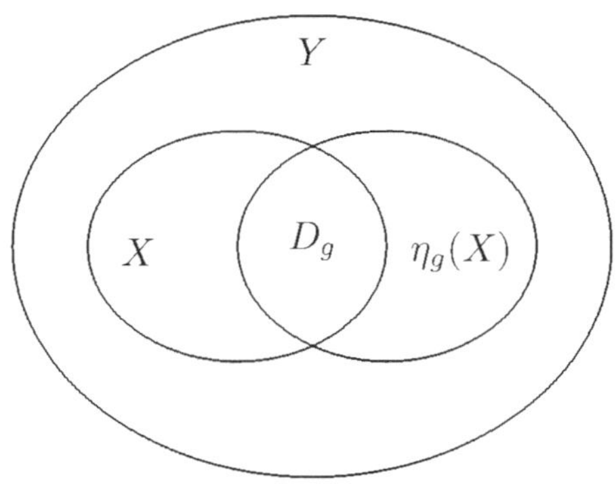

FIGURE 3.1. Restricting a global action to a partial action.

However, even when X is not η-invariant, we may restrict η to a *partial action* of G on X, by setting

$$D_g = \eta_g(X) \cap X,$$

and, after observing that $\eta_g(D_{g^{-1}}) = D_g$, we may let

$$\theta_g : D_{g^{-1}} \to D_g$$

be the restriction of η_g to $D_{g^{-1}}$, for each g in G. The easy verification that this indeed leads to a partial action is left to the reader.

Definition 3.2. The partial action θ of G on X defined above is called the *restriction* of the global action η to X.

It is readily seen that, under the above conditions, the inclusion of X into Y is a G-equivariant map.

Let us now discuss the relationship between the graph of a global action and that of its restriction to a subset.

Proposition 3.3. *Let η be a global action of a group G on a set Y and let θ be its restriction to a subset $X \subseteq Y$. Then*

$$\mathrm{Graph}(\theta) = \mathrm{Graph}(\eta) \cap (X \times G \times X).$$

PROOF. Left to the reader. □

Taking a different point of view, one may start with a partial action θ of G on X and wonder about the existence of a global action η on some set Y containing X, in such a way that θ is the restriction of η. Assuming that η does exist, the *orbit* of X in Y, namely

$$\mathrm{Orb}(X) := \bigcup_{g\in G} \eta_g(X)$$

will evidently be η-invariant and the restriction of η to $\mathrm{Orb}(X)$ will be another global action which, further restricted to X, will again produce the original partial action θ.

Definition 3.4. Let η be a global action of G on a set Y, and let θ be the partial action obtained by restricting η to a subset $X \subseteq Y$. If the orbit of X coincides with Y, we will say that η is a *globalization*, or an *enveloping action* for θ.

Theorem 3.5. *Every partial action admits a globalization, which is unique in the following sense: if θ is a partial action of G on X, and we are given globalizations η^i acting on sets Y^i, for $i = 1, 2$, then there exists an equivalence*

$$\phi : Y^1 \to Y^2,$$

such that ϕ coincides with the identity on X.

PROOF. Given a partial action $\theta = \big(\{D_g\}_{g\in G}, \{\theta_g\}_{g\in G}\big)$ of G on a set X, define an equivalence relation on $G \times X$ by

$$(g, x) \sim (h, y) \;\Leftrightarrow\; x \in D_{g^{-1}h}, \text{ and } \theta_{h^{-1}g}(x) = y.$$

That this is indeed a reflexive and symmetric relation is apparent, while the transitivity property may be verified as follows: assuming that

$$(g, x) \sim (h, y) \sim (k, z),$$

we have that $y \in D_{h^{-1}k}$, and $\theta_{k^{-1}h}(y) = z$. We then have that

$$x = \theta_{g^{-1}h}(y) \in \theta_{g^{-1}h}(D_{h^{-1}g} \cap D_{h^{-1}k}) = D_{g^{-1}h} \cap D_{g^{-1}k},$$

whence by (2.5.ii)

$$\theta_{k^{-1}g}(x) = \theta_{k^{-1}h}\big(\theta_{h^{-1}g}(x)\big) = \theta_{k^{-1}h}(y) = z,$$

showing that $(g, x) \sim (k, z)$.

Let $\tilde{X}$ be the quotient of $G \times X$ by this equivalence relation, and let us denote by $[g, x]$ the equivalence class of each $(g, x) \in G \times X$. The map

$$\iota : x \in X \mapsto [1, x] \in \tilde{X} \tag{3.5.1}$$

is clearly injective and hence we may identify X with its image X' in $\tilde{X}$ under ι. One then verifies that the expression

$$\tau_g([h, x]) = [gh, x], \quad \forall\, g, h \in G, \quad \forall\, x \in X,$$

gives a well defined global action τ of G on $\tilde{X}$.

We claim that

$$\tau_g(X') \cap X' = \iota(D_g).$$

In fact, an element of the form $[1, x]$ lies in $\tau_g(X')$ if and only if $[1, x] = [g, y]$, for some y in X, but this is so iff $x \in D_g$, and $\theta_{g^{-1}}(x) = y$, proving the claim.

In order to verify that θ indeed corresponds to the restriction of τ to X, let $x \in D_{g^{-1}}$, and observe that

$$\big(1, \theta_g(x)\big) \sim (g, x),$$

whence

$$\iota(\theta_g(x)) = [1, \theta_g(x)] = [g, x] = \tau_g([1, x]) = \tau_g\big(\iota(x)\big).$$

This is to say that $\iota \circ \theta_g = \tau_g \circ \iota$, which means that $\theta_g = \tau_g$, on $D_{g^{-1}}$, up to the above identification of X with X'.

Observing that $\tau_g([1, x]) = [g, x]$, we immediately see that the orbit of X coincides with $\tilde{X}$, so τ is indeed a globalization of θ.

If η is another globalization of θ, acting on the set $Y \supseteq X$, define a map

$$\phi : (g, x) \in G \times X \mapsto \eta_g(x) \in Y,$$

and observe that

$$\phi(g, x) = \phi(h, y) \;\Leftrightarrow\; \eta_g(x) = \eta_h(y) \;\Leftrightarrow\; x = \eta_{g^{-1}h}(y),$$

Under these conditions we have that $x \in \eta_{g^{-1}h}(X) \cap X = D_{g^{-1}h}$, and $y = \eta_{h^{-1}g}(x) = \theta_{h^{-1}g}(x)$, meaning that $(g, x) \sim (h, y)$.

Consequently the map ϕ, above, factors through the quotient $\tilde{X}$, providing an injective map

$$\tilde{\phi} : \tilde{X} \to Y.$$

Being a globalization, Y coincides with the orbit of X under η, from where one deduces that $\tilde{\phi}$ is surjective.

We leave it for the reader to prove that $\tilde{\phi}$ is G-equivariant and coincides with the identity on the respective copies of X in $\tilde{X}$ and Y. This proves that $\tilde{X}$ and Y are equivalent globalizations. □

The process of restricting a global action η to a non-invariant subset, studied above, could be generalized to the case in which η, itself, is a partial action. More precisely, suppose we are given a partial action

$$\eta = \big(\{Y_g\}_{g\in G}, \{\eta_g\}_{g\in G}\big)$$

of a group G on a set Y, and let $X \subseteq Y$ be a not necessarily invariant subset. Defining

$$D_g = Y_g \cap X \cap \eta_g\big(Y_{g^{-1}} \cap X\big), \quad \forall\, g \in G,$$

one may prove that $\eta_g(D_{g^{-1}}) = D_g$. If we then let θ_g be the restriction of η_g to $D_{g^{-1}}$, one has that

$$\theta = \big(\{D_g\}_{g\in G}, \{\theta_g\}_{g\in G}\big)$$

is a partial action of G on X, with respect to which the inclusion $X \hookrightarrow Y$ is a G-equivariant map.

Contrary to the case of restricting a global action, we have found no interesting applications of the above idea. Thus, until this situation changes, we will not bother to discuss it any further.

Notes and remarks. The notion of enveloping actions and the proof of Theorem (3.5) is due to Abadie and it first appeared in his PhD thesis [**1**], which led to the article [**2**].

CHAPTER 4

Inverse Semigroups

In the study of partial actions, the notion of inverse semigroups plays a predominant role. So let us briefly introduce this concept, referring the reader to [**79**] for proofs and further details.

Definition 4.1. An *inverse semigroup* is a nonempty set S equipped with a binary associative operation (i.e., S is a semigroup) such that, for every s in S, there exists a unique element s^* in S, such that

$$ss^*s = s, \quad \text{and} \quad s^*ss^* = s^*.$$

Given an inverse semigroup S, one may prove that the collection of idempotent elements in S, namely

$$E(S) = \{s \in S : s^2 = s\},$$

is a commutative sub-semigroup. Under the partial order relation defined in $E(S)$ by

$$e \leq f \iff ef = e, \quad \forall e, f \in E(S),$$

$E(S)$ becomes a *semilattice*, meaning that for every e and f in $E(S)$, there exists a largest element which is smaller than e and f, namely ef. One therefore often refers to $E(S)$ as the *idempotent semilattice* of S.

There is also a natural partial order relation defined on S itself by

$$s \leq t \iff ts^*s = s \iff ss^*t = s, \quad \forall s, t \in S. \tag{4.2}$$

This is compatible with the multiplication operation in the sense that

$$s \leq t, \ s' \leq t' \ \Rightarrow \ ss' \leq tt'. \tag{4.3}$$

One of the main examples of inverse semigroups is as follows: given a set X, two subsets $C, D \subseteq X$, and a bijective map

$$f : C \to D,$$

we will say that f is a *partial symmetry* of X. The set

$$\mathcal{I}(X) = \{f : f \text{ is a partial symmetry of } X\} \tag{4.4}$$

may be turned into a semigroup by equipping it with the operation of composition, where, as before, the composition of two partially defined maps is defined on the largest domain where it makes sense.

It may be easily proven that $\mathcal{I}(X)$ is an inverse semigroup, where, for every f in $\mathcal{I}(X)$, one has that f^* is the inverse of f.

The idempotent elements of $\mathcal{I}(X)$ are just the identity maps defined on subsets of X. The order among idempotents happens to be the same as the order of inclusion of their domains.

More generally, the natural order among general elements of $\mathcal{I}(X)$ is the order given by "extension", meaning that, for f and g in $\mathcal{I}(X)$, one has that $f \leq g$, if and only if g is an extension of f, in symbols

$$f \leq g \Leftrightarrow f \subseteq g, \quad \forall f, g \in \mathcal{I}(X).$$

The classical Wagner-Preston Theorem [**105**], [**92**] asserts that any inverse semigroup is isomorphic to a *-invariant sub-semigroup of $\mathcal{I}(X)$, for some X. This may be considered as the version for inverse semigroups of the well known Cayley Theorem for groups.

Given a partial action θ of a group G on a set X, notice that each θ_g is an element of $\mathcal{I}(X)$.

Proposition 4.5. *Let G be a group, X be a set, and*

$$\theta : G \to \mathcal{I}(X)$$

be a map. Then θ is a partial action of G on X if and only if, for every g and h in G, one has that

(i) *θ_1 is the identity map of X,*
(ii) *$\theta_{g^{-1}} = (\theta_g)^*$,*
(iii) *$\theta_g\theta_h\theta_{h^{-1}} = \theta_{gh}\theta_{h^{-1}}$,*
(iv) *$\theta_{g^{-1}}\theta_g\theta_h = \theta_{g^{-1}}\theta_{gh}$.*

PROOF. Assuming that θ is a partial action, one immediately checks (i) and (ii). As for (iii), observe that, following (2.2), the domain of $\theta_{gh}\theta_{h^{-1}}$ is

$$\theta_h(D_{h^{-1}} \cap D_{(gh)^{-1}}) \overset{(2.6)}{=} D_h \cap D_{g^{-1}},$$

which is clearly also the domain of $\theta_g\theta_h\theta_{h^{-1}}$. For any x in this common domain we have that

$$\theta_{h^{-1}}(x) \in \theta_{h^{-1}}(D_h \cap D_{g^{-1}}) = D_{h^{-1}} \cap D_{h^{-1}g^{-1}},$$

so (iii) follows immediately from (2.5.ii). Finally, (iv) follows from (iii) and (ii), in view of the fact that the star operation reverses multiplication.

Conversely, suppose that θ satisfies (i–iv) and denote the range of θ_g by D_g. Since $\theta_{g^{-1}}$ is the inverse of θ_g by (i), we have that the range of $\theta_{g^{-1}}$ is the domain of θ_g, meaning that θ_g is a map

$$\theta_g : D_{g^{-1}} \to D_g.$$

Since (2.1.i) is granted, let us verify (2.1.ii). For this, let us suppose we are given g and h in G, as well as an element x in the domain of $\theta_g\theta_h$. Then the element $y := \theta_h(x)$ clearly lies in the domain of $\theta_g\theta_h\theta_{h^{-1}}$, and hence (iii) implies that it is also in the domain of $\theta_{gh}\theta_{h^{-1}}$, which is to say that x is in the domain of θ_{gh}. Moreover

$$\theta_g\big(\theta_h(x)\big) = \theta_g\big(\theta_h(\theta_{h^{-1}}(y))\big) = \theta_{gh}\big(\theta_{h^{-1}}(y)\big) = \theta_{gh}(x),$$

proving that $\theta_g\theta_h \subseteq \theta_{gh}$. □

The language of inverse semigroups is especially well suited to the introduction of our next example. In order to describe it, let X be a set and let $\{f_\lambda\}_{\lambda\in\Lambda}$ be any collection of partial symmetries of X.

Letting $\mathbb{F} = \mathbb{F}(\Lambda)$ be the free group on the index set Λ, our plan is to construct a partial action of $\mathbb{F}$ on X in the form of a map

$$\theta : \mathbb{F} \to \mathcal{I}(X).$$

As a first step, let us define

$$\theta_\lambda := f_\lambda, \quad \text{and} \quad \theta_{\lambda^{-1}} := f_\lambda^{-1}, \quad \forall\, \lambda \in \Lambda.$$

Given any element $g \in \mathbb{F}$, write

$$g = x_1 x_2 \ldots x_n,$$

in *reduced form*, meaning that each $x_j \in \Lambda \cup \Lambda^{-1}$, and $x_{j+1} \neq x_j^{-1}$. It is well known that a unique such decomposition of g exists. We then put

$$\theta_g := \theta_{x_1}\theta_{x_2}\ldots\theta_{x_n}, \tag{4.6}$$

with the convention that, if $g = 1$, then its reduced form is "empty", and θ_1 is the identity function on X.

Proposition 4.7. *The map θ defined in* (4.6), *above, is a partial action of $\mathbb{F}$ on X.*

Proof. We leave it for the reader to prove (2.1.i). In order to verify (2.1.ii), pick g and h in $\mathbb{F}$, and write

$$g = x_n x_{n-1} \ldots x_2 x_1, \quad \text{and} \quad h = y_1 y_2 \ldots y_{m-1} y_m,$$

in reduced form, as above (for reasons which will soon become clear, we have chosen to reverse the indices in the reduced form of g).

Let p be the number of cancellations occurring when performing the multiplication gh, meaning that

$$x_i = y_i^{-1}, \quad \forall\, i = 1, \ldots, p,$$

and that p is maximal with this property. Define g', h', and k in $\mathbb{F}$, as follows

$$g = \overbrace{x_n \ldots x_{p+1}}^{g'}\ \overbrace{x_p \ldots x_1}^{k}, \quad \text{and} \quad h = \overbrace{y_1 \ldots y_p}^{k^{-1}}\ \overbrace{y_{p+1} \ldots y_m}^{h'},$$

so that

$$gh = g'h' = x_n \ldots x_{p+1}\ y_{p+1} \ldots y_m$$

is in reduced form. Denoting the identity map on X by id_X, we then clearly have that

$$\theta_k \theta_k^{-1} \subseteq id_X,$$

so

$$\theta_g\theta_h = \theta_{g'}\theta_k\theta_k^{-1}\theta'_h \overset{(4.3)}{\subseteq} \theta_{g'}\, id_X\, \theta_{h'} = \theta_{g'}\theta_{h'} = \theta_{g'h'} = \theta_{gh}, \tag{4.7.1}$$

concluding the proof. □

Observe that in (4.7.1) we were allowed to use that

$$\theta_{g'}\theta_{h'} = \theta_{g'h'}$$

because the juxtaposition of the reduced forms of g' and h' turned out to be the reduced form of $g'h'$. Equivalently, $|g'h'| = |g'| + |h'|$, where $|\cdot|$ refers to *word length*.

Definition 4.8. A *length function* on a group G is a function $\ell : G \to \mathbb{R}_+$ such that

(i) $\ell(1) = 0$, and
(ii) $\ell(gh) \leq \ell(g) + \ell(h)$, for all g and h in G.

Evidently $|\cdot|$ is a length function for the free group.

Definition 4.9. Let G be a group equipped with a length function ℓ. A partial action θ of G is said to be *semi-saturated* (with respect to the given length function ℓ) if

$$\ell(gh) = \ell(g) + \ell(h) \quad \Longrightarrow \quad \theta_g\theta_h = \theta_{gh}, \quad \forall\, g, h \in G.$$

In the free group, observe that the condition that $|gh| = |g| + |h|$ means that the juxtaposition of the reduced forms of g and h is precisely the reduced form of gh. So the partial action θ defined in (4.6) is easily seen to be semi-saturated.

Summarizing we have:

Proposition 4.10. *Let X be a set and let $\{f_\lambda\}_{\lambda\in\Lambda}$ be any collection of partial symmetries of X. Then there exists a unique semi-saturated partial action θ of $\mathbb{F}(\Lambda)$ on X such that*

$$\theta_\lambda = f_\lambda, \quad \forall\, \lambda \in \Lambda.$$

PROOF. The existence was proved above and we leave the proof of uniqueness as an easy exercise. □

Notes and remarks. Semi-saturated partial actions are related to Quigg and Raeburn's notion of *multiplicativity*, as defined in [**94**, Definition 5.1]. The relationship between partial actions and inverse semigroups is further discussed in [**101**], [**50**], [**55**], [**61**], [**22**], [**23**] and [**24**].

CHAPTER 5

Topological Partial Dynamical Systems

We shall now be concerned with the topological aspects of partial actions, so, ▶ throughout this chapter, we will fix a group G and topological space X.

Definition 5.1. A *topological partial action* of the group G on the topological space X is a partial action $\theta = \big(\{D_g\}_{g\in G}, \{\theta_g\}_{g\in G}\big)$ on the underlying set X, such that each D_g is *open* in X, and each θ_g is a *homeomorphism*. By a *topological partial dynamical system* we shall mean a partial dynamical system

$$\big(X,\ G,\ \{D_g\}_{g\in G},\ \{\theta_g\}_{g\in G}\big)$$

where X is a topological space and $\big(\{D_g\}_{g\in G}, \{\theta_g\}_{g\in G}\big)$ is a topological partial action of G on X.

When it is understood that we are working in the category of topological spaces and there is no chance for confusion we will drop the adjective *topological* and simply say *partial action* or *partial dynamical system*.

The notion of topological partial actions may also take into account topological groups (see [**48**] and [**2**]) but, for the sake of simplicity, we will only be concerned with discrete groups in this book.

Recall from (4.4) that $\mathcal{I}(X)$ denotes the inverse semigroup formed by all partial symmetries of X.

Definition 5.2. We will say that a partial symmetry $f \in \mathcal{I}(X)$ is a *partial homeomorphism* of X, if the domain and range of f are open subsets of X, and f is a homeomorphism from its domain to its range. We will denote by pHomeo(X) the collection of all partial homeomorphisms of X. It is evident that pHomeo(X) is an inverse sub-semigroup of $\mathcal{I}(X)$.

As an immediate consequence of (4.5) we have:

Proposition 5.3. *Let G be a group, X be a topological space, and*

$$\theta : G \to \text{pHomeo}(X)$$

be a map. Then θ is a topological partial action of G on X if and only if conditions (i–iv) of (4.5) *are fulfilled.*

We will say that two topological partial actions of G are *topologically equivalent* if they are equivalent in the sense of Definition (2.7) and, in addition, the map ϕ mentioned there is a homeomorphism.

The process of restriction may be applied in the context of topological spaces as follows: suppose we are given a global topological action η of a group G on a space Y, as well as an *open subset* $X \subseteq Y$.

Let θ be the restriction of η to X, as defined in (3.2) and recall that

$$D_g = \eta_g(X) \cap X.$$

Since η_g is a homeomorphism we have that $\eta_g(X)$ is open in Y, so D_g is an open subset of X. Also, since the restricted maps θ_g are obviously homeomorphisms, we conclude that θ is a topological partial action of G on X.

Definition 5.4. Let η be a topological global action of G on a space Y, and let θ be the topological partial action obtained by restricting η to an open subset $X \subseteq Y$. If the orbit of X coincides with Y, we will say that η is a *topological globalization* of θ.

Proposition 5.5. *Every topological partial action admits a topological globalization, unique up to topological equivalence.*

PROOF. Given a topological partial action θ of the group G on the space X, let τ and $\tilde{X}$ be as in the proof of (3.5).

Viewing G as a discrete space, consider the product topology on $G \times X$ and let us equip $\tilde{X}$ with the quotient topology. We then claim that τ is a topological globalization of θ. In other words we must prove that each τ_g is a homeomorphism, the map ι mentioned in (3.5.1) is a homeomorphism onto its image, and $\iota(X)$ is open.

The first assertion follows easily from the fact that, for $g \in G$, the map

$$(h, x) \in G \times X \mapsto (gh, x) \in G \times X$$

is a homeomorphism respecting the equivalence relation "$\sim$".

Observe that $\iota(x) = \pi(1, x)$, for every x in X, where

$$\pi : G \times X \to G \times X/\sim$$

is the quotient map, so ι is clearly continuous. In order to prove that ι is a homeomorphism onto its image it is therefore enough to show that it is an open map. With this purpose in mind let $U \subseteq X$ be an open set, and let us prove that $\iota(U)$ is open in $\tilde{X}$ (in fact, it would be enough to prove that $\iota(U)$ is open in $\iota(X)$).

By definition of the quotient topology, $\iota(U)$ is open in $\tilde{X}$ if and only if $\pi^{-1}\big(\iota(U)\big)$ is open in $G \times X$. Moreover observe that

$$(g, x) \in \pi^{-1}\big(\iota(U)\big)$$

if and only if $(g, x) \sim (1, y)$, for some $y \in U$, but this is clearly equivalent to saying that $x \in D_{g^{-1}}$, and $\theta_g(x) \in U$, or that $x \in \theta_g^{-1}(D_g \cap U)$. So

$$\pi^{-1}\big(\iota(U)\big) = \bigcup_{g \in G} \{g\} \times \theta_g^{-1}(D_g \cap U),$$

which is then seen to be open in $G \times X$, whence $\iota(U)$ is open in $\tilde{X}$, as claimed. This proves that ι is a homeomorphism onto its image, and it also implies that $\iota(X)$ is an open subset of $\tilde{X}$. Consequently τ is a topological globalization of θ, taking care of the existence part of the statement.

In order to prove uniqueness, let us be given another topological globalization of θ, say η, acting on the space Y. In particular η is a set-theoretical globalization of θ, so by the uniqueness part of (3.5), there exists an equivalence

$$\phi : Y \to \tilde{X},$$

coinciding with the identity map on X (or rather, its canonical copies within $\tilde{X}$ and Y). In order to conclude the proof it is now enough to prove that ϕ is a homeomorphism.

Given $y \in Y$, use the fact that the orbit of X coincides with Y to find some g in G such that $y \in \eta_g(X)$. For any other $y' \in \eta_g(X)$, we then have that $x' := \eta_g^{-1}(y') \in X$, so

$$\phi(y') = \phi(\eta_g(x')) = \tau_g(\phi(x')) = \tau_g(x') = \tau_g(\eta_g^{-1}(y')).$$

Since the map

$$y' \in \eta_g(X) \mapsto \tau_g(\eta_g^{-1}(y')) \in \tau_g(X)$$

is clearly continuous and coincides with ϕ on $\eta_g(X)$, which is an open neighborhood of y, we see that ϕ is continuous at y. Similarly one may prove that the inverse of ϕ is continuous, and hence that ϕ is a homeomorphism, as claimed. □

The reader should be warned that there is a catch in the above result: the quotient topology defined on $\tilde{X}$ may not be Hausdorff, even if X is Hausdorff. In fact it is easy to characterize the occurrence of this problem, as seen in the next:

Proposition 5.6. *A topological partial action of a group G on a Hausdorff space X admits a Hausdorff globalization if and only if its graph is closed in $X \times G \times X$, where G is given the discrete topology.*

PROOF. Assuming that η is a globalization of θ on a Hausdorff space Y, observe that

$$\text{Graph}(\eta) = \{(y, g, x) \in Y \times G \times Y : y = \eta_g(x)\}.$$

This should be contrasted with Definition (2.8), as here there is no need to require x to lie in the domain of η_g, a globally defined map! For this reason we see that $\text{Graph}(\eta)$ is closed in $Y \times G \times Y$ and it then follows from (3.3) that $\text{Graph}(\theta)$ is closed in $X \times G \times X$.

Conversely, assuming that $\text{Graph}(\theta)$ is closed, let η be any topological globalization of θ, acting on the space Y (recall from (5.5) that there is only one such).

We will prove that Y is Hausdorff. For this, let us be given two distinct points y_1 and y_2 in Y, and write $y_i = \eta_{g_i}(x_i)$, with $g_i \in G$, and $x_i \in X$. Noticing that

$$\eta_{g_1}(x_1) = y_1 \neq y_2 = \eta_{g_2}(x_2),$$

one sees that $\eta_{g_2^{-1}g_1}(x_1) \neq x_2$, whence $(x_2, g_2^{-1}g_1, x_1)$ is not in the graph of θ. From the hypothesis it follows that there are open subsets $U_1, U_2 \subseteq X$, with $x_i \in U_i$, such that

$$U_2 \times \{g_2^{-1}g_1\} \times U_1 \ \cap \ \text{Graph}(\theta) = \emptyset.$$

Since X is open in Y, we have that each U_i is open in Y and hence so is $\eta_{g_i}(U_i)$. Since $x_i \in U_i$, we have that

$$y_i = \eta_{g_i}(x_i) \in \eta_{g_i}(U_i).$$

In order to verify Hausdorff's axiom, it now suffices to prove that $\eta_{g_1}(U_1)$ and $\eta_{g_2}(U_2)$ are disjoint. Arguing by contradiction, suppose that

$$y \in \eta_{g_1}(U_1) \cap \eta_{g_2}(U_2),$$

so we may write $y = \eta_{g_1}(z_1) = \eta_{g_2}(z_2)$, with $z_i \in U_i$. It follows that $\eta_{g_2^{-1}g_1}(z_1) = z_2$, but since both z_1 and z_2 lie in X, we indeed have that

$$\theta_{g_2^{-1}g_1}(z_1) = z_2,$$

whence

$$(z_2, g_2^{-1}g_1, z_1) \in \text{Graph}(\theta) \ \cap \ U_2 \times \{g_2^{-1}g_1\} \times U_1 = \emptyset,$$

a contradiction. This proves that $\eta_{g_1}(U_1)$ and $\eta_{g_2}(U_2)$ are disjoint and hence that Y is Hausdorff. $\square$

We will see that in many interesting examples, the D_g are closed besides being open. In this case we have:

Proposition 5.7. *A topological partial dynamical system*

$$\big(X,\ G,\ \{D_g\}_{g\in G},\ \{\theta_g\}_{g\in G}\big),$$

such that each D_g is closed, always admits a Hausdorff globalization.

PROOF. If $\{(y_i, g_i, x_i)\}_i$ is a net in $\text{Graph}(\theta)$, converging to (y, g, x), then g_i is eventually constant, since G has the discrete topology. So we may assume, without loss of generality that $g_i = g$, for all i. Then every x_i lie in $D_{g^{-1}}$, and hence x is in $D_{g^{-1}}$, by hypothesis. Consequently

$$y = \lim_i y_i = \lim_i \theta_g(x_i) = \theta_g(x),$$

so (y, g, x) lies in $\text{Graph}(\theta)$, which is therefore proven to be closed. The conclusion then follows from (5.6). $\square$

Let us now give an important example of a topological partial dynamical system based on Bernoulli's action. For this, let G be a group and consider the compact topological space

$$\{0,1\}^G$$

relative to the product topology.

It is well known that $\{0,1\}^G$ is naturally equivalent to $\mathscr{P}(G)$, the power set of G, in such a way that, when a given $\omega \in \{0,1\}^G$ is seen as a subset of G, the Boolean value of the expression "$g \in \omega$" is given by the coordinate ω_g. In symbols

$$\omega_g = [g \in \omega], \tag{5.8}$$

where we are using brackets to denote Boolean value. In what follows we will often identify $\{0,1\}^G$ and $\mathscr{P}(G)$ based on this correspondence.

For each $g \in G$, and each $\omega \in \mathscr{P}(G)$, let us indicate by

$$g\omega = \{gh : h \in \omega\},$$

as usual, and let us consider the mapping

$$\eta_g : \omega \in \mathscr{P}(G) \mapsto g\omega \in \mathscr{P}(G). \tag{5.9}$$

It is easy to see that each η_g is a homeomorphism and that η is a topological global action of G on $\{0,1\}^G$.

Definition 5.10. The action η of G on $\{0,1\}^G$, defined above, will be called the *Bernoulli global action* of G.

The next concept to be defined is based on the observation that the set

$$\Omega_1 = \big\{\omega \in \{0,1\}^G : 1 \in \omega\big\} \tag{5.11}$$

is a compact open subset of $\{0,1\}^G$.

Definition 5.12. The *partial Bernoulli action* of a group G is the topological partial action β of G on Ω_1 obtained by restricting the global Bernoulli action to Ω_1, according to (3.2).

Denoting by

$$D_g = \Omega_1 \cap \eta_g(\Omega_1), \tag{5.13}$$

as demanded by (3.2), observe that an element ω in $\eta_g(\Omega_1)$ is characterized by the fact that $g \in \omega$. Thus

$$D_g = \{\omega \in \Omega_1 : g \in \omega\} = \big\{\omega \in \{0,1\}^G : 1, g \in \omega\big\}. \tag{5.14}$$

As already noted, besides being open, Ω_1 is a compact set, hence D_g is compact for every g in G.

The partial Bernoulli action will prove to be of utmost importance in our study of partial representations subject to relations.

Notes and remarks. Propositions (5.5) and (5.6) were proved by Abadie in his PhD thesis [**1**].

CHAPTER 6

Algebraic Partial Dynamical Systems

With this chapter we start our study of partial actions of groups on algebraic structures. Among these, we would eventually like to include rings, algebras, *-algebras and C*-algebras. We thus begin with a device meant to allow us to treat rings and algebras in the same footing.

► We will assume, throughout, that $\mathbb{K}$ is a unital commutative ring. Whenever convenient we will assume that $\mathbb{K}$ is equipped with a *conjugation*, that is, an involutive automorphism

$$r \in \mathbb{K} \mapsto \bar{r} \in \mathbb{K}, \tag{6.1}$$

which will be fixed from now on. In the absence of more interesting conjugations one could take the identity map by default. When $\mathbb{K}$ is the field of complex numbers we will always choose the standard complex conjugation.

Definition 6.2. By a $\mathbb{K}$-*algebra* we will mean a ring A equipped with the structure of a left $\mathbb{K}$-module, such that, for all $\lambda \in \mathbb{K}$, and all $a, b \in A$, one has

(i) $1a = a$,
(ii) $(\lambda a)b = a(\lambda b) = \lambda(ab)$.

When $\mathbb{K}$ is a field, this is nothing but the usual concept of an algebra over a field. On the other hand, any ring A may be seen as a $\mathbb{Z}$-algebra, as long as we equip it with the obvious left $\mathbb{Z}$-module structure. With this device we may therefore treat rings and algebras in the same footing.

Ideals in $\mathbb{K}$-algebras will always be assumed to be $\mathbb{K}$-sub-modules (that is, closed under multiplication by elements of $\mathbb{K}$) and homomorphisms between $\mathbb{K}$-algebras will always be assumed to be $\mathbb{K}$-linear.

Definition 6.3. By a **-algebra* over $\mathbb{K}$ we mean a $\mathbb{K}$-algebra equipped with an *involution*, namely a map

$$a \in A \mapsto a^* \in A,$$

such that, for all $a, b \in A$, and all $\lambda \in \mathbb{K}$, one has

(i) $(a^*)^* = a$,
(ii) $(a + \lambda b)^* = a^* + \bar{\lambda} b^*$,
(iii) $(ab)^* = b^* a^*$.

Ideals in *-algebras will always be assumed to be *self-adjoint* (that is, closed under *) and homomorphisms between *-algebras will always be assumed to be **-homomorphisms* (that is, preserving the involution).

Among the most important examples of *-algebras, we mention the group ring $\mathbb{K}(G)$, for any group G, equipped with the involution

$$\Big(\sum_{g\in G} \lambda_g \delta_g\Big)^* = \sum_{g\in G} \bar{\lambda}_g \delta_{g^{-1}}.$$

Definition 6.4. An *algebraic* (resp. **-algebraic*) partial action of a group G on a $\mathbb{K}$-algebra (resp. *-algebra) A is a partial action

$$\theta = \big(\{D_g\}_{g\in G}, \{\theta_g\}_{g\in G}\big)$$

on the underlying set A, such that each D_g is a *two-sided ideal* (resp. *self-adjoint two-sided ideal*) in A, and each θ_g is an *isomorphism* (resp. **-isomorphism*). Taking the point of view of (2.1) or (5.1), one likewise defines the notions of *algebraic* and **-algebraic partial dynamical systems.*

When the category one is working with is understood, be it $\mathbb{K}$-algebras or *-algebras, and when there is no chance for confusion, we will drop the adjectives *algebraic* or **-algebraic* and simply say *partial action* or *partial dynamical system.*

We will say that two algebraic (resp. *-algebraic) partial actions of G are *algebraically equivalent* (resp. **-algebraically equivalent*) if they are equivalent in the sense of Definition (2.7) and, in addition, the map ϕ mentioned there is a homomorphism (resp. *-homomorphism).

The process of restriction may also be applied in the algebraic context: given a (*-)algebraic global action η of G on a (*-)algebra B, and given a two-sided ideal $A \trianglelefteq B$ (supposed to be self-adjoint in the *-algebra case), let θ be the restriction of η to A, as defined in (3.2). Since

$$D_g = \eta_g(A) \cap A,$$

we have that D_g is a (self-adjoint) two-sided ideal of A and, since the restricted maps θ_g are obviously (*-)homomorphisms, we conclude that θ is a (*-)algebraic partial action of G on A.

Reversing this process we may likewise define the notion of (*-)algebraic globalization, except that we need to adapt the notion of orbit employed in (3.4) to the algebraic context. Before we spell out the appropriate definition of algebraic globalization let us discuss a minor technical point.

Proposition 6.5. *Let η be an algebraic global action of the group G on the $\mathbb{K}$-algebra B, and let A be a two-sided ideal of B. Then the smallest η-invariant subalgebra of B containing A is*

$$\sum_{g\in G} \eta_g(A).$$

PROOF. Since each $\eta_g(A)$ is a two-sided ideal in B, the sum of all such is also a two-sided ideal in B, and hence also a subalgebra, evidently η-invariant. That it is the smallest η-invariant subalgebra is also clear. □

If, in addition to the hypotheses of the result above, η is a *-algebraic global action and A is a self-adjoint ideal, then it is easily verified that $\sum_{g\in G} \eta_g(A)$ is the smallest η-invariant *-subalgebra of B.

The notion of algebraic globalization is thus adapted from (3.4) as follows:

Definition 6.6. Let η be a (*-)algebraic global action of G on a (*-)algebra B, and let A be a (self-adjoint) two-sided ideal of B. Also let θ be the partial action obtained by restricting η to A. If

$$B = \sum_{g\in G} \eta_g(A),$$

we will say that η is a (*-)*algebraic globalization* of θ.

► In order to avoid cluttering the exposition we will now restrict ourselves to the category of $\mathbb{K}$-algebras, even though most of our results from now on are valid for *-algebras as well.

Unlike the case of partial actions on sets or topological spaces, algebraic globalizations do not always exist, and when they do, uniqueness may fail. Postponing the existence question, let us present an example of a partial action admitting many non-equivalent algebraic globalizations.

Example 6.7. This is essentially an example of actions on $\mathbb{K}$-modules masqueraded as $\mathbb{K}$-algebras. Given a left $\mathbb{K}$-module M, we may view it as a $\mathbb{K}$-algebra by introducing the trivial multiplication operation, namely,

$$ab = 0, \quad \forall\, a, b \in M. \tag{6.7.1}$$

Notice that every $\mathbb{K}$-sub-module of M is an ideal and that any $\mathbb{K}$-linear map between two such $\mathbb{K}$-modules is automatically a homomorphism.

This said, let us assume that $\mathbb{K} = \mathbb{R}$, the field of real numbers, and let us choose a real number

$$\alpha \in \mathbb{R} \setminus \mathbb{Q}.$$

Consider the action η^α of $\mathbb{Z}$ on $\mathbb{R}^2$ defined in terms of the group homomorphism

$$\eta^\alpha : n \in \mathbb{Z} \mapsto \begin{bmatrix} \cos(2n\pi\alpha) & -\sin(2n\pi\alpha) \\ \sin(2n\pi\alpha) & \cos(2n\pi\alpha) \end{bmatrix} \in \mathrm{GL}_2(\mathbb{R}).$$

In other words, each η^α_n acts by rotating the plane by an angle of $2n\pi\alpha$.

If, as above, we make $\mathbb{R}^2$ into an algebra by introducing the trivial multiplication, we may view η^α as a global algebraic action of $\mathbb{Z}$ on $\mathbb{R}^2$.

Let A be the one-dimensional subspace of $\mathbb{R}^2$ spanned by the vector $(1, 0)$, hence a two-sided ideal, and let θ be the restriction of η^α to A. Since η^α_n has no real eigenvectors for nonzero n, we have that

$$D_n := \eta^\alpha_n(A) \cap A = \{0\},$$

and consequently θ_n is the zero map, except of course for θ_0, which is the identity map on $D_0 = A$. It is also easy to see that

$$\mathbb{R}^2 = \sum_{n\in\mathbb{Z}} \eta^\alpha_n(A),$$

so η^α is an algebraic globalization of θ.

Notice that the complex eigenvalues for η^α_1 are $e^{\pm 2\pi i\alpha}$ so, given two distinct irrational numbers α_1 and α_2 in $(0, 1/2)$, the corresponding global actions η^{α_1} and η^{α_2} are not equivalent. Therefore θ admits an uncountable number of non-equivalent algebraic globalizations.

Many aspects of partial actions become greatly simplified when the ideals involved are unital.

► In order to discuss these aspects, let us fix, for the time being, a partial action $\theta = \big(\{D_g\}_{g\in G}, \{\theta_g\}_{g\in G}\big)$ of a group G on an algebra A, such that D_g is unital, with unit

$$1_g \in D_g,$$

for all g in G.

We will now prove a few technical lemmas which will later be useful to study the globalization question under the present hypotheses.

Lemma 6.8. *For every $g, h \in G$, one has that*

$$\theta_g(1_{g^{-1}}1_h) = 1_g 1_{gh}.$$

PROOF. This follows immediately from the fact that $1_{g^{-1}}1_h$ is the unit for $D_{g^{-1}} \cap D_h$, while $1_g 1_{gh}$ is the unit for $D_g \cap D_{gh}$, and that θ_g is an isomorphism between these ideals by (2.6). □

Observe that 1_g is a central idempotent in A, as is the case for the unit of any two-sided ideal. Moreover, since $D_g = 1_g A$, the correspondence

$$\Theta_g : a \in A \mapsto \theta_g(1_{g^{-1}}a) \in A, \tag{6.9}$$

gives a well defined endomorphism of A.

The range of Θ_g is clearly D_g, so we have that

$$\Theta_g(a) = \Theta_g(a)1_g, \quad \forall a \in A. \tag{6.10}$$

Lemma 6.11. *For every $g, h \in G$, and every $a \in A$, one has that*

$$\Theta_g\big(\Theta_h(a)\big) = 1_g\Theta_{gh}(a).$$

PROOF. We have

$$\begin{aligned}
\Theta_g\big(\Theta_h(a)\big) &= \theta_g\big(1_{g^{-1}}\theta_h(1_{h^{-1}}a)\big)\\
&= \theta_g\big(1_{g^{-1}}1_h\theta_h(1_{h^{-1}}a)\big)\\
&\overset{(6.8)}{=} \theta_g\big(\theta_h(1_{h^{-1}g^{-1}}1_{h^{-1}})\,\theta_h(1_{h^{-1}}a)\big)\\
&= \theta_g\big(\theta_h(1_{h^{-1}g^{-1}}1_{h^{-1}}a)\big)\\
&\overset{(2.5.ii)}{=} \theta_{gh}(1_{h^{-1}g^{-1}}1_{h^{-1}}a)\\
&= \theta_{gh}(1_{h^{-1}g^{-1}}1_{h^{-1}})\cdot\theta_{gh}(1_{h^{-1}g^{-1}}a)\\
&\overset{(6.8)}{=} 1_{gh}1_g\Theta_{gh}(a)\\
&= 1_g\Theta_{gh}(a).
\end{aligned}$$

□

Lemma 6.12. *Suppose that A is an algebra which is a (not necessarily direct) sum of a finite number of unital ideals. Then A is a unital algebra.*

PROOF. By induction on the number of summands it is enough to consider two ideals, say $A = J + K$. Letting 1_J and 1_K denote the units of J and K, respectively, one proves that 1_J and 1_K are central idempotents in A and that

$$1_J + 1_K - 1_J \cdot 1_K$$

is the unit for A. □

We are now ready to tackle our first globalization result for algebraic partial actions. So far we will only treat the unital case, leaving the highly involved non-unital case for later.

Theorem 6.13. *Let $\theta = \big(\{D_g\}_{g\in G}, \{\theta_g\}_{g\in G}\big)$ be a partial action of a group G on a unital algebra A. Then a necessary and sufficient condition for θ to admit a globalization is that D_g be unital for every g in G. Moreover, in this case the globalization is unique up to algebraic equivalence.*

PROOF. Let η be a globalization of θ, acting on the algebra B. Denote by 1_A the unit of A and observe that, for each g in G, one has that

$$1_A \eta_g(1_A) \in A \cap \eta_g(A) = D_g.$$

One then easily checks that $1_A \eta_g(1_A)$ is the unit for D_g, which is therefore proven to be a unital ideal.

Conversely, supposing that each D_g is unital, let us build a globalization for θ. As a first step we consider the algebra A^G, formed by all functions

$$f : G \to A,$$

with pointwise operations. Consider the global action η of G on A^G defined by

$$\eta_g(f)|_h = f(g^{-1}h), \quad \forall\, g, h \in G, \quad \forall\, f \in A^G,$$

and let

$$\iota : A \to A^G$$

be the homomorphism defined by

$$\iota(a)|_g = \theta_{g^{-1}}(1_g a) = \Theta_{g^{-1}}(a), \quad \forall\, a \in A, \quad \forall\, g \in G.$$

Since $\iota(a)|_1 = a$, we have that ι is injective, so we may identify A with its image

$$A' := \iota(A) \subseteq A^G.$$

We have therefore embedded A into the larger algebra A^G, where a global action of G is defined but, unfortunately, A' is not necessarily an ideal in A^G, so we need to do some more work if we are to obtain the desired globalization of θ. The trick is to find an η-invariant subalgebra of A^G, containing A' as an ideal. In doing so the crucial technical ingredient is to check that for every $g, h \in G$, one has that

$$\eta_g(A') \cdot \eta_h(A') \subseteq \eta_g(A') \cap \eta_h(A'). \tag{6.13.1}$$

In order to prove this let $a, b \in A$, and notice that, for all $k \in G$, one has

$$\begin{aligned} \eta_g\big(\iota(a)\big)|_k \cdot \eta_h\big(\iota(b)\big)|_k & \\ &= \iota(a)|_{g^{-1}k} \cdot \iota(b)|_{h^{-1}k} \\ &= \Theta_{k^{-1}g}(a) \cdot \Theta_{k^{-1}h}(b) \\ &= (\star). \end{aligned}$$

This can be developed in two different ways. On the one hand

$$\begin{aligned} (\star) &\overset{(6.10)}{=} \Theta_{k^{-1}g}(a) 1_{k^{-1}h} \Theta_{k^{-1}h}(b) \\ &\overset{(6.11)}{=} \Theta_{k^{-1}h}\big(\Theta_{h^{-1}g}(a)\big) \cdot \Theta_{k^{-1}h}(b) \\ &= \Theta_{k^{-1}h}\big(\Theta_{h^{-1}g}(a) b\big) \\ &= \iota\big(\Theta_{h^{-1}g}(a) b\big)|_{h^{-1}k} \\ &= \eta_h\Big(\iota\big(\Theta_{h^{-1}g}(a) b\big)\Big)|_k. \end{aligned}$$

This shows that

$$\eta_g\big(\iota(a)\big) \cdot \eta_h\big(\iota(b)\big) = \eta_h\Big(\iota\big(\Theta_{h^{-1}g}(a) b\big)\Big) \in \eta_h(A').$$

On the other hand we may write

$$
\begin{aligned}
(\star) &\overset{(6.10)}{=} \Theta_{k^{-1}g}(a)1_{k^{-1}g}\Theta_{k^{-1}h}(b) \\
&\overset{(6.11)}{=} \Theta_{k^{-1}g}(a)\cdot\Theta_{k^{-1}g}\big(\Theta_{g^{-1}h}(b)\big) \\
&= \Theta_{k^{-1}g}\big(a\Theta_{g^{-1}h}(b)\big) \\
&= \iota\big(a\Theta_{g^{-1}h}(b)\big)|_{g^{-1}k} \\
&= \eta_g\Big(\iota\big(a\Theta_{g^{-1}h}(b)\big)\Big)|_k,
\end{aligned}
$$

whence

$$\eta_g\big(\iota(a)\big)\cdot\eta_h\big(\iota(b)\big) = \eta_g\Big(\iota\big(a\Theta_{g^{-1}h}(b)\big)\Big) \in \eta_g(A'),$$

thus proving (6.13.1).

Let B be the subset of A^G defined by

$$B = \sum_{g\in G}\eta_g(A').$$

Using (6.13.1) one immediately sees that B is a subalgebra of A^G. Moreover, by plugging $h = 1$ in (6.13.1) one concludes that A' is a left ideal of B, while with $g = 1$, (6.13.1) implies that A' is a right ideal of B, so A' is a two-sided ideal.

It is clear that B is invariant relative to the global action η, so we may restrict η to B, obtaining another global action which, by abuse of language, we still denote by η. The very definition of B then implies that it is the smallest η-invariant subalgebra of A^G containing A', as seen in (6.5).

In order to complete the existence part of the proof it is now enough to check that the restriction of η to A' is equivalent to θ. As a first step let us verify that

$$\eta_g(A') \cap A' = \iota(D_g). \tag{6.13.2}$$

Given $x \in \eta_g(A') \cap A'$, write $x = \iota(a) = \eta_g(\iota(b))$, for $a, b \in A$, and notice that

$$a = \iota(a)|_1 = \eta_g\big(\iota(b)\big)|_1 = \iota(b)|_{g^{-1}} = \Theta_g(b) \in D_g,$$

so $x \in \iota(D_g)$. Conversely, if $x = \iota(a)$, with $a \in D_g$, let $b = \theta_{g^{-1}}(a)$, and notice that, for all $h \in G$,

$$
\begin{aligned}
\eta_g\big(\iota(b)\big)|_h & \\
&= \iota(b)|_{g^{-1}h} \\
&= \Theta_{h^{-1}g}(b) \\
&= \Theta_{h^{-1}g}\big(\Theta_{g^{-1}}(a)\big) \\
&\overset{(6.11)}{=} \Theta_{h^{-1}}(a)1_{h^{-1}g}.
\end{aligned}
\tag{6.13.3}
$$

Observe, however that

$$\Theta_{h^{-1}}(a) = \theta_{h^{-1}}(1_h a) \in \theta_{h^{-1}}(D_h \cap D_g) \overset{(2.6)}{=} D_{h^{-1}} \cap D_{h^{-1}g},$$

so we deduce from (6.13.3) that

$$\eta_g\big(\iota(b)\big)|_h = \Theta_{h^{-1}}(a) = \iota(a)|_h,$$

whence

$$x = \iota(a) = \eta_g\big(\iota(b)\big) \in \eta_g(A') \cap A',$$

proving (6.13.2).

We must now prove that the restriction of each η_g to $\iota(D_{g^{-1}})$ corresponds to θ_g. For this, let $a \in D_{g^{-1}}$ and observe that, for all k in G we have

$$\eta_g\big(\iota(a)\big)|_k = \iota(a)|_{g^{-1}k} = \theta_{k^{-1}g}(1_{g^{-1}k}a) \overset{(2.6)}{=} \theta_{k^{-1}}\big(\theta_g(1_{g^{-1}k}a)\big) = (\dagger).$$

Notice that the argument of $\theta_{k^{-1}}$, above, may be rewritten as

$$\begin{aligned}\theta_g(1_{g^{-1}k}a) &= \theta_g(1_{g^{-1}k}1_{g^{-1}}a)\\ &= \theta_g(1_{g^{-1}k}1_{g^{-1}})\theta_g(a)\\ &\overset{(6.8)}{=} 1_k 1_g \theta_g(a)\\ &= 1_k\theta_g(a),\end{aligned}$$

so

$$(\dagger) = \theta_{k^{-1}}\big(1_k\theta_g(a)\big) = \Theta_{k^{-1}}\big(\theta_g(a)\big) = \iota(\theta_g(a))|_k.$$

This shows that $\eta_g\big(\iota(a)\big) = \iota\big(\theta_g(a)\big)$, which is to say that restriction of η_g to $\iota(D_g)$ corresponds to θ_g, as desired.

This takes care of existence, so let us now focus on proving uniqueness. Therefore we suppose we are given two globalizations of θ, say η and η', acting respectively on the algebras B and B'. As a first step we will prove that given $a_1, a_2, \ldots, a_n \in A$, and $g_1, g_2, \ldots, g_n \in G$, one has that

$$\sum_{i=1}^n \eta_{g_i}(a_i) = 0 \quad \Rightarrow \quad \sum_{i=1}^n \eta'_{g_i}(a_i) = 0. \tag{6.13.4}$$

We begin by claiming that, for all $g \in G$, and all a and b in A, one has that

$$b\eta_g(a) = b\Theta_g(a). \tag{6.13.5}$$

In fact observe that

$$b\eta_g(a) = \eta_g\big(\eta_{g^{-1}}(b)a\big),$$

so, since A is an ideal in B, we see that

$$b\eta_g(a) \in A \cap \eta_g(A) = D_g.$$

Therefore, recalling that η_g extends θ_g, we have that

$$b\eta_g(a) = b\eta_g(a)1_g = b\eta_g(a)\eta_g(1_{g^{-1}}) = b\eta_g(a1_{g^{-1}}) = b\theta_g(a1_{g^{-1}}) = b\Theta_g(a),$$

proving (6.13.5). If we now apply this for $h^{-1}g$, where h is another element of G, and then compute η_h on both sides of the resulting expression, we obtain

$$\eta_h(b)\eta_g(a) = \eta_h\big(b\Theta_{h^{-1}g}(a)\big). \tag{6.13.6}$$

Therefore, under the hypothesis of (6.13.4), we have for all $b \in A$, and $h \in G$, that

$$0 = \eta_h(b)\cdot\sum_{i=1}^n \eta_{g_i}(a_i) \overset{(6.13.6)}{=} \sum_{i=1}^n \eta_h\big(b\Theta_{h^{-1}g_i}(a_i)\big) = \eta_h\Big(\sum_{i=1}^n b\Theta_{h^{-1}g_i}(a_i)\Big).$$

As a consequence we deduce that

$$\sum_{i=1}^n b\Theta_{h^{-1}g_i}(a_i) = 0.$$

Let us denote by $Z = \sum_{i=1}^n \eta'_{g_i}(a_i)$, so that proving (6.13.4) amounts to proving that $Z = 0$. Employing (6.13.6), this time for η', it then follows that

$$0 = \eta'_h\Big(\sum_{i=1}^n b\Theta_{h^{-1}g_i}(a_i)\Big) = \eta'_h(b)\cdot\sum_{i=1}^n \eta'_{g_i}(a_i) = \eta'_h(b)Z.$$

Since $B' = \sum_{h\in G} \eta'_h(A)$, this shows that Z is in the annihilator ideal of B'. Notice that Z lies in

$$\sum_{i=1}^{n} \eta'_{g_i}(A),$$

which is a finite sum of unital ideals of B, and hence itself a unital ideal by (6.12). From this we easily see that $Z = 0$, hence concluding the proof of (6.13.4).

As a most important consequence, it follows that there exists a well defined $\mathbb{K}$-linear map $\phi : B \to B'$, such that

$$\phi\big(\sum_{i=1}^{n} \eta_{g_i}(a_i)\big) = \sum_{i=1}^{n} \eta'_{g_i}(a_i),$$

for all $a_1, a_2, \ldots, a_n \in A$, and $g_1, g_2, \ldots, g_n \in G$. By reversing the implication in (6.13.4), we see that ϕ admits an inverse, and so it is a bijection from B onto B'.

We leave it for the reader to prove the easy facts that ϕ is G-equivariant and restricts to the identity over A.

Finally, to see that ϕ is a homomorphism, we must prove that $\phi(xy) = \phi(x)\phi(y)$, for all $x, y \in B$, but it is clearly enough to consider $x = \eta_g(a)$, and $y = \eta_h(b)$, with $g, h \in G$, and $a, b \in A$. We have

$$\begin{aligned}
&\phi(xy)\\
&\qquad = \phi\big(\eta_g(a)\eta_h(b)\big)\\
&\qquad \overset{(6.13.6)}{=} \phi\big(\eta_g(a\Theta_{g^{-1}h}(b)\big)\\
&\qquad = \eta'_g\big(a\Theta_{g^{-1}h}(b)\big)\\
&\qquad \overset{(6.13.6)}{=} \eta'_g(a)\eta'_h(b)\\
&\qquad = \phi(x)\phi(y).
\end{aligned}$$

□

With this result it is easy to produce examples of algebraic partial actions possessing no algebraic globalizations:

Example 6.14. Take any unital algebra A containing a non-unital two-sided ideal J, and consider the partial action θ of the group $\mathbb{Z}_2 = \{0, 1\}$ on A, defined by setting $\theta_0 = id_A$, and $\theta_1 = id_J$, so that $D_0 = A$, and $D_1 = J$. By (6.13), the absence of a unit in D_1 precludes the existence of a globalization for θ.

Notes and remarks. Partial actions on C*-algebras were studied long before partial actions on rings and algebras. The reason for this is the fact that the first proofs of the associativity property for C*-algebraic partial crossed products (defined later in this book) uses the existence of approximate identities, which are absent in the purely algebraic case.

Having realized this difficulty early on, I have discussed it with many colleagues in the Algebra community but it took a while before anyone was convinced of the relevance of this question. I eventually managed to attract the attention of Misha Dokuchaev, and together we were able to understand exactly what was going on [**33**]. Since then, Misha has been able to convince a host of people to study the algebraic aspects of partial actions. Unfortunately the growing body of knowledge in that direction is poorly represented in this book and it might soon deserve a book of its own.

Theorem (6.13) was proved in [**33**]. It is an early attempt to replicate Abadie's previous work on enveloping actions in the realm of C*-algebras [**2**], which we will discuss later in this book. Further attempts were made in [**36**] and [**4**].

Although not covered by this work, the notion of *twisted* partial actions of groups on $\mathbb{K}$-algebras has also been considered [**48**], [**35**], [**36**].

CHAPTER 7

Multipliers

The notion of multiplier algebras is a well established concept among C*-algebraists, but its purely algebraic version has not appeared too often in the specialized literature. Since multipliers are an important tool in the treatment of algebraic partial actions we have included this brief chapter to subsidize our future work. See [**33**] for more details.

▶ Throughout this chapter we shall let A be a fixed $\mathbb{K}$-algebra, sometimes assumed to be a *-algebra.

Definition 7.1. Let L and R be $\mathbb{K}$-linear maps from A to itself. We shall say that the pair (L, R) is a *multiplier* of A if, for every a and b in A, one has that

(i) $L(ab) = L(a)b$,
(ii) $R(ab) = aR(b)$,
(iii) $R(a)b = aL(b)$.

If A is a unital algebra, and if (L, R) is a multiplier of A, notice that,

$$R(1) = R(1)1 \overset{(7.1.\text{iii})}{=} 1L(1) = L(1).$$

Moreover, letting $m := R(1) = L(1)$, we have for every a in A that

$$L(a) = L(1a) = L(1)a = ma,$$

while

$$R(a) = R(a1) = aR(1) = am,$$

so we see that $(L, R) = (L_m, R_m)$, where L_m is the operator of left multiplication by m, and R_m is the operator of right multiplication by m.

Definition 7.2. The *multiplier algebra* of A is the set $\mathcal{M}(A)$ consisting of all multipliers (L, R) of A. Given (L, R) and (L', R') in $\mathcal{M}(A)$, and $\lambda \in \mathbb{K}$, we define

$$\lambda(L, R) = (\lambda L, \lambda R),$$

$$(L, R) + (L', R') = (L + L', R + R'),$$

$$(L, R)(L', R') = (L \circ L', R' \circ R).$$

We leave it for the reader to prove that $\mathcal{M}(A)$ is an associative $\mathbb{K}$-algebra with the above operations. It is moreover a unital algebra with unit (id_A, id_A).

If A is a *-algebra, and $T : A \to A$ is a linear map, define a mapping $T^* : A \to A$ by the formula

$$T^*(a) = \big(T(a^*)\big)^*.$$

It is then easy to see that T^* is also linear.

Proposition 7.3. *If A is a *-algebra then so is $\mathcal{M}(A)$, when equipped with the involution*

$$(L,R)^* := (R^*, L^*), \quad \forall (L,R) \in \mathcal{M}(A).$$

PROOF. Left for the reader. □

Assuming that A is an ideal in some other algebra B, let m be a fixed element of B and consider the maps

$$L_m : a \in A \mapsto ma \in A, \quad \text{and} \quad R_m : a \in A \mapsto am \in A.$$

It is an easy exercise to show that the pair (L_m, R_m) is a multiplier of A and that the correspondence

$$\mu_B : m \in B \mapsto (L_m, R_m) \in \mathcal{M}(A), \tag{7.4}$$

is a homomorphism.

Definition 7.5. We shall say that A is an *essential ideal* in B if the map μ_B mentioned above is injective.

In general the kernel of μ_B is the intersection of the left and right annihilators of A in B. Therefore A is an essential ideal in B if and only if, for every nonzero element a in A, there exists some b in B such that either $ab \neq 0$ or $ba \neq 0$.

Definition 7.6. We shall say that A is a *non-degenerate* algebra if there is no nonzero element a in A such that $ab = ba = 0$, for all b in A. Equivalently, if A is an essential ideal in itself.

Proposition 7.7.

(i) *μ_A is an isomorphism onto $\mathcal{M}(A)$ if and only if A is unital.*
(ii) *The range of μ_A, henceforth denoted by A', is an ideal of $\mathcal{M}(A)$.*
(iii) *If A is non-degenerate then A' is isomorphic to A.*

PROOF. Left for the reader. □

Thus, when A is an essential ideal in some algebra B, a situation which is only possible when A is non-degenerate, we have that B is isomorphic to a subalgebra of $\mathcal{M}(A)$, namely the range of μ_B, containing A'. In other words, up to isomorphism, all algebras B containing A as an essential ideal are to be found among the subalgebras of $\mathcal{M}(A)$ containing A'.

Given two multipliers (L^1, R^1) and (L^2, R^2) in $\mathcal{M}(A)$ we shall be concerned with the validity of the formula

$$R^2 L^1 \overset{(?)}{=} L^1 R^2. \tag{7.8}$$

We will see that this is the crucial property governing the associativity of partial crossed products studied below.

To see the connection between associativity and property (7.8) above, notice that if A is an ideal in some algebra B, and if for all $i = 1, 2$ we let $(L^i, R^i) = \mu_B(m_i)$, for some $m_i \in B$, then for every a in A we have that

$$R^2\big(L^1(a)\big) = (m_1 a)m_2, \quad \text{while} \quad L^1\big(R^2(a)\big) = m_1(a m_2),$$

so that (7.8) holds as a consequence of the associativity property of B.

However (7.8) may just as well fail: take A to be any $\mathbb{K}$-module equipped with the trivial multiplication operation described in (6.7.1). In this case, observe that any pair (L, R) of linear operators on A would constitute a multiplier and one should clearly not expect (7.8) to hold in such a generality!

Proposition 7.9. *If A is either non-degenerate or idempotent then* (7.8) *holds for any pair of multipliers* (L^1, R^1) *and* (L^2, R^2).

PROOF. Given $a, b \in A$, we have that

$$R^2\big(L^1(a)\big)b = L^1(a)L^2(b) = L^1\big(aL^2(b)\big) = L^1\big(R^2(a)b\big) = L^1\big(R^2(a)\big)b.$$

This shows that $R^2\big(L^1(a)\big) - L^1\big(R^2(a)\big)$ lies in the left annihilator of A. With a similar argument one shows that this also lies in the right annihilator of A. So this proves (7.8) under the assumption that A is non-degenerate.

Next suppose we are given $a_1, a_2 \in A$. Letting $a = a_1a_2$, notice that

$$\begin{aligned} R^2\big(L^1(a)\big) & \\ &= R^2\big(L^1(a_1a_2)\big) \\ &= R^2\big(L^1(a_1)a_2\big) \\ &= L^1(a_1)R^2(a_2) \\ &= L^1\big(a_1R^2(a_2)\big) \\ &= L^1\big(R^2(a_1a_2)\big) \\ &= L^1\big(R^2(a)\big). \end{aligned}$$

Assuming that A is idempotent, we have that every element of A is a sum of terms of the form a_1a_2, whence the conclusion. □

CHAPTER 8

Crossed Products

The classical notion of crossed product[3] is a useful tool to construct interesting examples of algebras. Its basic ingredient is a global action η of a group G on an algebra A. One then defines the crossed product algebra $A \rtimes G$, sometimes also denoted $A \rtimes_\eta G$ when we want to make the action explicit, to consist of all finite formal linear combinations

$$\sum_{g \in G} a_g \delta_g, \tag{8.1}$$

(that is $a_g = 0$, except for finitely many g's), where the δ_g are seen as place markers[4].

A multiplication operation on $A \rtimes G$ is then defined by

$$(a\delta_g)(b\delta_h) = a\eta_g(b)\delta_{gh}, \tag{8.2}$$

for all a and b in A, and all g and h in G.

The intuitive idea behind the above definition of product is to think of the δ_g's as invertible elements implementing the given action. A somewhat imprecise but highly enlightening calculation motivating this definition is as follows

$$(a\delta_g)(b\delta_h) = \underbrace{a\delta_g b\delta_h}_{\text{eliminate parentheses}} = a\delta_g b \underbrace{\delta_g^{-1}\delta_g}_{\text{insert 1}} \delta_h = a \underbrace{\delta_g b \delta_g^{-1}}_{\text{view as conjugation}} \underbrace{\delta_g\delta_h}_{\text{apply the group law}} = a\eta_g(b)\delta_{gh}.$$

It is our intention to develop a similar construction for partial actions.

▶ So let us fix, for the remainder of this chapter, a group G, a $\mathbb{K}$-algebra A, and an algebraic partial action $\theta = \big(\{D_g\}_{g \in G}, \{\theta_g\}_{g \in G}\big)$ of G on A.

Evidently the main difficulty in generalizing the above construction of the crossed product is that, should η_g be a partially defined map, the term $\eta_g(b)$ appearing in (8.2) is only defined for b in the domain of η_g.

Definition 8.3. The *crossed product*[5] of the algebra A by the group G under the partial action $\theta = \big(\{D_g\}_{g \in G}, \{\theta_g\}_{g \in G}\big)$ is the algebra $A \rtimes G$, sometimes also

[3] Algebraists sometimes reserve the term *crossed product* for situations when, besides a group action, a cocycle is also involved. When only a group action is present, as above, the algebraic literature usually favors the expression *skew-group algebra*.

[4] Technically speaking, $A \rtimes G$ is the set of all finitely supported functions from G to A. Denoting by $a_g\delta_g$ the function supported on the singleton $\{g\}$, and whose value at g is the element a_g, the term in (8.1) corresponds simply to the function sending g to a_g. Note however that the expression "δ_g" has no meaning in itself according to the present convention.

[5] In order to emphasize that the action θ is partial, we will sometimes use the expression *partial crossed product* to refer to this construction.

denoted $A \rtimes_\theta G$ when we want to make θ explicit, consisting of all finite formal linear combinations

$$\sum_{g\in G} a_g\delta_g, \tag{8.3.1}$$

with $a_g \in D_g$, for all g in G. Addition and scalar multiplication are defined in the obvious way, while multiplication is determined by

$$(a\delta_g)(b\delta_h) = \theta_g\big(\theta_{g^{-1}}(a)b\big)\delta_{gh}, \tag{8.3.2}$$

for all g and h in G, and for all $a \in D_g$, and $b \in D_h$.

Here is another imprecise but likewise enlightening calculation, this time motivating the definition of the multiplication in (8.3.2):

$$
\begin{aligned}
(a\delta_g)(b\delta_h) &= \underbrace{a\delta_g b\delta_h}_{\substack{\text{eliminate}\\ \text{parentheses}}}\\
&= \underbrace{\delta_g\delta_g^{-1}}_{\text{insert 1}}\, a\delta_g b\, \underbrace{\delta_g^{-1}\delta_g}_{\text{insert 1}}\, \delta_h\\
&= \delta_g\, \underbrace{\delta_g^{-1}a\delta_g}_{\substack{\text{view as}\\ \text{conjugation}}}\, b\delta_g^{-1}\, \underbrace{\delta_g\delta_h}_{\substack{\text{apply the}\\ \text{group law}}}\\
&= \underbrace{\delta_g\theta_{g^{-1}}(a)b\delta_g^{-1}}_{\substack{\text{view again as}\\ \text{conjugation}}}\, \delta_{gh}\\
&= \theta_g\big(\theta_{g^{-1}}(a)b\big)\delta_{gh}.
\end{aligned}
$$

Although slightly longer than (8.2), the definition of the multiplication in (8.3.2) has the advantage of avoiding the temptation of applying a function to elements outside its domain. In fact, on the one hand notice that since $a \in D_g$, the reference to $\theta_{g^{-1}}(a)$ in (8.3.2) is perfectly legal. On the other hand, since $\theta_{g^{-1}}(a)$ belongs to the *ideal* $D_{g^{-1}}$, we see that $\theta_{g^{-1}}(a)b$ is in $D_{g^{-1}}$ as well, so the reference to $\theta_g\big(\theta_{g^{-1}}(a)b\big)$ is legal too.

As a final syntactic check, recall that in (8.3.1), each a_g must lie in D_g, so we need to ensure that $\theta_g\big(\theta_{g^{-1}}(a)b\big)$ lies in D_{gh}, if we are to allow it to stand besides δ_{gh}. But this is also granted because

$$\theta_{g^{-1}}(a)b \in D_{g^{-1}} \cap D_h,$$

so

$$\theta_g\big(\theta_{g^{-1}}(a)b\big) \in \theta_g(D_{g^{-1}} \cap D_h) \overset{(2.6)}{=} D_g \cap D_{gh}.$$

After extending the multiplication defined above to a bi-linear map on $A \rtimes G$, we must worry about the associativity property, but unfortunately it does not always hold.

In order to identify the origin of this problem let us begin with the trivial remark that $A \rtimes G$ is associative if and only if

$$(a\delta_g\ b\delta_h)\ c\delta_k = a\delta_g\ (b\delta_h\ c\delta_k), \tag{8.4}$$

for all $g, h, k \in G$, and all $a \in D_g$, $b \in D_h$, and $c \in D_k$. Focusing on the left-hand-side above we have

$$\begin{aligned}(a\delta_g\ b\delta_h)\ c\delta_k &= \theta_g\big(\theta_{g^{-1}}(a)b\big)\delta_{gh}\ c\delta_k \\ &= \theta_{gh}\Big(\theta_{h^{-1}g^{-1}}\Big[\theta_g\big(\theta_{g^{-1}}(a)b\big)\Big]c\Big)\delta_{ghk} \\ &= (\star_1).\end{aligned}$$

Observe that the term within square brackets above satisfies

$$\theta_g\big(\theta_{g^{-1}}(a)b\big) \in \theta_g(D_{g^{-1}} \cap D_h) = D_g \cap D_{gh},$$

which is precisely the set on which (2.5.ii) yields $\theta_{h^{-1}g^{-1}} = \theta_{h^{-1}}\theta_{g^{-1}}$. Therefore we may cancel out the composition "$\theta_{g^{-1}}\theta_g$", so that

$$(\star_1) = \theta_{gh}\Big(\ \underline{\theta_{h^{-1}}\big(\theta_{g^{-1}}(a)b\big)}\ c\Big)\delta_{ghk} = (\star_2).$$

Observe also that the underlined term above satisfies

$$\theta_{h^{-1}}\big(\theta_{g^{-1}}(a)b\big) \in \theta_{h^{-1}}(D_{g^{-1}} \cap D_h) = D_{h^{-1}g^{-1}} \cap D_{h^{-1}},$$

which is where $\theta_{gh} = \theta_g\theta_h$, again according to (2.5.ii). We thus finally obtain that

$$(\star_2) = \theta_g\Big[\theta_h\Big(\theta_{h^{-1}}\big(\theta_{g^{-1}}(a)b\big)c\Big)\Big]\delta_{ghk}.$$

On the other hand, the right-hand-side of (8.4) equals

$$a\delta_g\ (b\delta_h\ c\delta_k) = a\delta_g\Big(\theta_h\big(\theta_{h^{-1}}(b)c\big)\delta_{hk}\Big) = \theta_g\Big[\theta_{g^{-1}}(a)\theta_h\big(\theta_{h^{-1}}(b)c\big)\Big]\delta_{ghk}.$$

This said, we see that equation (8.4) is equivalent to

$$\theta_g\Big[\theta_{g^{-1}}(a)\theta_h\big(\theta_{h^{-1}}(b)c\big)\Big] = \theta_g\Big[\theta_h\Big(\theta_{h^{-1}}\big(\theta_{g^{-1}}(a)b\big)c\Big)\Big],$$

which is clearly the same as

$$\theta_{g^{-1}}(a)\theta_h\big(\theta_{h^{-1}}(b)c\big) = \theta_h\Big(\theta_{h^{-1}}\big(\theta_{g^{-1}}(a)b\big)c\Big).$$

Since the only occurrence of g and a, above, is in the term $\theta_{g^{-1}}(a)$, we may substitute a for $\theta_{g^{-1}}(a)$ without altering the logical content of this expression. We have therefore proven:

Lemma 8.5. *A necessary and sufficient condition for $A \rtimes G$ to be associative is that, for all $h \in G$, $b \in D_h$, and $a, c \in A$, one has*

$$a\theta_h\big(\theta_{h^{-1}}(b)c\big) = \theta_h\Big(\theta_{h^{-1}}\big(ab\big)c\Big). \tag{8.5.1}$$

The above condition might still look a bit messy at first, but it may be given a clean interpretation in terms of multipliers. With this goal in mind, consider the multiplier of D_h given by $(L^1, R^1) = \mu_A(a)$, as defined in (7.4).

There is another relevant multiplier of D_h given as follows: initially consider the multiplier

$$\mu_A(c) = (L_c, R_c) \in \mathcal{M}(D_{h^{-1}}).$$

Since θ_h is an isomorphism from $D_{h^{-1}}$ to D_h, we may transfer the above to a multiplier $(L^2, R^2) \in \mathcal{M}(D_h)$, by setting

$$L^2 = \theta_h L_c \theta_{h^{-1}}, \quad \text{and} \quad R^2 = \theta_h R_c \theta_{h^{-1}}.$$

Given $b \in D_h$, notice that

$$L^1(R^2(b)) = a\theta_h(\theta_{h^{-1}}(b)c), \quad \text{and} \quad R^2(L^1(b)) = \theta_h(\theta_{h^{-1}}(ab)c), \tag{8.6}$$

so that (8.5.1) is precisely expressing the commutativity of L^1 and R^2, as discussed in (7.8).

We thus arrive at the main result of this chapter:

Theorem 8.7. *Given an algebraic partial action* $\theta = (\{D_g\}_{g\in G}, \{\theta_g\}_{g\in G})$ *of the group* G *on the algebra* A, *a sufficient condition for* $A \rtimes G$ *to be an associative algebra is that each* D_g *be either non-degenerate or idempotent.*

Proof. Follows immediately from (8.6), (8.5), and (7.9). □

We refer the reader to [**33**, Proposition 3.6] for an example of a non-associative partial crossed product.

In virtually all of the examples of interest to us, sufficient conditions for associativity will be present, so we will not have to deal with non-associative algebras. However, the next few general results in this chapter do not need associativity, so we will temporarily be working with possibly non-associative algebras.

The following elementary fact should be noticed:

Proposition 8.8. *Given an algebraic partial action* θ *of the group* G *on the algebra* A, *the correspondence*

$$a \in A \mapsto a\delta_1 \in A \rtimes G$$

is an injective homomorphism.

We will therefore often identify A with its image in $A \rtimes G$ under the above map.

Let us now briefly study crossed products by *-algebraic partial actions.

Proposition 8.9. *Given a* **-algebraic partial action*

$$\theta = (\{D_g\}_{g\in G}, \{\theta_g\}_{g\in G})$$

of the group G *on the* **-algebra* A, *let*

$$* : A \rtimes G \to A \rtimes G,$$

be the unique conjugate-linear map such that

$$(a\delta_g)^* = \theta_{g^{-1}}(a^*)\delta_{g^{-1}}, \quad \forall\, g \in G, \quad \forall\, a \in D_g.$$

Then this operation satisfies (6.3.i)–(6.3.iii), *so* $A \rtimes G$ *is a (possibly non-associative)* **-algebra.*

Proof. We leave the easy proofs of (6.3.i)–(6.3.iii) to the reader and concentrate on (6.3.iii). We must then prove that

$$(a\delta_g\, b\delta_h)^* = (b\delta_h)^*(a\delta_g)^*, \quad \forall\, g, h \in G, \quad \forall\, a \in D_g, \quad \forall\, b \in D_h. \tag{8.9.1}$$

Observe that the left-hand-side above equals

$$\begin{aligned} (a\delta_g\, b\delta_h)^* &= \Big(\theta_g(\theta_{g^{-1}}(a)b)\delta_{gh}\Big)^* \\ &= \theta_{h^{-1}g^{-1}}\Big(\theta_g(b^*\theta_{g^{-1}}(a^*))\Big)\delta_{h^{-1}g^{-1}} \\ &= (\star). \end{aligned}$$

Notice that the term between the outermost pair of parenthesis above satisfies

$$\theta_g\big(b^*\theta_{g^{-1}}(a^*)\big) \in \theta_g(D_h \cap D_{g^{-1}}) \in D_{gh} \cap D_g,$$

where $\theta_{h^{-1}g^{-1}}$ coincides with $\theta_{h^{-1}}\theta_{g^{-1}}$, by (2.5.ii). So

$$(\star) = \theta_{h^{-1}}\big(b^*\theta_{g^{-1}}(a^*)\big)\delta_{h^{-1}g^{-1}}.$$

Speaking of the right-hand-side of (8.9.1), we have

$$\begin{aligned}(b\delta_h)^*(a\delta_g)^* &= \big(\theta_{h^{-1}}(b^*)\delta_{h^{-1}}\big)\ \big(\theta_{g^{-1}}(a^*)\delta_{g^{-1}}\big)\\ &= \theta_{h^{-1}}\big(b^*\theta_{g^{-1}}(a^*)\big)\delta_{h^{-1}g^{-1}},\end{aligned}$$

which coincides with $(\star)$ hence proving (8.9.1). This concludes the proof. □

Under the conditions of the above Proposition, it is clear that the canonical embedding of A into $A \rtimes G$ described in (8.8) is a *-homomorphism.

One of the main applications of partial actions is to the theory of graded algebras. Let us therefore formally introduce this important concept.

Definition 8.10. Let B be a (*-)algebra and let G be a group. By a *G-grading* of B we shall mean an independent collection $\{B_g\}_{g\in G}$ of $\mathbb{K}$-sub-modules of B, such that $B = \bigoplus_{g\in G} B_g$, and

$$B_g B_h \subseteq B_{gh}, \quad \forall\, g, h \in G.$$

In the *-algebra case we also require that

$$B_g^* \subseteq B_{g^{-1}}, \quad \forall\, g \in G.$$

Given such a G-grading, we say that B is a *G-graded algebra*, and each B_g is called a *grading space*, or a *homogeneous space*.

The next result states that a partial crossed product algebra has a canonical grading. Its proof is left as an easy exercise to the reader.

Proposition 8.11. *Given a (*-)algebraic partial action*

$$\theta = \big(\{D_g\}_{g\in G}, \{\theta_g\}_{g\in G}\big)$$

of G on the (*-)algebra A, let B_g be the subset of $A \rtimes G$ given by

$$B_g = D_g\delta_g.$$

Then $\{B_g\}_{g\in G}$ is a G-grading of $A \rtimes G$.

As already claimed, partial actions will often be used to describe graded algebras. Given a graded algebra $B = \bigoplus_{g\in G} B_g$, the plan is to look for sufficient conditions to ensure the existence of a partial action of G on[6] B_1 such that B is isomorphic to $B_1 \rtimes G$.

As an example let $M_n(\mathbb{K})$ denote the algebra of all $n \times n$ matrices with coefficients in $\mathbb{K}$, and consider the $\mathbb{Z}$-grading of $M_n(\mathbb{K})$ given by

$$B_k = \operatorname{span}\{e_{i,j} : i - j = k\}, \quad \forall\, k \in \mathbb{Z},$$

[6]Recall that, in the grading of $A \rtimes G$ provided by (8.11), the grading subspace corresponding to the unit group element is naturally isomorphic to A.

where the $e_{i,j}$ are the standard matrix units.

$$\begin{bmatrix} 0 & 0 & \star & 0 & 0 & 0 \\ 0 & 0 & 0 & \star & 0 & 0 \\ 0 & 0 & 0 & 0 & \star & 0 \\ 0 & 0 & 0 & 0 & 0 & \star \\ 0 & 0 & 0 & 0 & 0 & 0 \\ 0 & 0 & 0 & 0 & 0 & 0 \end{bmatrix}$$

An illustration of B_{-2} in $M_6(\mathbb{K})$.

One may immediately rule out the possibility that this grading comes from a global action since $B_k = \{0\}$, for all $|k| \geq n$. However we will soon see that it may be effectively described as the crossed product algebra for a partial action of $\mathbb{Z}$ on $\mathbb{K}^n$.

Let us now discuss a few elementary functorial properties of crossed products.

Proposition 8.12. *Let G be a group and suppose we are given $({}^*\text{-})$algebraic partial dynamical systems*

$$\theta^i = \big(A^i,\ G,\ \{D_g^i\}_{g\in G},\ \{\theta_g^i\}_{g\in G}\big),$$

for $i = 1, 2$. Suppose, in addition, that $\varphi : A^1 \to A^2$ is a G-equivariant $({}^*\text{-})$ homomorphism. Then there is a graded[7] *$({}^*\text{-})$ homomorphism*

$$\psi : A^1 \rtimes G \to A^2 \rtimes G,$$

such that

$$\psi(a\delta_g) = \varphi(a)\delta_g, \quad \forall\, a \in D_g^1$$

(we believe the context here is enough to determine the appropriate interpretations of the expressions "$a\delta_g$" and "$\varphi(a)\delta_g$", above, so as to dispense with heavier notation such as "$a\delta_g^1$" and "$\varphi(a)\delta_g^2$").

PROOF. Left to the reader. □

Injectivity and surjectivity of the induced map is discussed in the next Proposition. The proof is routine so we again leave it for the reader.

Proposition 8.13. *Under the conditions of* (8.12) *one has:*

(*i*) *If φ is injective, then so is ψ.*
(*ii*) *If $\varphi(D_g^1) = D_g^2$, for every g in G, then ψ is surjective.*

There are some recurring calculations with elements of $A \rtimes G$ which are useful to know in advance. So, before concluding this chapter, let us collect some of these for later reference.

Proposition 8.14. *In what follows, every time we write $a\delta_g$, it is assumed that g is an arbitrary element of G and a is an arbitrary element of D_g, satisfying explicitly stated conditions, if any. If an expression involves a "$*$", we assume we are speaking of a *-algebraic partial dynamical system.*

Wherever 1_g is mentioned, we tacitly assume that D_g is unital with unit 1_g. In this case recall that Θ_g was defined in (6.9).

(*a*) $(a\delta_1)(b\delta_h) = ab\delta_h,$
(*b*) $(a\delta_g)(b\delta_h) = a\theta_g(b)\delta_{gh},$ *provided* $b \in D_{g^{-1}} \cap D_h$,

[7] A map ψ between two graded algebras $B = \bigoplus_{g\in G} B_g$ and $C = \bigoplus_{g\in G} C_g$ is said to be *graded* if $\psi(B_g) \subseteq C_g$, for every g in G.

$$
\begin{array}{llcl}
(c) & \big(\theta_g(a)\delta_g\big)(b\delta_h) & = & \theta_g(ab)\delta_{gh}, \quad \textit{provided } a \in D_{g^{-1}}, \\
(d) & (u\delta_g)(a\delta_1)(v\delta_{g^{-1}}) & = & u\theta_g(av)\delta_1, \\
(e) & (a\delta_g)^*(b\delta_h) & = & \theta_{g^{-1}}(a^*b)\delta_{g^{-1}h}, \\
(f) & (a\delta_g)(b\delta_g)^* & = & ab^*\delta_1, \\
(g) & (1_g\delta_1)(a\delta_g) & = & a\delta_g, \\
(h) & (a\delta_g)(1_{g^{-1}}\delta_1) & = & a\delta_g, \\
(i) & (1_g\delta_g)(a\delta_1)(1_{g^{-1}}\delta_{g^{-1}}) & = & \theta_g(a)\delta_1, \quad \textit{provided } a \in D_{g^{-1}}, \\
(j) & (a\delta_g)(b\delta_h) & = & a\theta_g(1_{g^{-1}}b)\delta_{gh} \;=\; a\Theta_g(b)\delta_{gh}.
\end{array}
$$

PROOF. Left for the reader. □

Notes and remarks. Theorem (8.7) was first proved in [**33**]. It will be used again to give a proof of the associativity of C*-algebraic partial crossed products which, unlike the original proofs, does not use approximate identities, relying only on the fact that C*-algebras are non-degenerate.

Similar results on the associativity of *twisted* partial crossed products can be found in [**48**, Proposition 2.4] and [**35**, Theorem 2.4].

CHAPTER 9

Partial Group Representations

Given a partial action $\theta = (\{D_g\}_{g\in G}, \{\theta_g\}_{g\in G})$ of a group G on an algebra A, we have seen that $A \rtimes G$ consists of elements of the form

$$\sum_{g\in G} a_g\delta_g,$$

with $a_g \in D_g$. As already mentioned the δ_g appearing above are just place markers and thus have no meaning by themselves. However, in case θ is a global action and A is a unital algebra, the elements

$$v_g := 1\delta_g, \quad \forall\, g \in G,$$

make perfectly good sense and are indeed crucial in the study of skew products. In that case it is easy to see that each v_g is an invertible element in $A \rtimes G$ and, in addition, we have that

$$v_{gh} = v_g v_h, \quad \forall\, g, h \in G,$$

so v may be seen as a group representation of G in $A \rtimes G$. Identifying A with its canonical copy in $A \rtimes G$, we also have that

$$v_g a v_g^{-1} = \theta_g(a), \quad \forall\, g \in G, \quad \forall\, a \in A,$$

which is to say that θ_g coincides with the inner automorphism Ad_{v_g} on A.

In the case of a partial action the above definition of v_g does not make sense, unless we assume that each D_g is a unital ideal in which case we may redefine

$$u_g := 1_g\delta_g, \quad \forall\, g \in G,$$

where 1_g is the unit of D_g. With the appropriate modifications we will see that the u_g play as important a role relative to partial action as the v_g does in the global case.

Definition 9.1. Given a group G and a unital algebra B, we shall say that a map

$$u : G \to B$$

is a *partial representation* of G in B if, for all g and h in G, one has that

(i) $u_1 = 1$,
(ii) $u_g u_h u_{h^{-1}} = u_{gh} u_{h^{-1}}$,
(iii) $u_{g^{-1}} u_g u_h = u_{g^{-1}} u_{gh}$.
If B is a *-algebra and u moreover satisfies
(iv) $u_{g^{-1}} = (u_g)^*$,
we will say that u is a **-partial representation.*

The above definition of partial group representation may be generalized to maps from G to any unital semigroup (also called monoid), with or without an involution. Specializing this to the multiplicative semigroup of an algebra B, we evidently recover the above definition.

9.2. Given a unital *-algebra B, and a map $u : G \to B$ satisfying (9.1.iv), notice that (9.1.ii) is equivalent to (9.1.iii). Thus, for such a map to be proven a *-partial representation, one may choose to check only one axiom among (9.1.ii) and (9.1.iii).

For comparison purposes, let us recall a well known concept:

Definition 9.3. Given a group G and a unital algebra B, we shall say that a map $v : G \to B$ is a *group representation* of G in B if, for all g and h in G, one has that

(i) $v_1 = 1$, and
(ii) $v_g v_h = v_{gh}$.
If B is a *-algebra and v moreover satisfies
(iii) $v_{g^{-1}} = (v_g)^*$,
we will say that v is a *unitary representation*.

Needless to say, a (unitary) group representation is a (*-)partial group representation, but we shall encounter many examples of partial representations which are not group representations.

A rule of thumb to memorize condition (9.1.ii), above, is to think of it as the usual axiom for a group representation, namely "$u_g u_h = u_{gh}$", but which is being "observed" by the term $u_{h^{-1}}$ on the right. The observer apparently does not take any part in the computation and, in case it is invertible, one could effectively cancel it out and be left with the traditional "$u_g u_h = u_{gh}$". A similar comment could of course be made with respect to the left-hand observer $u_{g^{-1}}$ in (9.1.iii).

Returning to the discussion at the beginning of the present chapter, we now present an important source of examples of partial representations.

Proposition 9.4. *Let $\theta = \big(\{D_g\}_{g\in G}, \{\theta_g\}_{g\in G}\big)$ be a (*-)partial action of a group G on a (*-)algebra A, such that each D_g is unital, with unit denoted 1_g. Then the map*

$$u : g \in G \mapsto 1_g\delta_g \in A \rtimes G,$$

is a (*-)partial representation.

Proof. Condition (9.1.i) is evident. As for (9.1.iii), let $g, h \in G$, and observe that, using (8.14), we have

$$u_g u_h = (1_g\delta_g)(1_h\delta_h) = 1_g\theta_g(1_{g^{-1}}1_h)\delta_{gh} = 1_g 1_{gh}\delta_{gh},$$

so

$$\begin{aligned} u_{g^{-1}} u_g u_h &= (1_{g^{-1}}\delta_{g^{-1}})(1_g 1_{gh}\delta_{gh}) \\ &= 1_{g^{-1}}\theta_{g^{-1}}(1_g 1_{gh})\delta_h \\ &= 1_{g^{-1}}1_h\delta_h = (\star). \end{aligned}$$

On the other hand,

$$u_{g^{-1}} u_{gh} = (1_{g^{-1}}\delta_{g^{-1}})(1_{gh}\delta_{gh}) = 1_{g^{-1}}\theta_{g^{-1}}(1_g 1_{gh})\delta_h = 1_{g^{-1}}1_h\delta_h,$$

which coincides with $(\star)$, hence proving (9.1.iii). The proof of (9.1.ii) follows along similar lines.

In the *-algebra case, observe that 1_g^* is also a unit for D_g and, since an algebra has at most one unit, we must have $1_g^* = 1_g$. Therefore

$$(u_g)^* = (1_g\delta_g)^* = \theta_{g^{-1}}(1_g^*)\delta_{g^{-1}} = \theta_{g^{-1}}(1_g)\delta_{g^{-1}} = 1_{g^{-1}}\delta_{g^{-1}} = u_{g^{-1}}.$$

proving (9.1.iv). □

Let us now discuss another source of examples of partial representations, but let us first observe that, given an idempotent element p in an algebra B, then pBp is a unital subalgebra of B, with unit p. If B is moreover a *-algebra and p is self-adjoint (that is, $p^* = p$), then pBp is clearly also a *-algebra.

Proposition 9.5. *Let B be a unital (*-)algebra, and v be a (unitary) group representation of G in B. Suppose that p is a (self-adjoint) idempotent element of B such that $v_g p v_{g^{-1}}$ commutes with p, for every g in G. Then the formula*

$$u_g = p v_g p, \quad \forall\, g \in G$$

defines a (*-)partial representation of G in the unital (*-)algebra pBp.

PROOF. With respect to (9.1.i), we clearly have that $u_1 = p$, which is the unit of pBp. Given g and h in G we have

$$\begin{aligned} u_{g^{-1}} u_g u_h &= p\, \underline{v_{g^{-1}} p v_g}\ \underline{p}\, v_h p = p\, \underline{p}\ \underline{v_{g^{-1}} p v_g}\, v_h p \\ &= p v_{g^{-1}} p v_{gh} p = u_{g^{-1}} u_{gh}, \end{aligned}$$

proving (9.1.iii), while (9.1.ii) may be proved similarly. In the *-algebra case we have

$$u_{g^{-1}} = p v_{g^{-1}} p = p v_g^* p = (p v_g p)^* = (u_g)^*,$$

proving (9.1.iv). □

After developing the necessary tools we will come back to this construction proving that the above process may sometimes be reversed, so that certain partial representations may be *dilated* to unitary group representations.

Notice that a partial action may itself be viewed as a *-partial representation of G in the inverse semigroup $\mathcal{I}(X)$, according to (4.5). In what follows we will show that, conversely, partial representations have a characterization similar to the definition of partial actions given in (2.1).

Proposition 9.6. *Let G be a group, S be a unital inverse semigroup, and $u : G \to S$ be a map. Then u is a *-partial representation*[8] *if and only if, for all $g, h \in G$, one has that*

(*i*) $u_1 = 1$,
(*ii*) $u_{g^{-1}} = (u_g)^*$,
(*iii*) $u_g u_h \leq u_{gh}$.

PROOF. Assume conditions (i–iii) hold. Given g and h in G, we interpret (iii) from the point of view of (4.2) obtaining

$$u_g u_h = u_{gh} u_{h^{-1}} u_{g^{-1}} u_g u_h. \tag{9.6.1}$$

Focusing on (9.1.ii), we right multiply the above equation by $u_{h^{-1}}$ to obtain

$$\begin{aligned} u_g u_h u_{h^{-1}} &= u_{gh} u_{h^{-1}}\ u_{g^{-1}} u_g\ u_h u_{h^{-1}} \\ &= u_{gh} u_{h^{-1}}\ u_h u_{h^{-1}}\ u_{g^{-1}} u_g \\ &= u_{gh} u_{h^{-1}}\ u_{g^{-1}} u_g = \ldots \end{aligned} \tag{9.6.2}$$

[8]Here we are considering a generalized notion of partial representations, taking values in a semigroup with involution, rather than a *-algebra, as already commented after (9.1). In any case, when we say that u is a *-partial representation, all we mean is that (9.1.i)–(9.1.iv) are satisfied.

which is almost what we want, except for the extraneous term $u_{g^{-1}}u_g$ on the right-hand-side. Letting e denote the initial projection of $u_{gh}u_{h^{-1}}$, namely

$$e = u_h u_{h^{-1}g^{-1}} u_{gh} u_{h^{-1}},$$

we claim that

$$e \leq u_{g^{-1}}u_g. \tag{9.6.3}$$

Should the claim be verified, we could continue from (9.6.2) as follows:

$$\ldots = u_{gh}u_{h^{-1}}\; e\; u_{g^{-1}}u_g = u_{gh}u_{h^{-1}}\; e = u_{gh}u_{h^{-1}},$$

thus proving the desired axiom (9.1.ii). Unfortunately the proof we found for (9.6.3) is a bit cumbersome. It starts by applying (9.6.1) with gh and h^{-1} playing the roles of g and h, respectively, producing

$$u_{gh}u_{h^{-1}} = u_g u_h u_{h^{-1}g^{-1}} u_{gh} u_{h^{-1}}. \tag{9.6.4}$$

With an eye in (9.6.3) we then compute

$$eu_{g^{-1}}u_g = u_h u_{h^{-1}g^{-1}} u_{gh} u_{h^{-1}} u_{g^{-1}} u_g = \ldots$$

Replacing $u_{h^{-1}}$ by $u_{h^{-1}}u_h u_{h^{-1}}$, and using the fact that idempotents in S commute, the above equals

$$\ldots = u_h u_{h^{-1}g^{-1}} u_{gh} u_{h^{-1}} u_{g^{-1}} u_g u_h u_{h^{-1}} = \ldots$$

Repeating this procedure, but now replacing u_{gh} by $u_{gh}u_{h^{-1}g^{-1}}u_{gh}$, we get

$$\begin{aligned}
\cdots &= u_h u_{h^{-1}g^{-1}}\, \underbrace{u_{gh} u_{h^{-1}} u_{g^{-1}} u_g u_h}_{(9.6.1)}\, u_{h^{-1}g^{-1}} u_{gh} u_{h^{-1}} \\
&= u_h u_{h^{-1}g^{-1}}\, \underbrace{u_g u_h \qquad u_{h^{-1}g^{-1}} u_{gh} u_{h^{-1}}}_{(9.6.4)} \\
&= u_h u_{h^{-1}g^{-1}} u_{gh} u_{h^{-1}} \\
&= e.
\end{aligned}$$

This proves (9.6.3) and, as already mentioned, (9.1.ii) follows. Point (9.1.iii) is now an easy consequence of (9.1.ii) and (ii).

In order to prove the converse, given that u is a *-partial representation, we have, for all g and h in G, that

$$\begin{aligned}
u_{gh}u_{h^{-1}}u_{g^{-1}}u_g u_h &\overset{(9.1.ii)}{=} u_g\; u_h u_{h^{-1}}\; u_{g^{-1}}u_g\; u_h \\
&= u_g\; u_{g^{-1}}u_g\; u_h u_{h^{-1}}\; u_h \\
&= u_g u_h,
\end{aligned}$$

proving (iii). Points (i) and (ii) are evident from the hypothesis. □

Suggested by (4.9) we have the following important concept.

Definition 9.7. Let G be a group equipped with a length function ℓ. A partial representation u of G in a unital algebra B is said to be *semi-saturated* (with respect to the given length function ℓ) if

$$\ell(gh) = \ell(g) + \ell(h) \quad \Longrightarrow \quad u_g u_h = u_{gh}, \quad \forall\, g, h \in G.$$

Among the many interesting algebraic properties of partial representations we note the following:

Proposition 9.8. *Let u be a given $(^*\text{-})$partial representation of a group G in a unital $(^*\text{-})$algebra B. Denoting by*

$$e_g := u_g u_{g^{-1}}, \tag{9.8.1}$$

one has for every $g, h \in G$, that

(i) $u_g u_{g^{-1}} u_g = u_g$,
(ii) e_g *is a (self-adjoint) idempotent,*
(iii) $u_g e_h = e_{gh} u_g$,
(iv) $e_g e_h = e_h e_g$,
(v) $u_g u_h = e_g u_{gh}$.

PROOF. The first point follows easily from (9.1.i) and (9.1.ii). Point (ii) then follows immediately from (9.8.i) and the observation, in the *-algebra case, that

$$(e_g)^* = (u_g u_{g^{-1}})^* = (u_{g^{-1}})^*(u_g)^* = u_g u_{g^{-1}} = e_g.$$

As for (iii) we have

$$\begin{aligned} u_g e_h &= u_g u_h u_{h^{-1}} \\ &\overset{(9.1.\text{ii})}{=} u_{gh} u_{h^{-1}} \\ &\overset{(\text{i})}{=} u_{gh} u_{(gh)^{-1}} u_{gh} u_{h^{-1}} \\ &\overset{(9.1.\text{iii})}{=} u_{gh} u_{(gh)^{-1}} u_g \\ &= e_{gh} u_g. \end{aligned}$$

To prove (9.8.iv) we compute

$$e_g e_h = u_g u_{g^{-1}} e_h \overset{(\text{iii})}{=} u_g e_{g^{-1}h} u_{g^{-1}} \overset{(\text{iii})}{=} e_{gg^{-1}h} u_g u_{g^{-1}} = e_h e_g.$$

We finally have

$$u_g u_h \overset{(\text{i})}{=} u_g u_{g^{-1}} u_g u_h \overset{(9.1.\text{iii})}{=} e_g u_{gh}.$$

□

Even though a certain disregard for the group law is the defining feature of partial representations, sometimes the group law is duly respected:

Proposition 9.9. *Let u be a partial representation of the group G in a unital $\mathbb{K}$-algebra B. Also let h be a fixed element of G. Then the following are equivalent:*

(i) u_h *is left-invertible,*
(ii) $u_{h^{-1}} u_h = 1$,
(iii) $u_g u_h = u_{gh}$, *for all g in G,*
(iv) $u_{h^{-1}}$ *is right-invertible,*
(v) $u_{h^{-1}} u_g = u_{h^{-1}g}$, *for all g in G.*

PROOF. (i) ⇒ (ii). Let v be a left-inverse for u_h, meaning that $vu_h = 1$. Then

$$1 = vu_h \overset{(9.8.\text{i})}{=} vu_h u_{h^{-1}} u_h = u_{h^{-1}} u_h.$$

(ii) ⇒ (iii). For every g in G, we have that

$$u_g u_h = u_g u_h u_{h^{-1}} u_h \overset{(9.1.\text{ii})}{=} u_{gh} u_{h^{-1}} u_h = u_{gh}.$$

(iii) $\Rightarrow$ (i). Plugging $g = h^{-1}$ in (iii), we have

$$u_g u_h = u_{gh} = u_1 = 1,$$

so u_g is a left-inverse for u_h.

(ii) $\Rightarrow$ (v). For every g in G, we have that

$$u_{h^{-1}} u_g = u_{h^{-1}} u_h u_{h^{-1}} u_g \overset{(9.1.\text{iii})}{=} u_{h^{-1}} u_h u_{h^{-1}g} = u_{h^{-1}g}.$$

(v) $\Rightarrow$ (iv). Taking $g = h$ in (v), we have

$$u_{h^{-1}} u_g = u_{h^{-1}g} = u_1 = 1,$$

so u_g is a right-inverse for $u_{h^{-1}}$.

(iv) $\Rightarrow$ (ii). Let v be a right-inverse for $u_{h^{-1}}$, meaning that $u_{h^{-1}} v = 1$. Then

$$1 = u_{h^{-1}} v = u_{h^{-1}} u_h u_{h^{-1}} v = u_{h^{-1}} u_h.$$

□

One of the most important roles played by partial representations is as part of covariant representations of algebraic partial dynamical systems, a notion to be introduced next.

Definition 9.10. By a *covariant representation* of a (*-)algebraic partial action

$$\theta = \big(\{D_g\}_{g\in G},\ \{\theta_g\}_{g\in G}\big)$$

in a unital (*-)algebra B, we shall mean a pair (π, u), where $\pi : A \to B$ is a (*-)homomorphism, u is a (*-)partial representation of G in B, and,

$$u_g \pi(a) u_{g^{-1}} = \pi\big(\theta_g(a)\big), \quad \forall\, g \in G, \quad \forall\, a \in D_{g^{-1}}.$$

Under the assumptions of (9.4) we have seen that u is a partial representation of G in $A \rtimes G$. If, in addition, we denote by π the canonical inclusion of A in $A \rtimes G$, one may easily prove that the pair (π, u) is a covariant representation of θ in $A \rtimes G$. In doing so, the conclusions of (8.14) will make this task a lot easier.

Let us now present a few easy consequences of the above definition.

Proposition 9.11. *Let*

$$\theta = \big(\{D_g\}_{g\in G},\ \{\theta_g\}_{g\in G}\big)$$

be a (-)algebraic partial action of a group G on a (*-)algebra A. Also let (π, u) be a covariant representation of θ in a unital (*-)algebra B, and denote by $e_g = u_g u_{g^{-1}}$, as usual. Then, for every g in G, one has that:*

(i) $\pi(a) = e_g \pi(a) = \pi(a) e_g$, *for every* $a \in D_g$,
(ii) $u_g \pi(a) = \pi\big(\theta_g(a)\big) u_g$, *for every* $a \in D_{g^{-1}}$,
(iii) *the linear mapping* $\pi \times u : A \rtimes G \to B$ *determined by*

$$(\pi \times u)(a\delta_g) = \pi(a) u_g, \quad \forall\, g \in G, \quad \forall\, a \in D_g,$$

is a (-)homomorphism.*

PROOF. Given a in D_g, we have

$$\begin{aligned} e_g\pi(a)e_g &= u_g u_{g^{-1}}\pi(a)u_g u_{g^{-1}} \\ &= u_g\pi\big(\theta_{g^{-1}}(a)\big)u_{g^{-1}} \\ &= \pi\big(\theta_g(\theta_{g^{-1}}(a))\big) \\ &= \pi(a), \end{aligned}$$

from where one easily deduces (i). Now, if $a \in D_{g^{-1}}$, then

$$u_g\pi(a) \overset{(i)}{=} u_g\pi(a)e_{g^{-1}} = u_g\pi(a)u_{g^{-1}}u_g = \pi\big(\theta_g(a)\big)u_g,$$

proving (ii). As for (iii), we have for all $a \in D_g$ and $b \in D_h$, that

$$\begin{aligned} (\pi \times u)(a\delta_g)\ (\pi \times u)(b\delta_h) &= \pi(a)u_g\pi(b)u_h \\ &\overset{(ii)}{=} u_g\pi\big(\theta_{g^{-1}}(a)\big)\pi(b)u_h \\ &= u_g\pi\big(\theta_{g^{-1}}(a)b\big)u_h \\ &= \pi\Big(\theta_g\big(\theta_{g^{-1}}(a)b\big)\Big)u_g u_h \\ &\overset{(9.8.v)}{=} \pi\Big(\theta_g\big(\theta_{g^{-1}}(a)b\big)\Big)e_g u_{gh} \\ &\overset{(i)}{=} (\pi\times u)\Big(\theta_g\big(\theta_{g^{-1}}(a)b\big)\delta_{gh}\Big) \\ &= (\pi \times u)(a\delta_g\ b\delta_h), \end{aligned}$$

so we see that $\pi \times u$ is multiplicative. In the *-algebraic case, we have

$$\begin{aligned} \big((\pi \times u)(a\delta_g)\big)^* &= (\pi(a)u_g)^* \\ &= u_g^*\pi(a)^* \\ &\overset{(i)}{=} u_{g^{-1}}\pi(a^*)u_g u_{g^{-1}} \\ &= \pi\big(\theta_{g^{-1}}(a^*)\big)u_{g^{-1}} \\ &= (\pi \times u)\big(\theta_{g^{-1}}(a^*)\delta_{g^{-1}}\big) \\ &= (\pi \times u)\big((a\delta_g)^*\big), \end{aligned}$$

proving that $\pi \times u$ is a *-homomorphism. □

Therefore we see that covariant representations lead to *representations of* (simply meaning homomorphisms defined on) the partial crossed product algebra. One could then ask to what extent does every homomorphism defined on $A \rtimes G$ arise from a covariant representation. We will later see that this is always the case in the category of C*-algebras but for now we simply note that, if every D_g is unital, then a homomorphism

$$\phi : A \rtimes G \to B$$

into a unital algebra B gives rise to a covariant representation (π, u) in a trivial way: set $\pi(a) = \phi(a\delta_1)$, and define $u_g = \phi(1_g\delta_g)$. One would then have for all a in D_g, that

$$(\pi \times u)(a\delta_g) = \pi(a)u_g = \phi(a\delta_1)\phi(1_g\delta_g) = \phi(a\delta_1\ 1_g\delta_g) \overset{(8.14)}{=} \phi(a\delta_g),$$

hence proving that $\phi = (\pi \times u)$.

In case one is attempting to provide a concrete model for a given partial crossed product algebra, perhaps trying to prove it to be isomorphic to some well known algebra, it is useful to know when are homomorphisms defined on the crossed product injective. The following result may therefore come in handy.

Proposition 9.12. *Let B be a G-graded $(^*\text{-})$algebra and let (π, u) be a covariant representation of a $(^*\text{-})$algebraic partial dynamical system*

$$\big(A,\ G,\ \{D_g\}_{g\in G},\ \{\theta_g\}_{g\in G}\big),$$

in B, such that $\pi(A) \subseteq B_1$, and $u_g \in B_g$, for all $g \in G$. Then $\pi \times u$ is a graded homomorphism (meaning that it takes the G-grading subspace $D_g\delta_g$ into B_g), and moreover the following are equivalent:

(i) π is one-to-one,

(ii) $\pi \times u$ is one-to-one.

Proof. For all $a \in D_g$, notice that

$$(\pi \times u)(a\delta_g) = \pi(a_g)u_g \in B_1 B_g \subseteq B_g, \tag{9.12.1}$$

so $\pi \times u$ is a graded homomorphism, as desired.

Since the restriction of $\pi \times u$ to the canonical copy of A in $A \rtimes G$ coincides with π, it is obvious that (ii) implies (i).

Conversely, assuming that (i) holds, let $a = \sum_{g\in G} a_g\delta_g \in A \rtimes G$, be in the kernel of $\pi \times u$. We then have

$$0 = (\pi \times u)(a) = \sum_{g\in G} \pi(a_g)u_g. \tag{9.12.2}$$

Since each $\pi(a_g)u_g \in B_g$, by (9.12.1), and since the B_g are independent, we deduce that $\pi(a_g)u_g = 0$, for all g in G. Recalling that the $\pi(a_g) \in D_g$, we then have

$$\pi(a_g) \overset{(9.11.i)}{=} \pi(a_g)e_g = \pi(a_g)u_g u_{g^{-1}} = 0.$$

The assumed injectivity of π then implies that all $a_g = 0$, so $a = 0$, proving that $\pi \times u$ is injective. □

Notes and remarks. A first rudimentary notion of partial representations was introduced by McClanahan in [**81**], as part of the ingredients of covariant representations for partial dynamical systems. A refinement of this idea was subsequently presented by Quigg and Raeburn in [**94**], where the expression *partial representation* was coined. The current set of axioms, as described in (9.1), was introduced in [**50**], under the presence of an involution, and in [**34**], otherwise. The characterization of partial representations given in (9.6) is closely related to the original definition given by Quigg and Raeburn.

CHAPTER 10

Partial Group Algebras

Given a group G, one of the main reasons why one studies the classical group algebra $\mathbb{K}(G)$ is that its representation theory is equivalent to that of G. Having expanded the notion of group representations to include partial ones, we will now introduce the partial group algebra of G whose role relative to partial representations of G will be shown to parallel that of $\mathbb{K}(G)$.

▶ To begin, let us temporarily fix a partial representation u of a given group G in a unital algebra B. Recall from (9.8.iv) that the e_g (defined to be $u_g u_{g^{-1}}$) form a commutative set of idempotents, so the subalgebra A of B generated by all of the e_g's is a commutative algebra. Denote by D_g the ideal of A generated by each e_g, that is

$$D_g = Ae_g.$$

Since e_g is idempotent, we see that D_g is a unital ideal, with e_g playing the role of the unit. For each g in G, let us also consider the map

$$\theta_g : D_{g^{-1}} \to D_g,$$

defined by $\theta_g(a) = u_g a u_{g^{-1}}$, for all a in $D_{g^{-1}}$.

Proposition 10.1. *Under the above conditions one has that*

$$\big(A,\ G,\ \{D_g\}_{g\in G},\ \{\theta_g\}_{g\in G}\big)$$

is an algebraic partial dynamical system.

Proof. For $a, b \in D_{g^{-1}}$, notice that

$$\theta_g(a)\theta_g(b) = u_g a u_{g^{-1}} u_g b u_{g^{-1}} = u_g a e_{g^{-1}} b u_{g^{-1}} = u_g a b u_{g^{-1}} = \theta_g(ab),$$

so θ is a homomorphism. We next need to check condition (2.1.i), which is evident, as well as condition (2.1.ii). So let us be given g and h in G, and let a be in the domain of $\theta_g \circ \theta_h$, which we have seen in (2.2) to be $\theta_h^{-1}(D_h \cap D_{g^{-1}})$. This means that $a = ae_{h^{-1}}$, and also that

$$\begin{aligned} u_h a u_{h^{-1}} &= \theta_h(a) \\ &= \theta_h(a) e_h e_{g^{-1}} \\ &= u_h a u_{h^{-1}} e_h e_{g^{-1}} \\ &= u_h a u_{h^{-1}} e_{g^{-1}} \\ &\overset{(9.8.iii)}{=} u_h a e_{h^{-1}g^{-1}} u_{h^{-1}} \\ &= u_h a e_{(gh)^{-1}} u_{h^{-1}}. \end{aligned}$$

It follows that

$$\begin{aligned} a &= e_{h^{-1}} a e_{h^{-1}} \\ &= u_{h^{-1}} u_h a u_{h^{-1}} u_h \\ &= u_{h^{-1}} u_h a e_{(gh)^{-1}} u_{h^{-1}} u_h \\ &= a e_{(gh)^{-1}} e_{h^{-1}} \in D_{(gh)^{-1}} \cap D_{h^{-1}}, \end{aligned}$$

so we see that a lies in the domain of θ_{gh}. Moreover,

$$\theta_g\big(\theta_h(a)\big) = u_g u_h a u_{h^{-1}} u_{g^{-1}} = u_g u_h e_{h^{-1}} a e_{h^{-1}} u_{h^{-1}} u_{g^{-1}} = (\star),$$

but

$$u_g u_h e_{h^{-1}} = u_g u_h u_{h^{-1}} u_h \overset{(9.1.\text{ii})}{=} u_{gh} u_{h^{-1}} u_h = u_{gh} e_{h^{-1}},$$

and similarly $e_{h^{-1}} u_{h^{-1}} u_{g^{-1}} = e_{h^{-1}} u_{(gh)^{-1}}$. Therefore

$$(\star) = u_{gh} e_{h^{-1}} a e_{h^{-1}} u_{(gh)^{-1}} = u_{gh} a u_{(gh)^{-1}} = \theta_{gh}(a),$$

proving that $\theta_g \circ \theta_h \subseteq \theta_{gh}$, as desired. □

Denoting by ι the inclusion of A in B, it is then evident that (ι, u) is a covariant representation of the above partial dynamical system in B, so we may use (9.11.iii) to conclude that

$$\iota \times u : A \rtimes G \to B \tag{10.2}$$

is a homomorphism. The range of $\iota \times u$ is clearly the subalgebra B_0 of B generated by the range of u, so B_0 is a quotient of the partial crossed product. In general we cannot assert that B_0 is isomorphic to the partial crossed product, but under certain conditions this may be guaranteed.

Theorem 10.3. *Let G be a group and let $B = \bigoplus_{g\in G} B_g$ be a unital G-graded algebra which is generated by the range of a partial representation u of G, such that $u_g \in B_g$, for all g in G. Then B_1 is commutative and there exists a partial action of G on B_1 such that*

$$B \simeq B_1 \rtimes G,$$

as graded algebras.

Proof. Let A be the subalgebra of B generated by the e_g. For every g in G one has that

$$e_g = u_g u_{g^{-1}} \in B_g B_{g^{-1}} \subseteq B_1,$$

so $A \subseteq B_1$. Considering the algebraic partial dynamical system arising from u as in (10.1), we have that the homomorphism $\iota \times u$ described in (10.2) is one-to-one by (9.12). As already seen, the range of $\iota \times u$ is the algebra generated by the range of u, which is assumed to be B, whence $\iota \times u$ is also onto, thus proving that

$$A \rtimes G \simeq B.$$

Being a surjective graded homomorphism it is then clear that $\iota \times u$ satisfies

$$(\iota \times u)(D_g \delta_g) = B_g, \quad \forall\, g \in G,$$

and, in particular $B_1 = (\iota \times u)(A\delta_1)$. So B_1 is a commutative algebra isomorphic to A. We may then transfer the partial action on A over to B_1 via this isomorphism, so that

$$B_1 \rtimes G \simeq A \rtimes G \simeq B.$$

□

As an example let us return to the grading $\{B_n\}_{n\in\mathbb{Z}}$ of $M_n(\mathbb{K})$ discussed at the end of chapter (8). Consider the element

$$v = \begin{bmatrix} 0 & 0 & 0 & \cdots & 0 & 0 \\ 1 & 0 & 0 & \cdots & 0 & 0 \\ 0 & 1 & 0 & \cdots & 0 & 0 \\ \vdots & \vdots & \ddots & & \vdots & \vdots \\ 0 & 0 & 0 & \cdots & 1 & 0 \end{bmatrix} \in B_1 \subseteq M_n(\mathbb{K}),$$

and let $u : \mathbb{Z} \to M_n(\mathbb{K})$ be given by

$$u_k = \begin{cases} v^k, & \text{if } k \geq 0, \\ (v^*)^{|k|}, & \text{if } k < 0, \end{cases}$$

where v^* refers to the transpose of v. It is relatively easy to check that u is a partial representation of $\mathbb{Z}$ in $M_n(\mathbb{K})$ satisfying the conditions of (10.3) and, since B_0 is isomorphic to $\mathbb{K}^n$, we conclude that

$$M_n(\mathbb{K}) \simeq \mathbb{K}^n \rtimes \mathbb{Z}.$$

The above partial action of $\mathbb{Z}$ on $\mathbb{K}^n$ may be shown to be precisely the semi-saturated partial action of $\mathbb{Z}$ given by (4.10) in terms of the partial automorphism f of $\mathbb{K}^n$ defined by

$$f(\lambda_1, \lambda_2, \ldots, \lambda_{n-1}, 0) = f(0, \lambda_1, \lambda_2, \ldots, \lambda_{n-1}),$$

whose domain is, of course, the set of vectors in $\mathbb{K}^n$ with zero in the last coordinate.

Definition 10.4. Let G be a group. The *partial group algebra* of G, denoted $\mathbb{K}_{\text{par}}(G)$, is the universal unital $\mathbb{K}$-algebra generated by a set of symbols $\{[g] : g \in G\}$, subject to the relations

$$[1] = 1, \qquad [g][h][h^{-1}] = [gh][h^{-1}], \quad \text{and} \quad [g^{-1}][g][h] = [g^{-1}][gh].$$

for all g and h in G.

The universal property of $\mathbb{K}_{\text{par}}(G)$ may be phrased as follows:

Proposition 10.5. *The correspondence*

$$g \in G \mapsto [g] \in \mathbb{K}_{par}(G)$$

is a partial representation, which we will call the universal partial representation. In addition, for any partial representation u of G in a unital $\mathbb{K}$-algebra B, there exists a unique homomorphism

$$\phi : \mathbb{K}_{par}(G) \to B,$$

such that $u_g = \phi([g])$, for all $g \in G$.

In case $\mathbb{K}$ is equipped with a conjugation, as in (6.1), we have the following:

Proposition 10.6. *For every group G, one has that $\mathbb{K}_{par}(G)$ is a *-algebra under a unique involution satisfying*

$$[g]^* = [g^{-1}], \quad \forall g \in G,$$

whence the universal partial representation is a *-representation.

PROOF. Let $B = \mathbb{K}_{\text{par}}(G)^*$, meaning the conjugate-opposite algebra, in which the multiplication operation is replaced by

$$x \star y = yx, \quad \forall x, y \in \mathbb{K}_{\text{par}}(G),$$

and the scalar multiplication is replaced by

$$\lambda \cdot x = \bar{\lambda} x, \quad \forall \lambda \in \mathbb{K}, \quad \forall x \in \mathbb{K}_{\text{par}}(G).$$

One may then easily prove that the mapping

$$u : g \in G \mapsto [g^{-1}] \in B$$

is a partial representation, so the universal property (10.5) implies the existence of a homomorphism $\sigma : \mathbb{K}_{\text{par}}(G) \to B$ such that $\sigma([g]) = [g^{-1}]$, for all g in G. The required involution is then defined by

$$a^* := \sigma(a), \quad \forall a \in \mathbb{K}_{\text{par}}(G).$$

□

It is our next immediate goal to analyze the partial dynamical system arising from the universal partial representation of G, as in (10.1). Our method will be to first construct an abstract partial dynamical system and later prove it to be the one we are looking for. In particular we will be able to describe $\mathbb{K}_{\text{par}}(G)$ as a partial crossed product algebra.

The tip we will follow is that the idempotents $e_g = [g][g^{-1}]$ in $\mathbb{K}_{\text{par}}(G)$ are not likely to satisfy any algebraic relation other than the fact that they commute and $e_1 = 1$.

Definition 10.7. Let $A_{\text{par}}(G)$ be the universal unital $\mathbb{K}$-algebra generated by a set of symbols

$$\mathcal{E} := \{\varepsilon_g : g \in G\},$$

subject to the relations stating that the ε_g are commuting idempotents, and that $\varepsilon_1 = 1$.

Notice that the only ingredient of the group structure of G which is relevant for the above definition is the singling out of the unit element. This definition would therefore make sense for any set with a distinguished element.

Observing that $A_{\text{par}}(G)$ is a commutative algebra, consider, for each $g \in G$, the ideal of $A_{\text{par}}(G)$ generated by ε_g, namely

$$D_g = A_{\text{par}}(G)\varepsilon_g.$$

Noticing that

$$\mathcal{E}_g := \{\varepsilon_{gh}\varepsilon_g : h \in G\}$$

is a set of commuting idempotents in D_g, and that for $h = 1$, the corresponding idempotent there is the unit of D_g, we may invoke the universal property of $A_{\text{par}}(G)$ to conclude that there exists a homomorphism $\Theta_g : A_{\text{par}}(G) \to D_g$, such that

$$\Theta_g(\varepsilon_h) = \varepsilon_{gh}\varepsilon_g, \quad \forall h \in G.$$

Proposition 10.8. *For each g in G, denote by θ_g the restriction of Θ_g to $D_{g^{-1}}$. Then $\big(\{D_g\}_{g\in G}, \{\theta_g\}_{g\in G}\big)$ is a partial action of G on $A_{par}(G)$.*

PROOF. Initially observe that, for all $h \in G$, one has

$$\theta_g(\varepsilon_h \varepsilon_{g^{-1}}) = \Theta_g(\varepsilon_h)\Theta_g(\varepsilon_{g^{-1}}) = \varepsilon_{gh}\varepsilon_g \ \varepsilon_1\varepsilon_g = \varepsilon_{gh}\varepsilon_g,$$

which the reader might want to compare with (6.8).

To see that each θ_g is an isomorphism, let $h \in G$, and notice that, using the calculation above, we have

$$\theta_{g^{-1}}\big(\theta_g(\varepsilon_h\varepsilon_{g^{-1}})\big) = \theta_{g^{-1}}\big(\varepsilon_{gh}\varepsilon_g\big) = \varepsilon_h\varepsilon_{g^{-1}}.$$

Since $D_{g^{-1}}$ is generated, as an algebra, by the set $\{\varepsilon_h\varepsilon_{g^{-1}} : h \in G\}$, we conclude that $\theta_{g^{-1}}\theta_g$ is the identity on $D_{g^{-1}}$, whence $\theta_{g^{-1}}$ is the inverse of θ_g, showing that θ_g is invertible.

It is evident that θ_1 is the identity of $A_{\text{par}}(G)$, so the statement will be proved once we check (2.1.ii). For this, let $g, h \in G$, and recall from (2.2) that the domain of $\theta_g\theta_h$ is given by

$$\begin{aligned}\theta_h^{-1}(D_h \cap D_{g^{-1}}) &= \theta_{h^{-1}}(A_{\text{par}}(G)\varepsilon_h\varepsilon_{g^{-1}})\\ &= A_{\text{par}}(G)\varepsilon_{h^{-1}}\varepsilon_{h^{-1}g^{-1}}\\ &= D_{h^{-1}} \cap D_{h^{-1}g^{-1}},\end{aligned}$$

which we then see is a subset of the domain of θ_{gh}, as needed. For any $k \in G$, we have

$$\begin{aligned}\theta_g\big(\theta_h(\varepsilon_k\varepsilon_{h^{-1}}\varepsilon_{h^{-1}g^{-1}})\big) &= \theta_g(\varepsilon_{hk}\varepsilon_h\varepsilon_{g^{-1}})\big)\\ &= \varepsilon_{ghk}\varepsilon_{gh}\varepsilon_g\\ &= \theta_{gh}(\varepsilon_k\varepsilon_{h^{-1}}\varepsilon_{h^{-1}g^{-1}}).\end{aligned}$$

Since the elements considered above generate $D_{h^{-1}} \cap D_{h^{-1}g^{-1}}$, as an algebra, we see that $\theta_g\theta_h$ coincides with θ_{gh} on the domain of the latter, so $\theta_g\theta_h \subseteq \theta_{gh}$, concluding the proof. □

We may then form the crossed product $A_{\text{par}}(G) \rtimes G$.

Theorem 10.9. *For every group G, there exists an isomorphism*

$$\Phi : \mathbb{K}_{par}(G) \to A_{par}(G) \rtimes G,$$

such that $\Phi([g]) = \varepsilon_g\delta_g$, for all $g \in G$.

PROOF. Being under the hypothesis of (9.4), we have that the map

$$g \in G \mapsto \varepsilon_g\delta_g \in A_{\text{par}}(G) \rtimes G$$

is a partial representation of G. By the universal property of $\mathbb{K}_{\text{par}}(G)$ we then obtain a homomorphism

$$\Phi : \mathbb{K}_{\text{par}}(G) \to A_{\text{par}}(G) \rtimes G,$$

such that $\phi([g]) = \varepsilon_g\delta_g$, for all $g \in G$, and it now suffices to prove that Φ is bijective.

In order to produce an inverse of Φ, we will build a covariant representation (π, u) of our partial dynamical system in $\mathbb{K}_{\text{par}}(G)$. For π we take the homomorphism from $A_{\text{par}}(G)$ to $\mathbb{K}_{\text{par}}(G)$ obtained via the universal property of the former, such that

$$\pi(\varepsilon_g) = [g][g^{-1}], \quad \forall\, g \in G,$$

while, for u, we take the universal partial representation, namely

$$u_g = [g], \quad \forall\, g \in G.$$

To check that (π, u) is in fact a covariant representation, let $g, h \in G$, and notice that

$$\begin{aligned} u_g\pi(\varepsilon_h\varepsilon_{g^{-1}})u_{g^{-1}} &= [g][h][h^{-1}][g^{-1}] \\ &\overset{(9.8.\text{iii})}{=} [gh][(gh)^{-1}][g][g^{-1}] \\ &= \pi(\varepsilon_{gh}\varepsilon_g) \\ &= \pi\big(\theta_g(\varepsilon_h\varepsilon_{g^{-1}})\big), \end{aligned}$$

as desired. By (9.11.iii) we obtain a homomorphism $\pi \times u$ from $A_{\text{par}}(G) \rtimes G$ to $\mathbb{K}_{\text{par}}(G)$, such that

$$(\pi \times u)(a\delta_g) = \pi(a)[g], \quad \forall\, g \in G, \quad \forall\, a \in D_g.$$

On the one hand we have, for all $g, h \in G$, that

$$\Phi\big((\pi \times u)(\varepsilon_h\varepsilon_g\delta_g)\big) = \Phi\big([h][h^{-1}][g]\big) = (\varepsilon_h\delta_h)(\varepsilon_{h^{-1}}\delta_{h^{-1}})(\varepsilon_g\delta_g) = \varepsilon_h\varepsilon_g\delta_g,$$

from where we conclude that $\Phi \circ (\pi \times u)$ is the identity. On the other hand

$$(\pi \times u)\big(\Phi([g])\big) = (\pi \times u)(\varepsilon_g\delta_g) = [g][g^{-1}][g] = [g],$$

so $(\pi \times u) \circ \Phi$ is also the identity, whence Φ is an isomorphism. □

One may now easily verify that the partial dynamical system provided by (10.1) in terms of the universal partial representation is equivalent to the one given in (10.8).

Before concluding this chapter, let us mention, without proofs, another interesting feature of the partial group algebra. Recall that if G is a finite abelian group and $\mathbb{K}$ is the field of complex numbers, then the classical group algebra $\mathbb{K}(G)$ is isomorphic to $\mathbb{K}^{|G|}$. In particular, the only feature of G retained by its complex group algebra is the number of elements in G. When it comes to partial group algebras the situation is completely different.

Theorem 10.10. [**34**, Corollary 4.5] *Let G and H be two finite abelian groups and let $\mathbb{K}$ be an integral domain whose characteristic does not divide $|G|$. If the partial group algebras $\mathbb{K}_{par}(G)$ and $\mathbb{K}_{par}(H)$ are isomorphic, then G and H are isomorphic groups.*

Notes and remarks. The concept of partial group algebra was introduced in [**34**, Definition 2.4]. It is the purely algebraic version of the corresponding notion of *partial group C^*-algebra*, previously introduced in [**50**, Definition 6.4].

CHAPTER 11

C*-Algebraic Partial Dynamical Systems

In this chapter we will adapt the construction of the partial crossed product to the category of C*-algebras. We begin with a quick review of basic concepts.

Definition 11.1. A *C*-algebra* is a *-algebra A over the field of complex numbers, equipped with a norm $\|\cdot\|$, with respect to which it is a Banach space, and such that for all a and b in A, one has that

(i) $\|ab\| \leq \|a\|\|b\|$,
(ii) $\|a^*\| = \|a\|$,
(iii) $\|a^*a\| = \|a\|^2$.

There are many references for the basic theory of C*-algebras where the interested reader will find the basic results, such as [**89**], [**9**], [**84**] and [**32**].

Of special relevance to us is Gelfand's Theorem [**84**, Theorem 2.1.10] which asserts that there is an equivalence between the category of locally compact Hausdorff (*LCH* for short) topological spaces, with proper[9] continuous maps, on the one hand, and the category of abelian C*-algebras, with *non-degenerate*[10] *-homomorphisms, on the other hand. This equivalence is implemented by the contravariant functor

$$X \rightsquigarrow C_0(X),$$

where $C_0(X)$ refers to the C*-algebra formed by all continuous complex valued functions f defined on X, vanishing[11] at ∞.

If X and Y are LCH spaces then any proper continuous map $h : X \to Y$ induces a non-degenerate *-homomorphism

$$\phi_h : f \in C_0(Y) \mapsto f \circ h \in C_0(X),$$

and conversely, any non-degenerate *-homomorphism from $C_0(Y)$ to $C_0(X)$ is induced, as above, by a unique proper continuous map from X to Y. Moreover, ϕ_h is an isomorphism if and only if h is a homeomorphism.

Ideals (always assumed to be norm-closed and two-sided) in C*-algebras are automatically self-adjoint. Every ideal in a C*-algebra is both non-degenerate and idempotent, and hence the conclusion of (7.9) holds for them.

If X is a LCH space then there is a one-to-one correspondence between open subsets of X and ideals of $C_0(X)$ given as follows: to an open set $U \subseteq X$ we attach

---

[9] A map between locally compact topological spaces is said to be *proper* if the inverse image of every compact set is compact.

[10] A *-homomorphism φ from a C*-algebra A to another C*-algebra B, or perhaps even into the multiplier algebra $\mathcal{M}(B)$, is said to be *non-degenerate* if $B = [\varphi(A)B]$ (brackets denoting closed linear span). By taking adjoints, this is the same as saying that $B = [B\varphi(A)]$.

[11] We say that a map f *vanishes at* ∞, if for every real number $\varepsilon > 0$, the set $\{x \in X : |f(x)| \geq \varepsilon\}$ is compact.

the ideal given by

$$C_0(U) := \{f \in C_0(X) : f = 0 \ \text{ on } \ X \setminus U\}.$$

The reader should be aware that any open set $U \subseteq X$ may also be seen as a LCH space in itself, so this notation has a potential risk of confusion since, besides the above meaning of $C_0(U)$, one could also think of $C_0(U)$ as the set of all continuous complex valued functions *defined on* U, and vanishing at ∞.

However the two meanings of $C_0(U)$ give rise to naturally isomorphic C*-algebras, the isomorphism taking any function defined on U to its extension to the whole of X, declared zero outside of U. The very slight distinction between the two interpretations of this notation will fortunately not cause us any problems.

Recall from (4.4) that $\mathcal{I}(X)$ denotes the inverse semigroup formed by all partial symmetries of a set X.

Definition 11.2. Given a C*-algebra A, we will say that a partial symmetry $\phi \in \mathcal{I}(A)$ is a *partial automorphism* of A, if the domain and range of ϕ are closed two-sided ideals of A, and ϕ is a *-isomorphism from its domain to its range. We will denote by $\mathrm{pAut}(A)$ the collection of all partial automorphisms of A. It is evident that $\mathrm{pAut}(A)$ is an inverse sub-semigroup of $\mathcal{I}(A)$.

Given any partial homeomorphism of a LCH space X, say $h : U \to V$, where U and V are open subsets of X, the map

$$\phi_h : C_0(V) \to C_0(U)$$

is a *-isomorphism between ideals of $C_0(X)$, and hence may be seen as a partial automorphism of $C_0(X)$. It follows from what was said above that the correspondence

$$h \in \mathrm{pHomeo}(X) \ \mapsto \ \phi_{h^{-1}} \in \mathrm{pAut}\big(C_0(X)\big) \tag{11.3}$$

is a semigroup isomorphism.

Definition 11.4. A *C*-algebraic partial action* of the group G on the C*-algebra A is a partial action $\theta = \big(\{D_g\}_{g\in G}, \{\theta_g\}_{g\in G}\big)$ on the underlying set A, such that each D_g is a *closed* two-sided ideal of A, and each θ_g is a *-isomorphism from $D_{g^{-1}}$ to D_g. By a *C*-algebraic partial dynamical system* we shall mean a partial dynamical system

$$\big(A,\ G,\ \{D_g\}_{g\in G},\ \{\theta_g\}_{g\in G}\big)$$

where A is a C*-algebra and $\big(\{D_g\}_{g\in G}, \{\theta_g\}_{g\in G}\big)$ is a C*-algebraic partial action of G on A.

When it is understood that we are working in the category of C*-algebras and there is no chance for confusion we will drop the adjective *C*-algebraic* and simply say *partial action* or *partial dynamical system.*

As an immediate consequence of (4.5) we have:

Proposition 11.5. *Let G be a group, A be a C*-algebra, and*

$$\theta : G \to \mathrm{pAut}(A)$$

be a map. Then θ is a C-algebraic partial action of G on A if and only if conditions (i–iv) of* (4.5) *are fulfilled.*

Putting together (5.3), (11.5) and (11.3), one concludes:

Corollary 11.6. *If G is a group and X is a LCH space, then* (11.3) *induces a natural equivalence between topological partial actions of G on X and C*-algebraic partial actions of G on $C_0(X)$.*

► We now fix, for the time being, a C*-algebraic partial action

$$\theta = \big(\{D_g\}_{g\in G},\ \{\theta_g\}_{g\in G}\big)$$

of a group G on a C*-algebra A.

Since θ is in particular a *-algebraic partial action, we may apply the construction of the crossed product described in (8.3) to θ. However, the resulting algebra, which we will temporarily denote by

$$A \rtimes_{\text{alg}} G,$$

will most certainly not be a C*-algebra, so we will modify the construction a bit in order to stay in the category of C*-algebras. Meanwhile we observe that $A \rtimes_{\text{alg}} G$ is an associative algebra by (8.7), as well as a *-algebra by (8.9).

Definition 11.7. A *C*-seminorm* on a complex *-algebra B is a seminorm $p : B \to \mathbb{R}_+$, such that, for all $a, b \in B$, one has that

(i) $p(ab) \leq p(a)p(b)$,
(ii) $p(a^*) = p(a)$,
(iii) $p(a^*a) = p(a)^2$.

If B is a C*-algebra and p is a C*-seminorm on B, it is well known that

$$p(b) \leq \|b\|, \tag{11.8}$$

for all $b \in B$.

Proposition 11.9. *Let p be a C*-seminorm on $A \rtimes_{alg} G$. Then, for every $a = \sum_{g\in G} a_g\delta_g$ in $A \rtimes_{alg} G$, one has that*

$$p(a) \leq \sum_{g\in G} \|a_g\|.$$

PROOF. Notice that $A\delta_1$ is isomorphic to the C*-algebra A, so by what was said above we have that $p(a\delta_1) \leq \|a\|$, for all $a \in A$. We then have that

$$p(a_g\delta_g)^2 = p\big((a_g\delta_g)(a_g\delta_g)^*\big) \overset{(8.14)}{=} p(a_g a_g^*\delta_1) \leq \|a_g a_g^*\| = \|a_g\|^2,$$

so the statement follows from the triangle inequality. □

Let us therefore define a seminorm on $A \rtimes_{\text{alg}} G$, by

$$\|a\|_{\max} = \sup\big\{p(a) : p \text{ is a C*-seminorm on } A \rtimes_{\text{alg}} G\big\}. \tag{11.10}$$

By (11.9) we see that $\|a\|_{\max}$ is always finite and it is not hard to see that $\|\cdot\|_{\max}$ is a C*-seminorm on $A \rtimes_{\text{alg}} G$ (we will later prove that it is in fact a norm).

Definition 11.11. The *C*-algebraic crossed product* of a C*-algebra A by a group G under a C*-algebraic partial action $\theta = \big(\{D_g\}_{g\in G}, \{\theta_g\}_{g\in G}\big)$ is the C*-algebra $A \rtimes G$ obtained by completing $A \rtimes_{\text{alg}} G$ relative to the seminorm $\|\cdot\|_{\max}$ defined above.

The process of completing a semi-normed space involves first modding out the subspace formed by all vectors of zero length. However, as already mentioned, $\|\cdot\|_{\max}$ will be shown to be a norm on $A \rtimes_{\mathrm{alg}} G$, so the modding out part will be seen to be unnecessary.

Definition 11.12. From now on, for any $a \in D_g$, we will let $a\delta_g^{alg}$ denote the element of $A \rtimes_{\mathrm{alg}} G$ we have so far been denoting by $a\delta_g$, while we will reserve the notation $a\delta_g$ for the canonical image of $a\delta_g^{alg}$ in $A \rtimes G$.

As we will be mostly working with the C*-algebraic crossed product, the notation $a\delta_g^{alg}$ will only rarely be used in the sequel.

Definition 11.13. We will denote by

$$\iota : A \to A \rtimes G,$$

the mapping defined by $\iota(a) = a\delta_1$, for every $a \in A$.

As already mentioned, we will later prove that the natural map from $A \rtimes_{\mathrm{alg}} G$ to $A \rtimes G$ is injective and consequently ι will be seen to be injective as well.

The following is a useful device in producing *-homomorphisms defined on crossed product algebras:

Proposition 11.14. *Let B be a C*-algebra and let*

$$\varphi_0 : A \rtimes_{alg} G \to B$$

*be a *-homomorphism. Then there exists a unique *-homomorphism φ from $A \rtimes G$ to B, such that the diagram*

$$\begin{array}{ccc} A \rtimes_{alg} G & \xrightarrow{\varphi_0} & B \\ \downarrow & \nearrow_{\varphi} & \\ A \rtimes G & & \end{array}$$

commutes, where the vertical arrow is the canonical mapping arising from the completion process.

Proof. It is enough to notice that $p(x) := \|\varphi_0(x)\|$ defines a C*-seminorm on $A \rtimes_{\mathrm{alg}} G$, which is therefore bounded by $\|\cdot\|_{\max}$. Thus φ_0 is continuous for the latter, and hence extends to the completion. □

Notes and remarks. As already mentioned, partial actions on C*-algebras were introduced in [**45**] for the case of the group of integers, and in [**81**] for general groups. Although not covered by this book, the notion of *continuous* partial actions of *topological groups* on C*-algebras, *twisted* by a *cocycle* or otherwise, has also been considered [**48**].

CHAPTER 12

Partial Isometries

When working in the category of C*-algebras, we will often consider *-partial representations

$$u : G \to A \tag{12.1}$$

of a given group G in a given C*-algebra A, according to Definition (9.1). In other words, the definition of *-partial representations given in (9.1) needs no further adaptation to the C*-algebraic case, considering that C*-algebras are special cases of *-algebras. Incidentally, many *-partial representations studied in this book will take place in $\mathscr{L}(H)$, the C*-algebra of all bounded linear operators on a Hilbert space H.

Given a *-partial representation u, as in (12.1), observe that, by (9.1.iv) and (9.8.i), one has

$$u_g u_g^* u_g = u_g, \quad \forall g \in G. \tag{12.2}$$

Elements satisfying this equation are crucial for the present work, so we shall dedicate this entire chapter to their study. We begin by giving them a well deserved name:

Definition 12.3. Let A be a *-algebra.

(i) An element s in A is said to be a *partial isometry*, if $ss^*s = s$.
(ii) An element p in A is said to be a *projection*, if $p = p^* = p^2$.

A useful characterization of partial isometries in terms of projections is as follows:

Proposition 12.4. *Let A be a C*-algebra and let $s \in A$. Then the following are equivalent:*

(i) s is a partial isometry,
(ii) s^*s is a projection,
(iii) ss^* is a projection.

PROOF. The implications (i)⇒(ii & iii) are evident. On the other hand, we have

$$\begin{aligned}(ss^*s - s)^*(ss^*s - s) &= s^*ss^*ss^*s - s^*ss^*s - s^*ss^*s + s^*s \\ &= (s^*s)^3 - 2(s^*s)^2 + s^*s.\end{aligned}$$

Thus, if s^*s is a projection, the above vanishes and hence

$$\|ss^*s - s\|^2 = \|(ss^*s - s)^*(ss^*s - s)\| = 0,$$

showing that $ss^*s = s$, and hence that s is a partial isometry. This proves that (ii)⇒(i). The proof of (iii)⇒(i) is obtained by replacing s by s^*. □

Given a projection p, we have by (11.1.iii) that

$$\|p\|^2 = \|p^*p\| = \|p^2\| = \|p\|,$$

so, unless $p = 0$, we have that $\|p\| = 1$. If s is a partial isometry, then s^*s is a projection by (12.4.ii), so unless $s = 0$, we have

$$\|s\|^2 = \|s^*s\| = 1.$$

This shows that nonzero projections, as well as nonzero partial isometries have norm exactly 1.

Definition 12.5. Let A be a C*-algebra and let s be a partial isometry in A. Then the projections s^*s and ss^* are called the *initial* and *final* projections of s, respectively.

Given a Hilbert space H, it is well known that a bounded linear operator s in $\mathscr{L}(H)$ is a partial isometry if and only if s is isometric when restricted to the orthogonal complement of its kernel, a space which is known as the *initial space* of s. On the other hand, the range of s is known as its *final space*. It is easy to see that the range of the initial projection of s coincides with its initial space, and similarly for the final projection and the final space.

If we denote the initial space of s by H_0 and its final space by H_1, then the effect of applying s to a vector $\xi \in H$ consists in projecting ξ orthogonally onto H_0, followed by the application of an isometric linear transformation from H_0 onto H_1 (namely the restriction of s to H_0).

The adjoint s^* of a partial isometry is easily seen to be a partial isometry, while the roles of the initial and final spaces of s are interchanged with those of s^*.

We thus see that, given a *-partial representation u of a group G in a C*-algebra, each u_g is a partial isometry and the element

$$e_g = u_g u_{g^{-1}} = u_g u_g^*,$$

which has already played an important role, is nothing but the final projection of u_g. On the other hand, the initial projection of u_g is clearly $e_{g^{-1}}$.

Let us now discuss some simple facts about partial isometries and projections in C*-algebras.

Lemma 12.6. *Let A be a C*-algebra and let $p \in A$ be such that $p^2 = p$, and $\|p\| \leq 1$. Then $p = p^*$.*

PROOF. By [**9**, Theorem 1.7.3] we may assume that A is a closed *-subalgebra of operators on some Hilbert space H. In that case, notice that for every $\xi \in p(H)^\perp$ and every $\lambda \in \mathbb{R}$, we have

$$|1+\lambda|\, \|p(\xi)\| = \|p(\xi) + \lambda p(\xi)\| = \|p\big(\xi + \lambda p(\xi)\big)\| \leq \|\xi + \lambda p(\xi)\|,$$

so, by the Pythagoras Theorem,

$$(1+\lambda)^2 \|p(\xi)\|^2 \leq \|\xi\|^2 + \lambda^2 \|p(\xi)\|^2,$$

which is easily seen to imply that

$$(1+2\lambda)\|p(\xi)\|^2 \leq \|\xi\|^2,$$

and since λ is arbitrary, we must have that $p(\xi) = 0$. This says that p vanishes on $p(H)^\perp$ and, since p is the identity on $p(H)$, it must coincide with the orthogonal projection onto $p(H)$. Hence $p = p^*$. □

Lemma 12.7. *Let p and q be projections in a C^*-algebra A. Then pq is idempotent if and only if p and q commute.*

PROOF. If pq is idempotent, then, since $\|pq\| \leq 1$, we have, by (12.6), that $pq = (pq)^* = qp$. The converse is trivial. □

The product of two partial isometries is not always a partial isometry. However we have:

Proposition 12.8. *Let s and t be partial isometries in a C^*-algebra A. Then st is a partial isometry if and only if s^*s and tt^* commute.*

PROOF. By definition st is a partial isometry if and only if

$$\begin{aligned} st(st)^*st = st &\Leftrightarrow stt^*s^*st = st \\ &\Leftrightarrow s^*stt^*s^*stt^* = s^*stt^* \\ &\Leftrightarrow (s^*stt^*)^2 = s^*stt^*, \end{aligned}$$

which, by (12.7), is equivalent to the commutativity of s^*s and tt^*. □

If we are given a set S of partial isometries in a C*-algebra A, we may always consider the multiplicative semigroup $\langle S\rangle$ generated by S. However, as seen above, unless there is enough commutativity among range and source projections, it is likely that $\langle S\rangle$ will include elements which are not partial isometries.

Definition 12.9. A set S of partial isometries in a C*-algebra A is said to be *tame* if the multiplicative sub-semigroup of A generated by $S \cup S^*$, henceforth denoted by $\langle S \cup S^*\rangle$, consists exclusively of partial isometries.

Given a tame set S, we then have that $\langle S \cup S^*\rangle$ is a multiplicative semigroup formed by partial isometries, which is easily seen to be a self-adjoint set.

Proposition 12.10. *Let $\mathfrak{S}$ be a self-adjoint multiplicative sub-semigroup of a C^*-algebra consisting of partial isometries. Then $\mathfrak{S}$ is an inverse semigroup.*

PROOF. Given s in $\mathfrak{S}$, we must prove that any element t in $\mathfrak{S}$ such that $sts = s$, and $tst = t$, necessarily coincides s^*.

Observe that both ts and st are idempotent elements with norm no bigger than 1. So ts and st are self-adjoint by (12.6). Therefore

$$ts = (ts)^* = (tss^*s)^* = (s^*s)^*(ts)^* = s^*sts = s^*s,$$

and

$$st = (st)^* = (ss^*st)^* = (st)^*(ss^*)^* = stss^* = ss^*.$$

Hence

$$t = tst = tss^*st = s^*ss^*ss^* = s^*.$$

□

As a consequence of the above result and our discussion just before it, we have:

Corollary 12.11. *For any tame set S of partial isometries in a C^*-algebra, one has that $\langle S \cup S^*\rangle$ is an inverse semigroup.*

Recall from (9.8.iv) that the projections e_g arising from a partial representation always commute. This is the basis for the close relationship between tame sets of partial isometries and *-partial representations, as we shall now see.

Proposition 12.12. *Let u be a *-partial representation of a group G in a unital C^*-algebra A. Then the range of u is a tame set of partial isometries.*

PROOF. Let s be an element in the multiplicative semigroup generated by the range of u (which is a self-adjoint set). Thus $s = u_{g_1} \cdots u_{g_n}$, for suitable elements $g_1, \ldots, g_n$ of G. By induction on n we then have

$$\begin{aligned} ss^*s &= u_{g_1} \cdots u_{g_{n-1}} e_{g_n} u^*_{g_{n-1}} \cdots u^*_{g_1} u_{g_1} \cdots u_{g_n} \\ &\overset{(9.8.\text{iii})}{=} e_{g_1 \cdots g_n} \underline{u_{g_1} \cdots u_{g_{n-1}}}\ \underline{u^*_{g_{n-1}} \cdots u^*_{g_1}}\ \underline{u_{g_1} \cdots u_{g_{n-1}}}\ u_{g_n} \\ &= e_{g_1 \cdots g_n} \underline{u_{g_1} \cdots u_{g_{n-1}}} u_{g_n} \\ &= u_{g_1} \cdots u_{g_{n-1}} e_{g_n} u_{g_n} \\ &= u_{g_1} \cdots u_{g_{n-1}} u_{g_n} \\ &= s. \end{aligned}$$

So s is a partial isometry. □

Roughly speaking, the following result includes a converse of the previous one.

Proposition 12.13. *Let $S = \{s_\lambda\}_{\lambda \in \Lambda}$ be a family of partial isometries in a unital C^*-algebra A, and denote by $\mathbb{F}$ the free group on the index set Λ. Then the following are equivalent:*

(*i*) *There exists a semi-saturated partial representation u of $\mathbb{F}$ in A, such that $u_\lambda = s_\lambda$, for every $\lambda \in \Lambda$.*

(*ii*) *There exists a partial representation u of $\mathbb{F}$ in A, such that $u_\lambda = s_\lambda$, for every $\lambda \in \Lambda$.*

(*iii*) *S is tame.*

PROOF. (i) ⇒ (ii): Obvious.

(ii) ⇒ (iii): Follows from (12.12).

(iii) ⇒ (i): For all $\lambda \in \Lambda$, we define $u_\lambda = s_\lambda$, and $u_{\lambda^{-1}} = s^*_\lambda$. If $g = g_1 \cdots g_n$, with $g_i \in \Lambda \cup \Lambda^{-1}$, is in reduced form, we put

$$u_g = u_{g_1} \cdots u_{g_n}.$$

This defines a map $u : \mathbb{F} \to A$, which we claim is a semi-saturated partial representation. Adopting the usual convention that the reduced form of the unit group element of $\mathbb{F}$ is the empty string, and also that a product involving zero factors equals one, we see that u satisfies (9.1.i). The easy verification of (9.1.iv) is left to the reader.

Before concluding the verification of the remaining axioms in (9.1), we observe that the condition for semi-saturatedness, namely that $u_g u_h = u_{gh}$, whenever g and h satisfy $|gh| = |g| + |h|$, is evidently satisfied simply because, in this case, the reduced form gh is the juxtaposition of the corresponding reduced forms of g and h.

In order to prove (9.1.ii), namely

$$u_g u_h u_{h^{-1}} = u_{gh} u_{h^{-1}}, \quad \forall\, g, h \in \mathbb{F}, \tag{12.13.1}$$

we use induction on $|g| + |h|$. If either $|g|$ or $|h|$ is zero, there is nothing to prove. So, assuming that $|g|, |h| \geq 1$, we may write

$$g = a\lambda, \quad \text{and} \quad h = \mu b,$$

where $a, b \in \mathbb{F}$, $\lambda, \mu \in \Lambda \cup \Lambda^{-1}$ and, moreover, $|g| = |a| + 1$ and $|h| = |b| + 1$.

In case $\lambda^{-1} \neq \mu$, we have $|gh| = |g| + |h|$, so $u_{gh} = u_g u_h$, as seen above. If, on the other hand, $\lambda^{-1} = \mu$, we have

$$u_g u_h u_{h^{-1}} = u_{a\lambda} u_{\lambda^{-1}b} u_{b^{-1}\lambda} = u_a u_\lambda u_{\lambda^{-1}} u_b u_{b^{-1}} u_\lambda = \cdots$$

By (iii) and the induction hypothesis, we conclude that the above equals

$$\begin{aligned} \cdots &= u_a u_b u_{b^{-1}} u_\lambda u_{\lambda^{-1}} u_\lambda \\ &= u_{ab} u_{b^{-1}} u_\lambda \\ &= u_{gh} u_{b^{-1}\lambda} \\ &= u_{gh} u_{h^{-1}}, \end{aligned}$$

proving (12.13.1). As already remarked, (9.1.iv) and (9.1.ii) together imply (9.1.iii), so u is indeed a semi-saturated partial representation and the proof is concluded. □

The study of partial isometries in a C*-algebra resembles the theory of inverse semigroups in the sense that every *tame* set of partial isometries is contained in an inverse semigroup by (12.11). Thus, as long as we are focusing on partial isometries lying in a single tame set, we may apply many of the tools of the theory of inverse semigroups.

This should be compared to quantum theory in the sense that, when we are working with a set of *commuting* self-adjoint operators, we are allowed to apply results from function theory, since our set of operators generates a commutative C*-algebra which, by Gelfand's Theorem, is necessarily of the form $C_0(X)$, for some locally compact Hausdorff space X. On the down side, if our operators do not commute, function theory becomes unavailable and we must face true quantum phenomena.

In what follows we will develop some elementary results about partial isometries in a C*-algebra which are not always explicitly required to lie in a tame set. Since wild (as opposed to tame) sets of partial isometries are very hard to handle, our guiding principle will be to stay as close as possible to the theory of inverse semigroups.

We begin with a result supporting a subsequent definition of an order relation among partial isometries. This should be compared with (4.2).

Proposition 12.14. *Let s and t be partial isometries in a C*-algebra A. Then the following are equivalent*

(*i*) $ts^*s = s$,
(*ii*) $ss^*t = s$.

In this case $s^*s \leq t^*t$, and $ss^* \leq tt^*$.

PROOF. Given that (i) holds, we have

$$ts^*s = s = ss^*s = ts^*s(ts^*s)^*ts^*s = ts^*st^*ts^*s.$$

Multiplying on the left by t^* gives

$$t^*ts^*s = t^*ts^*st^*ts^*s,$$

so, if we let $p = t^*t$, and $q = s^*s$, we see that $pq = pqpq$, so pq is idempotent. It then follows from (12.7) that p and q commute. Consequently

$$t^*s = t^*ts^*s = pq,$$

so t^*s is self-adjoint, and then

$$(\diamond) \qquad t^*s = (t^*s)^* = s^*t.$$

From (i) we also deduce that

$$st^* = ts^*st^*,$$

so st^* is self-adjoint as well, and hence

$$(\star) \qquad st^* = (st^*)^* = ts^*.$$

Focusing on (ii) we then have

$$ss^*t \overset{(\diamond)}{=} st^*s \overset{(\star)}{=} ts^*s \overset{(i)}{=} s.$$

The converse is verified by applying the part that we have already proved for $s' := s^*$ and $t' := t^*$.

With respect to the final sentence in the statement, we have

$$s^*s \overset{(i)}{=} (ts^*s)^*ts^*s = s^*st^*ts^*s = s^*st^*t,$$

where the last equality is a consequence of the commutativity of p and q. So $s^*s \leq t^*t$, and one similarly proves that $ss^* \leq tt^*$. □

The above result should be compared with (4.2), hence motivating the following:

Definition 12.15. Given two partial isometries s and t in a C*-algebra, we will say that s is *dominated* by t, or that $s \preceq t$, if the equivalent conditions of (12.14) are satisfied.

It is elementary to verify that "$\preceq$" is a reflexive and antisymmetric relation. We also have:

Proposition 12.16. *The order "$\preceq$" defined above is transitive.*

PROOF. Suppose that r, s and t are partial isometries with $r \preceq s \preceq t$. Then $r = sr^*r$ and $s = ts^*s$. Therefore

$$tr^*r = t(sr^*r)^*sr^*r = tr^*rs^*sr^*r \overset{(12.14)}{=} ts^*sr^*r = sr^*r = r.$$

So $r \preceq t$. □

The following provides a useful alternative characterization of the order relation "$\preceq$".

Proposition 12.17. *Given two partial isometries s and t in a C^*-algebra, the following are equivalent:*

(*i*) $s \preceq t$,
(*ii*) $ts^* = ss^*$,
(*iii*) $s^*t = s^*s$.

PROOF. (i)⇒(ii):

$$ts^* = ts^*ss^* = ss^*.$$

(ii)⇒(i):

$$ts^*s = ss^*s = s.$$

The proof that (i)⇔(iii) follows along similar lines. □

Let us now prove invariance of "$\preceq$" under multiplication:

Lemma 12.18. *Let s_1, s_2, t_1 and t_2 be partial isometries in a C^*-algebra, such that $s_1 \preceq s_2$, and $t_1 \preceq t_2$. If $s_i^* s_i$ commutes with $t_i t_i^*$, for all $i = 1, 2$, then $s_1 t_1 \preceq s_2 t_2$.*

PROOF. Recall that each $s_i t_i$ is a partial isometry by (12.8). We have

$$\begin{aligned} s_2 t_2 (s_1 t_1)^* &= s_2 t_2 t_1^* s_1^* \\ &\overset{(12.17.\text{ii})}{=} s_2 t_1 t_1^* s_1^* s_1 s_1^* \\ &= s_2 s_1^* s_1 t_1 t_1^* s_1^* \\ &= s_1 t_1 t_1^* s_1^* \\ &= s_1 t_1 (s_1 t_1)^*. \end{aligned}$$

This verifies (12.17.ii), so $s_1 t_1 \preceq s_2 t_2$, as desired. □

If a net $\{s_i\}_i$ of partial isometries on a Hilbert space strongly converges to a partial isometry s, then the final projections of the s_i might not converge to the final projection of s. An example of this is obtained by taking $s_n = (s^*)^n$, where s is the unilateral shift on $\ell^2(\mathbb{N})$. In this case the s_n converge strongly to zero, but the final projections of the s_n all coincide with the identity operator. On the bright side we have:

Lemma 12.19. *Let $\{s_i\}_{i \in I}$ be an <u>increasing</u> net of partial isometries on a Hilbert space H. Then*

(i) $\{s_i\}_{i \in I}$ strongly converges to a partial isometry s,
(ii) $\{s_i^*\}_{i \in I}$ strongly converges to s^*,
(iii) $\{s_i^* s_i\}_{i \in I}$ strongly converges to $s^* s$,
(iv) $\{s_i s_i^*\}_{i \in I}$ strongly converges to $s s^*$.

PROOF. We will first show that

$$\exists \lim_i s_i(\xi), \tag{12.19.1}$$

for all ξ in H.

By the last sentence of (12.14) we have that the corresponding initial projections, say

$$e_i = s_i^* s_i,$$

form an increasing net in the usual order of projections. Letting H_i be the range of e_i, also known as the initial space of s_i, we then see that the H_i form an increasing family of subspaces of H, so

$$H_0 := \bigcup_{i \in I} H_i$$

is a linear subspace of H. We will next prove (12.19.1) for every ξ in H_0. Given such ξ, let i be such that $\xi \in H_i$. Consequently, for all $j \geq i$, we have

$$s_j(\xi) = s_j e_i(\xi) = s_j s_i^* s_i(\xi) = s_i(\xi),$$

so we see that the net $\{s_j(\xi)\}_j$ is eventually constant, hence convergent. Observing that our net is uniformly bounded, we then have that (12.19.1) also holds for all ξ in the closure of H_0. On the other hand, if $\xi \in H_0^\perp$, then $e_i(\xi) = 0$, for all i, hence

$$s_i(\xi) = s_i e_i(\xi) = 0,$$

so (12.19.1) is verified for ξ in $H_0^\perp$ as well, hence also for all ξ in H. So we may define

$$s(\xi) = \lim_i s_i(\xi), \quad \forall \xi \in H,$$

and it is easy to see that s is isometric on $\overline{H}_0$, while s vanishes on $H_0^\perp$, so s is a partial isometry. This proves (i).

The initial space of s is easily seen to coincide with $\overline{H}_0$, whence s^*s is the orthogonal projection onto $\overline{H}_0$. On the other hand, it is clear that the orthogonal projection onto $\overline{H}_0$ is the strong limit of the e_i, so

$$s^*s = \lim_i e_i = \lim_i s_i^* s_i,$$

hence (iii) follows.

The order "$\preceq$" being evidently invariant under conjugation, we have that $\{s_i^*\}_{i\in I}$ is an increasing net of partial isometries, hence by the above reasoning it strongly converges to some partial isometry t, and moreover the corresponding net of initial projections $\{s_i s_i^*\}_{i\in I}$ (sic) converges to t^*t.

Unfortunately the operation of conjugation is not strongly continuous, but it is well known to be weakly continuous, hence $s_i^* \overset{i\to\infty}{\longrightarrow} s^*$ weakly. Since the weak limit is unique, we deduce that $t = s^*$, from where (ii) and (iv) follow. □

In the following we present another important relation involving partial isometries.

Definition 12.20. Let A be a C*-algebra. Given two partial isometries s and t in A, we will say that s and t are *compatible*, if

$$st^*t = ts^*s, \quad \text{and} \quad tt^*s = ss^*t.$$

If S is a subset of A consisting of partial isometries, we will say that S is a *compatible set* when the elements in S are pairwise compatible.

We will soon give a geometric interpretation of this concept, but let us first prove a useful result.

Proposition 12.21. *Given partial isometries s and t in a C*-algebra A, the following are equivalent:*

(a) s and t are compatible,

(b) st^* and s^*t are positive elements of A. In this case one also has that

- *(i) the final projections ss^* and tt^* commute,*
- *(ii) the initial projections s^*s and t^*t commute,*
- *(iii) $st^* = ts^* = ss^*tt^*$,*
- *(iv) $s^*t = t^*s = s^*st^*t$.*
- *(v) $ts^*s = tt^*s$, and consequently all of the four terms involved in the definition of compatibility coincide.*

Proof. Assuming that s and t are compatible, we have

$$st^* = st^*tt^* = ts^*st^* \geq 0,$$

and

$$s^*t = s^*ss^*t = s^*tt^*s \geq 0,$$

proving (b). Conversely, assuming (b), observe that, since positive elements are necessarily self-adjoint, we have

$$(\star) \qquad st^* = (st^*)^* = ts^*,$$

and

$$(\diamond) \qquad s^*t = (s^*t)^* = t^*s,$$

thus verifying the first identities in (iii) and (iv). Therefore

$$tt^*ss^* \overset{(\diamond)}{=} ts^*ts^* \overset{(\star)}{=} st^*st^* \overset{(\diamond)}{=} ss^*tt^*,$$

and

$$t^*ts^*s \overset{(\star)}{=} t^*st^*s \overset{(\diamond)}{=} s^*ts^*t \overset{(\star)}{=} s^*st^*t,$$

proving (i) and (ii).

Repeating an earlier calculation we have

$$(\dagger) \qquad (st^*)^2 = st^*st^* \overset{(\diamond)}{=} ss^*tt^*.$$

This implies that $(st^*)^2$ is a projection, whose spectrum is therefore contained in $\{0,1\}$. By the Spectral Mapping Theorem we have

$$\big(\sigma(st^*)\big)^2 = \sigma\big((st^*)^2\big) \subseteq \{0,1\},$$

so the spectrum of st^* is contained in $\{-1,0,1\}$, but since st^* is assumed to be positive, its spectrum must in fact be a subset of $\{0,1\}$. Consequently st^* is a projection, and then the last identity in (iii) follows from (†). On the other hand,

$$(s^*t)^2 = s^*ts^*t \overset{(\star)}{=} s^*st^*t,$$

so $(s^*t)^2$ is also a projection and the same reasoning adopted above leads to the proof of the last identity in (iv).

We may now prove the first of the two conditions in (12.20), namely

$$ts^*s \overset{(\star)}{=} st^*s \overset{(\diamond)}{=} ss^*t = ss^*tt^*t \overset{(\diamond)}{=} st^*st^*t = st^*t.$$

The second condition in (12.20) is proved in a similar way, so we deduce that s and t are compatible, as desired. Finally, with respect to (v), we have

$$ts^*s \overset{\text{(iii)}}{=} ss^*tt^*s \overset{\text{(i)}}{=} tt^*ss^*s = tt^*s.$$

□

12.22. As promised, let us give a geometric interpretation for the notion of compatibility of partial isometries. For this let S and T be compatible partially isometric linear operators on a Hilbert space H. By (12.21) we have that the initial projections of S and T commute, so we may decompose H as an orthogonal direct sum

$$H = K \oplus H_S \oplus H_T \oplus L,$$

such that $K \oplus H_S$ is the initial space of S, and $K \oplus H_T$ is the initial space of T. K is therefore the intersection of the initial spaces of S and T, and the orthogonal projection onto K is thus the product of the initial projections of S and T, namely S^*ST^*T. If k is a vector in K we then have that

$$S(k) = ST^*T(k) \overset{(12.20)}{=} TS^*S(k) = T(k),$$

so $S = T$ on K. We may then define an operator $S \vee T$ on H by

$$\begin{aligned}(S \vee T)(k, x_S, x_T, l) &= S(k) + S(x_S) + T(x_T)\\ &= T(k) + S(x_S) + T(x_T),\end{aligned}$$

for all k in K, x_S in H_S, x_T in H_T, and l in L, and it may be proved that $S \vee T$ is a partial isometry which coincides with S on the initial space of S, and with T on the initial space of T. In our next result we will generalize this idea for partial isometries in any C*-algebra.

Proposition 12.23. *Let A be a C^*-algebra and let s and t be partial isometries in A. Then the following are equivalent*

(a) s and t are compatible

(b) s^*s and t^*t commute, ss^* and tt^* commute, and there exists a partial isometry dominating both s and t.

In this case, defining

$$\begin{aligned}u &= s + t - st^*t\\ &= s + t - ts^*s,\end{aligned}$$

we have that u is a partial isometry such that:

(i) $s \preceq u$, and $t \preceq u$,

(ii) if v is a partial isometry such that $s \preceq v$, and $t \preceq v$, then $u \preceq v$,

(iii) the initial projection of u coincides with[12] *$s^*s \vee t^*t$,*

(iv) the final projection of u coincides with $ss^* \vee tt^*$.

PROOF. Supposing that s and t are compatible, we compute

$$\begin{aligned}u^*u &= (s^* + t^* - t^*ts^*)(s + t - st^*t)\\ &= s^*s + s^*t - s^*st^*t + t^*s + t^*t - t^*st^*t - t^*ts^*s - t^*ts^*t + t^*ts^*st^*t\\ &= s^*s + s^*t - s^*st^*t + t^*s + t^*t - s^*tt^*t \qquad\qquad - t^*tt^*s\\ &= s^*s \qquad - s^*st^*t \qquad + t^*t\\ &= s^*s \vee t^*t, \qquad (12.23.1)\end{aligned}$$

where we have used (12.21) to conclude that s^*s and t^*t commute. This proves that u^*u is a projection, and by (12.4) we deduce that u is a partial isometry. This also proves (iii), and the proof of (iv) is done along similar lines. In order to prove (i) we compute

$$us^*s = (s + t - ts^*s)s^*s = ss^*s + ts^*s - ts^*s = s,$$

while

$$ut^*t = (s + t - st^*t)t^*t = st^*t + tt^*t - st^*t = t,$$

so $s, t \preceq u$. Notice that this also proves that (a)⇒(b).

Next suppose that v is a partial isometry dominating s and t. Then

$$vu^*u \overset{(12.23.1)}{=} v(s^*s \vee t^*t) = vs^*s + vt^*t - vs^*st^*t = s + t - st^*t = u,$$

so $u \preceq v$, taking care of (ii).

[12] If p and q are commuting projections in an algebra A, one denotes by $p \vee q = p + q - pq$. It is well known that $p \vee q$ is again a projection, which is the least upper bound of p and q among the projections in A.

In order to prove that (b)⇒(a), assume that the commutativity conditions in (b) hold, and let v be a partial isometry dominating both s and t. Then

$$st^*t = us^*st^*t = ut^*ts^*s = ts^*s,$$

while

$$tt^*s = tt^*ss^*v = ss^*tt^*v = ss^*t.$$

□

The result above shows that, when two partial isometries are compatible, their least upper bound exists. This motivates and justifies the introduction of the following notation:

Definition 12.24. Given a set S of partial isometries in a C*-algebra A, suppose that there exists a partial isometry u that dominates every element of S, and such that $u \preceq v$, for every other partial isometry v dominating all elements of S. Observing that such a u is necessarily unique, we denote it by $\vee S$. If S is a two-element set, say $S = \{s, t\}$, and $\vee S$ exists, we also denote $\vee S$ by $s \vee t$.

Of course this is nothing but the usual notion of least upper bounds, meaningful in any ordered set. We have spelled it out just for emphasis.

Given two compatible partial isometries s and t in a C*-algebra A, observe that (12.23) implies that $s \vee t$ exists, and moreover

$$\begin{aligned} s \vee t &= s + t - st^*t \\ &= s + t - ts^*s. \end{aligned}$$

Projections being special cases of partial isometries, the above notions may also be applied to the former. Given two projections p and q in a C*-algebra, notice that $p \le q$ if and only if $p \preceq q$. Also, p and q are compatible if and only if they commute. In this case the two meanings of the expression $p \vee q$ so far defined are easily seen to coincide.

Here is a sort of a distributivity property mixing compatibility of partial isometries and the notion of least upper bounds just mentioned:

Proposition 12.25. *Let r, s and t be compatible partial isometries in a C*-algebra. Then r is compatible with $s \vee t$.*

PROOF. We have

$$\begin{aligned} r(s \vee t)^*(s \vee t) &\overset{(12.23.iii)}{=} r(s^*s \vee t^*t) \\ &= rs^*s + rt^*t - rs^*st^*t \\ &= sr^*r + tr^*r - sr^*rt^*t \\ &= (s + t - st^*t)r^*r \\ &= (s \vee t)r^*r. \end{aligned}$$

On the other hand

$$\begin{aligned}(s \vee t)(s \vee t)^* r &\overset{(12.23.\text{iv})}{=} (ss^* \vee tt^*)r \\ &= ss^* r + tt^* r - ss^* tt^* r \\ &= rr^* s + rr^* t - ss^* rr^* t \\ &= rr^*(s + t - ss^* t) \\ &= rr^*(s + t - st^* s) \\ &= rr^*(s + t - ts^* s) \\ &= rr^*(s \vee t).\end{aligned}$$

□

Proposition 12.26. *Let S be a compatible set of partial isometries in a C^*-algebra A. In addition we assume that either A is a von Neumann algebra, or that S is finite. Then*

(i) $\vee S$ exists,
(ii) the initial (resp. final) projection of $\vee S$ coincides with the least upper bound of the initial (resp. final) projections of the members of S,
(iii) every partial isometry in A which is compatible with the members of S, is also compatible with $\vee S$.

PROOF. Let us first deal with the case of a finite S, and so we assume that $S = \{s_1, s_2, \ldots, s_n\}$. Our proof will be by induction on n.

In case $n = 0$, then S is the empty set and $\vee S = 0$. If $n = 1$, then $\vee S = s_1$, and in both cases (ii–iii) hold trivially.

In case $n = 2$, both the existence of $\vee S$ and (ii) follow from (12.23), while (iii) follows from (12.25).

Assuming now that $n > 2$, let

$$S' = \{s_1, s_2, \ldots, s_{n-1}\}.$$

By the induction hypothesis we have that $\vee S'$ exists, and it is compatible with every partial isometry r, which in turn is compatible with the members of S'. This evidently includes s_n. We then apply the already verified case $n = 2$ to the set

$$S'' = \{\vee S', s_n\}.$$

It is then elementary to prove that

$$\vee S'' = (\vee S') \vee s_n = \vee S.$$

By induction we have that

$$(\vee S')^*(\vee S') = \vee\{s^* s : s \in S'\},$$

and

$$(\vee S')(\vee S')^* = \vee\{ss^* : s \in S'\},$$

so (ii) follows from (12.23.iii) & (12.23.iv).

If the partial isometry r is compatible with all of the members of S, then it is also compatible with $\vee S'$, by induction, and obviously also with s_n. Hence r is compatible with $\vee S$ by (12.25).

We next tackle the case that S is infinite, assuming that A is a von Neumann algebra of operators on a Hilbert space H. For each finite set $F \subseteq S$, let us denote by

$$s_F = \vee F.$$

We may then view $\{s_F\}_F$ as an increasing net, indexed by the set of all finite subsets of S, ordered by inclusion. By (12.19) the strong limit of this net exists, and we shall denote it by u.

Given $s \in S$, and a finite set $F \subseteq S$, with $s \in F$, we have that $s \preceq s_F$, whence, for all ξ in H, we have

$$s(\xi) = s_F s^* s(\xi).$$

Taking the limit as $F \to \infty$, we deduce that $s(\xi) = us^*s(\xi)$, so $s \preceq u$, hence u is an upper bound for S.

In order to prove that u is in fact the least upper bound, let v be an upper bound for S. In particular v is also an upper bound for all finite $F \subseteq S$, whence $s_F \preceq v$, which is to say that for all ξ in H one has

$$s_F(\xi) = vs_F^* s_F(\xi).$$

Taking the limit as $F \to \infty$, we have

$$u(\xi) = \lim_F s_F(\xi) = \lim_F vs_F^* s_f(\xi) \overset{(12.19.\text{iii})}{=} vu^*u(\xi),$$

so $u \preceq v$.

Point (ii) follows easily from the first part of the proof and (12.19.iii) & (12.19.iv). With respect to (iii), if the partial isometry r is compatible with all of the members of S, then r is compatible with s_F, for all finite $F \subseteq S$, by the first part of the proof. Thus

$$s_F r^* r = rs_F^* s_F, \quad \text{and} \quad rr^* s_F = s_F s_F^* r.$$

Again taking the limit as $F \to \infty$, we deduce from (12.19) that

$$ur^*r = ru^*u, \quad \text{and} \quad rr^*u = uu^*r,$$

so u is compatible with r. □

While the ideas of the above proof are still hot, let us observe that we have also proven:

Proposition 12.27. *Let S be a compatible set of partial isometries in a von Neumann algebra A. Then $\vee S$ is the strong limit of the net $\{\vee F\}_F$, where F ranges in the directed set consisting of all finite subsets of S.*

Proposition 12.28. *Let A be a C^*-algebra (resp. von-Neumann algebra) and let S and T be compatible sets of partial isometries in A such that s^*s commutes with tt^* for every s in S, and every t in T. In the C^*-case we moreover assume that S and T are finite. Then*

$$ST := \{st : s \in S,\ t \in T\}$$

is a compatible set of partial isometries and

$$\vee(ST) = (\vee S)(\vee T).$$

PROOF. That ST consists of partial isometries is a consequence of (12.8).

We will next prove that ST is a compatible set. For this choose any pair of elements in ST, say s_1t_1 and s_2t_2, where $s_1, s_2 \in S$, and $t_1, t_2 \in T$. Notice that the set

$$\{s_1^*s_1,\ s_2^*s_2,\ t_1t_1^*,\ t_2t_2^*\}$$

is commutative because the $s_i^*s_i$ commute with the $t_jt_j^*$ by hypothesis, while the $t_jt_j^*$ commute amongst themselves by (12.21.i), and the $s_i^*s_i$ commute with each other by (12.21.ii). We then have

$$\begin{aligned} s_1t_1(s_2t_2)^* &= s_1t_1t_2^*s_2^* \\ &\overset{(12.21.\text{iii})}{=} s_1\, t_1t_1^*\ t_2t_2^*\ s_2^*s_2\, s_2^* \\ &= s_1\, s_2^*s_2\ t_1t_1^*\ t_2t_2^*\, s_2^* \\ &= s_2\, s_1^*s_1\ t_1t_1^*\ t_2t_2^*\, s_2^* \\ &\geq 0. \end{aligned}$$

We also have

$$\begin{aligned} (s_1t_1)^*s_2t_2 &= t_1^*s_1^*s_2t_2 \\ &\overset{(12.21.\text{iv})}{=} t_1^*\, t_1t_1^*\ s_1^*s_1\ s_2^*s_2\, t_2 \\ &= t_1^*\ s_1^*s_1\ s_2^*s_2\ t_1t_1^*\, t_2 \\ &= t_1^*\ s_1^*s_1\ s_2^*s_2\ t_2t_2^*\, t_1 \\ &\geq 0. \end{aligned}$$

By (12.21) we then have that s_1t_1 is compatible with s_2t_2, proving that ST is indeed a compatible set. Employing (12.26.ii) we have

$$(\vee S)^*(\vee S) = \vee\{\, s^*s : s \in S\}, \quad \text{and} \quad (\vee T)(\vee T)^* = \vee\{\, tt^* : t \in T\},$$

and hence the initial projection of $\vee S$ commutes with the final projection of $\vee T$, so we deduce from (12.8) that $(\vee S)(\vee T)$ is a partial isometry.

We will next prove that

$$\vee(ST) = (\vee S)(\vee T), \tag{12.28.1}$$

under the assumption that S and T are finite sets. We begin by analyzing several possibilities for the number of elements in S and T. For example, when $S = \{s_1, s_2\}$, and $T = \{t\}$, we have

$$\begin{aligned} \vee(ST) &= \vee\{s_1t, s_2t\} \\ &= (s_1t) \vee (s_2t) \\ &= s_1t + s_2t - s_1t(s_2t)^*s_2t \\ &= s_1t + s_2t - s_1tt^*s_2^*s_2t \\ &= s_1t + s_2t - s_1s_2^*s_2t \\ &= (s_1 + s_2 - s_1s_2^*s_2)t \\ &= (s_1 \vee s_2)t \\ &= (\vee S)(\vee T). \end{aligned}$$

By induction it then easily follows that (12.28.1) holds when S has an arbitrary finite number of elements and T a singleton. Suppose now that $S = \{s\}$, and $T = \{t_1, t_2\}$. Then

$$\begin{aligned}\vee(ST) &= (st_1) \vee (st_2) \\ &\overset{(12.21.v)}{=} st_1 + st_2 - st_1(st_1)^* st_2 \\ &= st_1 + st_2 - st_1 t_1^* s^* st_2 \\ &= st_1 + st_2 - st_1 t_1^* t_2 \\ &= s(t_1 + t_2 - t_1 t_1^* t_2) \\ &= s(t_1 \vee t_2) \\ &= (\vee S)(\vee T).\end{aligned}$$

Again by induction it follows that (12.28.1) holds for S a singleton and T finite.

We will now prove (12.28.1) in the general finite case using induction on the number of elements of T. Writing $T = \{t_1, t_2, \ldots, t_n\}$, consider the set $T' = \{t_2, \ldots, t_n\}$, and observe that

$$\begin{aligned}(\vee S)(\vee T) &= (\vee S)\big(t_1 \vee (\vee T')\big) \\ &= (\vee S)t_1 \ \vee \ (\vee S)(\vee T') \\ &= (\vee St_1) \ \vee \ (\vee ST') \\ &= \vee(St_1 \cup ST') \\ &= \vee ST.\end{aligned}$$

Finally, let us tackle the general infinite case, assuming that A is a von-Neumann algebra. For this recall from (12.27) that

$$\vee S = \lim_{F\to\infty} \vee F, \quad \text{and} \quad \vee T = \lim_{G\to\infty} \vee G,$$

where F and G range in the directed sets consisting of all finite subsets of S and T, respectively. Since multiplication is doubly continuous on bounded sets for the strong operator topology, we have

$$\begin{aligned}(\vee S)(\vee T) &= (\lim_{F\to\infty} \vee F)(\lim_{G\to\infty} \vee G) \\ &= \lim_{F,G\to\infty} (\vee F)(\vee G) \\ &= \lim_{F,G\to\infty} (\vee FG). \qquad (12.28.2)\end{aligned}$$

Again by (12.27) we have that $\vee(ST)$ is the strong limit of the net

$$\{\vee H\}_{\substack{H\subseteq ST \\ H \text{ finite}}}, \qquad (12.28.3)$$

and we observe that the collection formed by all $H = FG$, where F and G are finite subsets of S and T, respectively, is co-final in the collection of finite subsets of ST. This gives rise to a subnet of (12.28.3), which therefore also converges to $\vee(ST)$. In other words, the last limit in (12.28.2) coincides with $\vee(ST)$, thus proving (12.28.1) in the general case. □

Definition 12.29. Let S be a set of partial isometries in a C*-algebra A.

(i) We will say that S is *finitely-$\vee$-closed* if, whenever T is a *finite* compatible subset of S, one has that $\vee T$ lies in S.

(ii) In case A is a von-Neumann algebra, we will say that S is $\vee$-*closed* if the condition above holds for every compatible subset $T \subseteq S$, regardless of whether it is finite or infinite.

The set of all partial isometries in a C*-algebra A is evidently finitely-$\vee$-closed. Moreover, the intersection of an arbitrary family of finitely-$\vee$-closed sets is again finitely-$\vee$-closed. Thus we may speak of the *finite-$\vee$-closure* of a given set S of partial isometries in A, namely the intersection of all finitely-$\vee$-closed sets of partial isometries containing S.

All that was said in the above paragraph clearly remains true if we remove all references to finiteness, as long as we take A to be a von-Neumann algebra. In particular we may also speak of the $\vee$-*closure* of a set of partial isometries in a von-Neumann algebra.

Lemma 12.30. *Let A be a C^*-algebra (resp. von-Neumann algebra) and let S be a self-adjoint multiplicative sub-semigroup of A consisting of partial isometries. Then the finite-$\vee$-closure (resp. $\vee$-closure) of S is also a self-adjoint multiplicative sub-semigroup of A, hence an inverse semigroup by* (12.10).

PROOF. Letting

$$\mathfrak{S} = \{ \vee T : T \text{ is a (finite) compatible subset of } S \},$$

we will first prove that $\mathfrak{S}$ is (finitely-)$\vee$-closed. For this let us be given a (finite) compatible subset of $\mathfrak{S}$, say

$$\mathfrak{T} = \{\vee T_j : j \in J\},$$

where each T_j is a (finite) compatible subset of S, and J is a (finite) set of indices. We then claim that

$$T := \cup_j T_j$$

is a compatible subset of S. To see this, we must prove that any two elements t and s in T are compatible, and we will do this by verifying (12.23.b). Initially notice that, thanks to S being a self-adjoint multiplicative semigroup, both s^*t and st^* are partial isometries. So we deduce from (12.8) that ss^* commutes with tt^*, and that s^*s commutes with t^*t.

On the other hand, since $\mathfrak{T}$ is a compatible set, we may use (12.26) to obtain a partial isometry u in A with

$$\vee T_j \preceq u, \quad \forall j \in J.$$

It is then evident that u dominates every element of every T_j, and consequently u dominates every element of T. In particular, u dominates s and t, thus verifying all of the conditions in (12.23.b), and hence we conclude that s and t are compatible. This proves that T is a compatible set.

It is now easy to see that

$$\vee \mathfrak{T} = \bigvee_{j \in J} \big(\vee T_j \big) = \vee T \in \mathfrak{S},$$

thus showing that $\mathfrak{S}$ is (finitely-)$\vee$-closed[13].

[13]It is worth observing that the assumption that S is a self-adjoint multiplicative semigroup, hence tame, was used above in an essential way in order to prove that $\mathfrak{S}$ is $\vee$-closed. In order to get to the $\vee$-closure of an arbitrary (wild) set of partial isometries it might be necessary to iterate the above construction more than once.

In order to prove that $\mathfrak{S}$ is closed under multiplication, we pick two elements in $\mathfrak{S}$, say $\vee T$ and $\vee U$, where T and U are (finite) compatible subsets of S. Recalling that S is assumed to be closed under multiplication, we have that $TU \subseteq S$, so clearly TU consists of partial isometries, which is equivalent to saying that the initial projections of the members of T commute with the final projections of the members of U. From (12.28) we then conclude that

$$(\vee T)(\vee U) = \vee(TU) \in \mathfrak{S},$$

proving that $\mathfrak{S}$ is closed under multiplication.

In order to prove that $\mathfrak{S}$ is also closed under adjoints, pick a generic element $\vee T \in \mathfrak{S}$, where T is a (finite) compatible subset of S. Then it is easy to see that T^* is also a compatible set, and

$$(\vee T)^* = \vee(T^*) \in \mathfrak{S}.$$

□

An easy consequence of the above result is in order:

Corollary 12.31. *The finite-$\vee$-closure (resp. $\vee$-closure) of any tame set of partial isometries in a C^*-algebra (resp. von-Neumann algebra) is tame.*

PROOF. Given a tame set T of partial isometries in a C*-algebra (resp. von-Neumann algebra) A, let S be the multiplicative sub-semigroup of A generated by $T \cup T^*$. Then S satisfies the hypothesis of (12.30), hence the finite-$\vee$-closure (resp. $\vee$-closure) of S, which we denote by $\mathfrak{S}$, is a self-adjoint multiplicative semigroup of partial isometries, therefore necessarily tame.

The finite-$\vee$-closure (resp. $\vee$-closure) of T is then clearly a subset of $\mathfrak{S}$, hence it is also tame. □

Notes and remarks. Partial isometries have been a topic of interest for a long time. Among other things it appears in the polar decomposition of bounded operators. A thorough study of single partial isometries is to be found in [**67**].

Proposition (12.8) has been reproved in the literature many times. The first reference for this result we know of is [**44**].

The notion of a tame set of partial isometries and its relationship to partial representations, as discussed in Proposition (12.13), first appeared in [**51**]. The order relation among partial isometries defined in (12.15) has been considered in [**67**] in the case of operators on Hilbert's space. As mentioned above, it is inspired by the usual order relation on inverse semigroups.

CHAPTER 13

Covariant Representations of C*-Algebraic Dynamical Systems

In (9.11.iii) we have seen that a covariant representation of an algebraic partial dynamical system gives rise to a representation of the crossed product. In the paragraph following (9.11) we have also noticed that a converse of this result is easily obtained in the case of unital ideals. In the present chapter we will prove a similar converse for C*-algebras, despite the fact that ideals are not always unital.

► Let us therefore fix a C*-algebraic partial dynamical system

$$\big(A,\ G,\ \{D_g\}_{g\in G},\ \{\theta_g\}_{g\in G}\big),$$

for the duration of this chapter.

The appropriate definition of covariant representations of a C*-algebraic partial dynamical system is identical to Definition (9.10), with the understanding that the target algebra B mentioned there will always be a C*-algebra. In respect to this, it is noteworthy that every *-homomorphism between C*-algebras is automatically continuous.

Quite often the target algebra for our covariant representations will be taken to be the C*-algebra $\mathscr{L}(H)$ consisting of all bounded linear operators on a Hilbert space H.

Proposition 13.1. *Given a covariant representation (π, u) of a C*-algebraic partial dynamical system $\big(A, G, \{D_g\}_{g\in G}, \{\theta_g\}_{g\in G}\big)$ in a unital C*-algebra B, there exists a unique *-homomorphism*

$$\pi\times u : A\rtimes G\to B,$$

such that $(\pi\times u)(a\delta_g)=\pi(a)u_g$, for all g in G, and all a in D_g.

Proof. Let us use $\pi\times^{alg} u$ to denote the *-homomorphism from $A\rtimes_{\text{alg}} G$ to B, provided by (9.11.iii). The conclusion then follows immediately by applying (11.14) to $\pi\times^{alg} u$. □

The following is the main result of this chapter:

Theorem 13.2. *Given a C*-algebraic partial action*

$$\theta=\big(\{D_g\}_{g\in G},\ \{\theta_g\}_{g\in G}\big)$$

of a group G on a C-algebra A, and a non-degenerate[14] *-representation*

$$\rho : A\rtimes G\to\mathscr{L}(H),$$

[14] A *-representation $\rho : B\to\mathscr{L}(H)$, of a C*-algebra B on a Hilbert space H is said to be *non-degenerate* if $[\rho(B)H]=H$ (closed linear span). This is the same as saying that ρ is a non-degenerate *-homomorphism into the multiplier algebra of $\mathscr{K}(H)$, but it is not equivalent to the much stronger property stating that $\mathscr{L}(H)=[\varphi(A)\mathscr{L}(H)]$.

where H is a Hilbert space, there exists a unique covariant representation (π, u) of θ in $\mathscr{L}(H)$, such that:

(i) *π is a non-degenerate representation of A,*
(ii) *$u_g u_{g^{-1}}$ is the orthogonal projection onto $[\pi(D_g)H]$ (brackets meaning closed linear span),*
(iii) *$\rho = \pi \times u$.*

PROOF. Denote by π the representation of A on H given by

$$\pi(a) = \rho(a\delta_1), \quad \forall\, a \in A.$$

To see that π is non-degenerate, observe that, since ρ is non-degenerate and since $A \rtimes_{\text{alg}} G$ is dense in $A \rtimes G$, the restriction of ρ to the former is non-degenerate. Thus, the set

$$X := \{\rho(a\delta_g)\eta : g \in G,\ a \in D_g,\ \eta \in H\}$$

spans a dense subset of H. Given any element ξ in X, say $\xi = \rho(a\delta_g)\eta$, as above, use the Cohen-Hewitt Theorem [**68**, 32.22] to write $a = bc$, with $b, c \in D_g$. Then

$$\xi = \rho(a\delta_g)\eta \overset{(8.14)}{=} \rho(b\delta_1 c\delta_g)\eta = \rho(b\delta_1)\rho(c\delta_g)\eta = \pi(b)\rho(c\delta_g)\eta \in \pi(A)H.$$

This shows that $X \subseteq \pi(A)H$, so $\pi(A)H$ also spans a dense subset of H, meaning that π is non-degenerate, as desired. For each g in G, let

$$H_g = [\pi(D_g)H].$$

We should remark that, again by the Cohen-Hewit Theorem, every element ξ in H_g may be written as

$$\xi = \pi(a)\eta,$$

for some a in D_g, and η in H, which means that

$$H_g = \pi(D_g)H, \tag{13.2.1}$$

even without taking closed linear span.

Letting $\{v_i\}_{i\in I}$ be an approximate identity[15] for D_g, we claim that the orthogonal projection onto H_g, which we henceforth denote by e_g, is given by

$$e_g(\xi) = \lim_{i\to\infty} \pi(v_i)\xi, \quad \forall\, \xi \in H. \tag{13.2.2}$$

In fact, given ξ in H, write $\xi = \xi_1 + \xi_2$, with ξ_1 in H_g, and ξ_2 in $H_g^\perp$. For every a in D_g, and every η in H, observe that

$$\langle \pi(a)\xi_2, \eta\rangle = \langle \xi_2, \underbrace{\pi(a^*)\eta}_{\in H_g}\rangle = 0,$$

from where we see that $\pi(a)\xi_2 = 0$. In particular $\pi(v_i)\xi_2 = 0$, for all $i \in I$. On the other hand, let us use (13.2.1) in order to write $\xi_1 = \pi(a)\eta$, where $a \in D_g$, and

[15] An *approximate identity* for a C*-algebra B is a net $\{v_i\}_{i\in I} \subseteq B$ of positive elements, with $\|v_i\| \leq 1$, such that $b = \lim_{i\to\infty} bv_i = \lim_{i\to\infty} v_i b$, for every b in B. Every C*-algebra (and hence also every ideal in a C*-algebra) is known to admit an approximate identity.

$\eta \in H$. Then

$$\begin{aligned}\lim_{i\to\infty} \pi(v_i)\xi &= \lim_{i\to\infty} \pi(v_i)\big(\pi(a)\eta + \xi_2\big)\\ &= \lim_{i\to\infty} \pi(v_i a)\eta\\ &= \pi\big(\lim_{i\to\infty} v_i a\big)\eta\\ &= \pi(a)\eta = \xi_1 = e_g(\xi),\end{aligned}$$

proving (13.2.2).

It is clear that each H_g is invariant under π, so one may easily prove that e_g commutes with $\pi(a)$, for every a in A. We further claim that, given another group element $h \in G$, one has

$$e_g e_h = e_h e_g. \tag{13.2.3}$$

Indeed, given ξ in H, observe that

$$\begin{aligned}e_h\big(e_g(\xi)\big) &= e_h\big(\lim_{i\to\infty} \pi(v_i)\xi\big)\\ &= \lim_{i\to\infty} e_h\big(\pi(v_i)\xi\big)\\ &= \lim_{i\to\infty} \pi(v_i)\big(e_h(\xi)\big)\\ &= e_g\big(e_h(\xi)\big).\end{aligned}$$

Another fact we will need later is

$$H_g \cap H_h = [\pi(D_g \cap D_h)H], \quad \forall\, g, h \in G. \tag{13.2.4}$$

In order to prove it, we note that the inclusion "$\supseteq$" is evident, while the reverse inclusion may be proved as follows: pick any ξ in $H_g \cap H_h$, and emphasizing the role of h, write $\xi = \pi(a)\eta$, with a in D_h, and η in H, by (13.2.1). Once more employing our approximate identity $\{v_i\}_{i\in I}$ for D_g, we then have

$$\begin{aligned}\xi &= e_g(\xi)\\ &= e_g\big(\pi(a)\eta\big)\\ &= \lim_{i\to\infty} \pi(v_i)\pi(a)\eta\\ &= \lim_{i\to\infty} \pi(v_i a)\eta \in [\pi(D_g \cap D_h)H],\end{aligned}$$

proving that

$$H_g \cap H_h \subseteq [\pi(D_g \cap D_h)H].$$

As already observed, Cohen-Hewit implies that $\pi(D_g \cap D_h)H$ (without taking closed linear span) is a closed linear subspace of H, so (13.2.4) is verified.

In order to kick-start the construction of the partial representation u referred to in the statement, let us now construct, for any given g in G, an isometric linear operator $v_g : H_{g^{-1}} \to H$, satisfying

$$v_g\big(\pi(a)\xi\big) = \rho\big(\theta_g(a)\delta_g\big)\xi, \quad \forall\, a \in D_{g^{-1}}, \quad \forall\, \xi \in H.$$

With this goal in mind we claim that, for all $a_1, \ldots, a_n \in D_{g^{-1}}$, and all $\xi_1, \ldots, \xi_n \in H$, one has that

$$\Big\| \sum_{i=1}^{n} \pi(a_i)\xi_i \Big\| = \Big\| \sum_{i=1}^{n} \rho\big(\theta_g(a_i)\delta_g\big)\xi_i \Big\|. \tag{13.2.5}$$

Indeed, starting from the right-hand-side above, we have

$$\Big\| \sum_{i=1}^{n} \rho\big(\theta_g(a_i)\delta_g\big)\xi_i\Big\|^2 = \Big\langle \sum_{i,j=1}^{n} \rho\Big(\big(\theta_g(a_j)\delta_g\big)^*\big(\theta_g(a_i)\delta_g\big)\Big)\xi_i, \xi_j\Big\rangle$$
$$\overset{(8.14)}{=} \Big\langle \sum_{i,j=1}^{n} \pi(a_j^* a_i)\xi_i, \xi_j\Big\rangle$$
$$= \Big\langle \sum_{i=1}^{n} \pi(a_i)\xi_i, \sum_{j=1}^{n} \pi(a_j)\xi_j\Big\rangle$$
$$= \Big\| \sum_{i=1}^{n} \pi(a_i)\xi_i\Big\|^2,$$

proving (13.2.5). This said, the correspondence

$$\sum_{i=1}^{n} \pi(a_i)\xi_i \mapsto \sum_{i=1}^{n} \rho\big(\theta_g(a_i)\delta_g\big)\xi_i,$$

may now be shown to provide a well defined isometric linear map on the linear span of $\pi(D_{g^{-1}})H$, which we have seen coincides with $H_{g^{-1}}$ by (13.2.1), providing an isometric linear operator v_g on $H_{g^{-1}}$ satisfying the desired properties.

Next we show that the range of v_g is precisely H_g. For this, given $a \in D_{g^{-1}}$, write $a = bc$, with $b, c \in D_{g^{-1}}$, by the Cohen-Hewitt Theorem. So, for all $\xi \in H$, we have that

$$v_g\big(\pi(a)\xi\big) = \rho\big(\theta_g(bc)\delta_g\big)\xi = \rho\big(\theta_g(b)\delta_1\big)\ \rho\big(\theta_g(c)\delta_g\big)\ \xi \in \pi\big(\theta_g(b)\big)H \subseteq H_g,$$

proving that $v_g(H_{g^{-1}}) \subseteq H_g$. On the other hand, given any a in D_g and ξ in H, use Cohen-Hewit again to write $\theta_{g^{-1}}(a) = bc$, with $b, c \in D_{g^{-1}}$. Then

$$a\delta_1 = \theta_g(bc)\delta_1 \overset{(8.14)}{=} \big(\theta_g(b)\delta_g\big)\ (c\delta_{g^{-1}}),$$

so

$$\pi(a)\xi = \rho(a\delta_1)\xi = \rho\big(\theta_g(b)\delta_g\big)\underbrace{\rho(c\delta_{g^{-1}})\xi}_{\eta} = v_g\big(\pi(b)\eta\big) \in v_g(H_{g^{-1}}),$$

where η is as indicated. It follows that $H_g \subseteq v_g(H_{g^{-1}})$, thus showing that v_g is indeed an isometric operator from $H_{g^{-1}}$ onto H_g.

Taking one step closer to our desired partial representation, for each g in G, we let

$$u_g : H \to H$$

be the linear operator defined on the whole of H by extending v_g to be zero on the orthogonal complement of $H_{g^{-1}}$. It is then clear that u_g is a partial isometry in $\mathscr{L}(H)$, with initial space $H_{g^{-1}}$ and final space H_g. Consequently the orthogonal projection onto H_g satisfies

$$e_g = u_g u_g^*. \tag{13.2.6}$$

If a is in $D_{g^{-1}}$ we have by definition that $u_g\pi(a)\xi = \rho\big(\theta_g(a)\delta_g\big)\xi$, for all ξ in H, which means that

$$u_g\pi(a) = \rho\big(\theta_g(a)\delta_g\big), \quad \forall\, a \in D_{g^{-1}}. \tag{13.2.7}$$

Having shown that π is non-degenerate, it is evident that $H_1 = H$, and it is easy to see that u_1 is the identity operator on H, as required by (9.1.i).

We next claim that

$$u_g\big(u_{g^{-1}}(\xi)\big) = \xi, \quad \forall\, g \in G, \quad \forall\, \xi \in H_g. \tag{13.2.8}$$

Assuming, as we may, that $\xi = \pi(a)\eta$, with $a \in D_g$, and $\eta \in H$, write $\theta_{g^{-1}}(a) = bc$, with $b, c \in D_{g^{-1}}$, by Cohen-Hewit, and observe that

$$\begin{aligned} u_{g^{-1}}(\xi) &= u_{g^{-1}}\big(\pi(a)\eta\big) \\ &= \rho\big(\theta_{g^{-1}}(a)\delta_{g^{-1}}\big)\eta \\ &= \rho(b\delta_1\ c\delta_{g^{-1}})\eta \\ &= \pi(b)\ \underbrace{\rho(c\delta_{g^{-1}})\eta}_{\zeta} \\ &= \pi(b)\zeta. \end{aligned}$$

Therefore the left-hand-side of (13.2.8) equals

$$\begin{aligned} &u_g\big(u_{g^{-1}}(\xi)\big) \\ &\qquad = u_g\big(\pi(b)\zeta\big) \\ &\qquad = \rho\big(\theta_g(b)\delta_g\big)\zeta \\ &\qquad = \rho\big(\theta_g(b)\delta_g\big)\ \rho(c\delta_{g^{-1}})\eta \\ &\qquad = \rho\big(\theta_g(bc)\delta_1\big)\eta \\ &\qquad = \pi(a)\eta = \xi, \end{aligned}$$

proving (13.2.8), from where we deduce that

$$u_{g^{-1}} = u_g^*, \quad \forall\, g \in G, \tag{13.2.9}$$

hence verifying (9.1.iv) and, in view of (13.2.6), this also proves condition (ii) in the statement.

We next focus on proving (9.1.ii), namely

$$u_{gh}u_{h^{-1}} = u_g u_h u_{h^{-1}},$$

for all $g, h \in G$. As a first step we claim that

$$u_{gh}u_{h^{-1}}(\eta) = u_g(\eta), \quad \forall\, \eta \in H_h \cap H_{g^{-1}}. \tag{13.2.10}$$

In order to do this we use (13.2.4) to write $\eta = \pi(a)\xi$, with a in $D_h \cap D_{g^{-1}}$. Observing that

$$\theta_{h^{-1}}(a) \in D_{h^{-1}} \cap D_{h^{-1}g^{-1}},$$

we may write $\theta_{h^{-1}}(a) = bc$, with b and c in $D_{h^{-1}} \cap D_{h^{-1}g^{-1}}$. Therefore

$$\begin{aligned} &u_{h^{-1}}(\eta) \\ &\qquad = u_{h^{-1}}\big(\pi(a)\xi\big) \\ &\qquad = \rho\big(\theta_{h^{-1}}(a)\delta_{h^{-1}}\big)\xi \\ &\qquad = \rho(bc\delta_{h^{-1}})\xi \\ &\qquad = \rho(b\delta_1)\ \rho(c\delta_{h^{-1}})\xi \\ &\qquad = \pi(b)\ \rho(c\delta_{h^{-1}})\xi. \end{aligned}$$

So the left-hand-side of (13.2.10) equals

$$\begin{aligned} &u_{gh}\big(u_{h^{-1}}(\eta)\big) \\ &\qquad = u_{gh}\big(\pi(b)\,\rho(c\delta_{h^{-1}})\xi\big) \\ &\qquad = \rho\big(\theta_{gh}(b)\delta_{gh}\big)\,\rho(c\delta_{h^{-1}})\xi \\ &\qquad = \rho\big(\theta_{gh}(bc)\delta_g\big)\xi \\ &\qquad = \rho\big(\theta_g(a)\delta_g\big)\xi \\ &\qquad = u_g\big(\pi(a)\xi\big) \\ &\qquad = u_g(\eta), \end{aligned}$$

taking care of (13.2.10).

Observing that the right-hand-side of (9.1.ii) equals $u_g e_h$, our next claim is that

$$u_g e_h = u_{gh} u_{h^{-1}} e_{g^{-1}}, \quad \forall g, h \in G. \tag{13.2.11}$$

In order to prove this we have

$$\begin{aligned} u_g e_h &= u_g e_{g^{-1}} e_h \\ &\overset{(13.2.10)}{=} u_{gh} u_{h^{-1}} e_{g^{-1}} e_h \\ &\overset{(13.2.3)}{=} u_{gh} u_{h^{-1}} e_h e_{g^{-1}} \\ &= u_{gh} u_{h^{-1}} e_{g^{-1}}, \end{aligned}$$

proving (13.2.11). Taking adjoints in (13.2.11), using (13.2.9), and changing variables appropriately leads to

$$e_h u_g = e_g u_h u_{h^{-1}g}. \tag{13.2.12}$$

Focusing now on the left-hand-side of (9.1.ii), observe that

$$u_{gh} u_{h^{-1}} = u_{gh} e_{(gh)^{-1}} u_{h^{-1}} \overset{(13.2.12)}{=} u_{gh} e_{h^{-1}} u_{(gh)^{-1}} u_g.$$

Since $u_g = u_g e_{g^{-1}}$, the right-hand-side above is unaffected by right multiplication by $e_{g^{-1}}$, and so is the left-hand-side. Thus

$$u_{gh} u_{h^{-1}} = u_{gh} u_{h^{-1}} e_{g^{-1}} \overset{(13.2.11)}{=} u_g e_h = u_g u_h u_{h^{-1}},$$

proving (9.1.ii). It has already been observed that (9.1.iii) is implied by (9.1.ii) and (9.1.iv), so the verification that u is a partial representation is complete. We now verify that (π, u) is a covariant representation, which is to say that

$$u_g \pi(a) u_{g^{-1}} = \pi\big(\theta_g(a)\big), \tag{13.2.13}$$

for any given g in G, and a in $D_{g^{-1}}$. To prove it, let $\xi \in H$, and let us first suppose that ξ is in $H_g^{\perp}$. Then, evidently $u_{g^{-1}}(\xi) = 0$, so the operator on the left-hand-side of (13.2.13) vanishes on ξ. We claim that the same is true with respect to the operator on the right-hand-side. Indeed, given any η in H, we have that

$$\big\langle \pi\big(\theta_g(a)\big)\xi, \eta\big\rangle = \big\langle \xi, \underbrace{\pi\big(\theta_g(a^*)\big)\eta}_{\in H_g} \big\rangle = 0.$$

Since η is arbitrary, we have that $\pi\big(\theta_g(a)\big)\xi = 0$, as claimed.

Suppose now that ξ is in H_g. Then we may write $\xi = \pi(b)\eta$, for $b \in D_g$, and $\eta \in H$, by (13.2.1), and

$$\begin{aligned} u_g\pi(a)u_{g^{-1}}(\xi) &= u_g\pi(a)u_{g^{-1}}\big(\pi(b)\eta\big) \\ &\overset{(13.2.7)}{=} \rho\big(\theta_g(a)\delta_g\big)\rho\big(\theta_{g^{-1}}(b)\delta_{g^{-1}}\big)\eta \\ &= \rho\Big(\theta_g(a)\delta_g\ \theta_{g^{-1}}(b)\delta_{g^{-1}}\Big)\eta \\ &\overset{(8.14)}{=} \rho\big(\theta_g(a)b\delta_1\big)\eta \\ &= \pi\big(\theta_g(a)\big)\pi(b)\eta \\ &= \pi\big(\theta_g(a)\big)\xi. \end{aligned}$$

This concludes the proof of (13.2.13), and hence that (π, u) is a covariant representation as needed. In order to show point (iii) in the statement, it clearly suffices to prove that

$$(\pi \times u)(a\delta_g) = \rho(a\delta_g), \tag{13.2.14}$$

for every g in G, and $a \in D_g$, which we will now do. Writing $a = bc$, with b and c in D_g, we have

$$\begin{aligned} \rho(a\delta_g) &\overset{(8.14)}{=} \rho\big(b\delta_g\,\theta_{g^{-1}}(c)\delta_1\big) \\ &= \rho(b\delta_g)\pi\big(\theta_{g^{-1}}(c)\big) \\ &\overset{(13.2.13)}{=} \rho(b\delta_g)u_{g^{-1}}\pi\big(c\big)u_g \\ &\overset{(13.2.7)}{=} \rho(b\delta_g)\rho\big(\theta_{g^{-1}}(c)\delta_{g^{-1}}\big)u_g \\ &\overset{(8.14)}{=} \rho(bc\delta_1)u_g \\ &= \pi(a)u_g \\ &= (\pi \times u)(a\delta_g), \end{aligned}$$

so (13.2.14) is verified.

We are therefore left with the only remaining task of proving uniqueness of the (π, u). With this goal in mind, assume that (π', u') is another covariant representation of our system in $\mathscr{L}(H)$, such that $\pi' \times u' = \rho$. For every a in A we then have that $\pi'(a) = \rho(a\delta_1) = \pi(a)$, so π and π' must coincide.

Given g in G, and ξ in $H_{g^{-1}}$, write $\xi = \pi(a)\eta$, with a in $D_{g^{-1}}$, and η in H, by (13.2.1). Then

$$\begin{aligned} &u'_g(\xi) \\ &\quad= u'_g\pi(a)\eta \\ &\quad= u'_g\pi'(a)\eta \\ &\quad= \big(\pi'(a^*)u'_{g^{-1}}\big)^*\eta \\ &\quad= \big((\pi' \times u')(a^*\delta_{g^{-1}})\big)^*\eta \\ &\quad= \big((\pi \times u)(a^*\delta_{g^{-1}})\big)^*\eta \\ &\quad= u_g\pi(a)\eta = u_g(\xi), \end{aligned}$$

which says that u'_g coincides with u_g on $H_{g^{-1}}$. By (ii) we have that both u'_g and u_g vanish on $H_{g^{-1}}^{\perp}$, so u'_g coincides with u_g on the whole of H. This proves that $u' = u$, and hence completes the proof of the uniqueness part. □

Notes and remarks. Theorem (13.2) is due to McClanahan [**81**, Proposition 2.8]. The above proof is inspired by [**59**, Theorem 1.3][16], except that we have avoided the use of approximate identities whenever possible, basing the arguments on the Cohen-Hewit Theorem instead. Even though the proof presented here turned out to be a bit long, we have chosen it because we believe a possible generalization of this result to algebraic partial actions will most likely use arguments based on idempotency rather than convergence of limits.

[16]Please note that [**59**] is the preprint version of [**60**].

CHAPTER 14

Partial Representations Subject to Relations

In the present chapter we will describe one of the most efficient ways to show a given C*-algebra to have a partial crossed product description.

In order to motivate our method, suppose that B is a unital C*-algebra defined in terms of generators and relations. Suppose also that the given relations imply that the generators are partial isometries or even, as it often happens, explicitly state this fact. Let us moreover suppose that these partial isometries form a tame set so, by (12.13), we may find a *-partial representation u of a group G (quite often a free group) whose range also generates B. Under all of these favorable circumstances B may then be given another presentation in which the set of generators turns out to be the range of a *-partial representation, and we may then consider the partial dynamical system described in (10.1).

Having the same goal as Theorem (10.3), although with different hypothesis, and in a different category, the main result of this chapter is intended to specify conditions under which the homomorphism given in (10.2) is in fact an isomorphism. Another major objective of this chapter is to give a very concrete picture of the spectrum of the commutative algebra involved in this system.

Given a group G, we will be dealing with universal[17] unital C*-algebras on the set of generators

$$\mathscr{G} = \{u_g : g \in G\},$$

subject to a given set of relations. In all instances below this set of relations will split as

$$\mathcal{R}' \cup \mathcal{R},$$

where $\mathcal{R}'$ consists precisely of relations (9.1.i)–(9.1.iv), which is to say that the correspondence $g \mapsto u_g$ is a *-partial representation of G in B. With respect to the remaining set of relations, namely $\mathcal{R}$, we will always make the assumption that it consists of algebraic relations involving only the e_g (defined to be $e_g = u_g u_{g^{-1}}$). By this we mean relations of the form

$$p(e_{g_1}, e_{g_2}, \ldots, e_{g_n}) = 0, \tag{14.1}$$

where p is a complex polynomial in n variables, and the g_i are in G. Recall that the e_g commute by (9.8.iv), so it is OK to apply a polynomial in n commuting variables to them.

Definition 14.2. Given a set $\mathcal{R}$ of relations of the form (14.1), we will denote by $C^*_{\text{par}}(G, \mathcal{R})$ the universal unital C*-algebra generated by the set $\mathscr{G} = \{u_g : g \in G\}$, subject to the set of relations $\mathcal{R}' \cup \mathcal{R}$, where $\mathcal{R}'$ consists of the relations (9.1.i)–(9.1.iv).

[17] See [**13**] for a definition of universal C*-algebras given by generators and relations.

Notice that, since the set $\mathcal{R}'$ is always involved by default, we have decided not to emphasize it in the notation introduced above. The subscript "$_{\rm par}$" should be enough to remind us that the axioms for *-partial representations, namely $\mathcal{R}'$, is also being taken into account.

Definition 14.3. Let v be a *-partial representation of a group G in a unital C*-algebra B, and let $\mathcal{R}$ be a set of relations of the form (14.1). We will say that v *satisfies* $\mathcal{R}$ if, for every relation "$p(e_{g_1}, e_{g_2}, \ldots, e_{g_n}) = 0$" in $\mathcal{R}$, one has that

$$p(v_{g_1}v_{g_1^{-1}}, v_{g_2}v_{g_2^{-1}}, \ldots, v_{g_n}v_{g_n^{-1}}) = 0.$$

The universal property of $C^*_{\rm par}(G, \mathcal{R})$ may then be expressed as follows.

Proposition 14.4. *For every *-partial representation v of G in a unital C*-algebra B, satisfying $\mathcal{R}$, there exists a unique *-homomorphism*

$$\varphi : C^*_{par}(G, \mathcal{R}) \to B,$$

such that $\varphi(u_g) = v_g$, for every $g \in G$.

PROOF. The proof follows immediately from the universality of $C^*_{\rm par}(G, \mathcal{R})$, once we realize that, besides satisfying $\mathcal{R}$, the v_g also satisfy $\mathcal{R}'$, as a consequence of it being a *-partial representation. □

As an example of relations of the form (14.1) one is allowed to express that a certain u_g is an isometry, since this reads as "$(u_g)^*u_g = 1$", and may therefore be expressed as "$e_{g^{-1}} - 1 = 0$." As another example, one may express that two partial isometries u_g and u_h have orthogonal ranges by writing down the relation "$e_g e_h = 0$". On the other hand we are ruling out relations such as "$u_g + u_h = u_k$", since these are not of the above form.

We should however remark that certain relations which do not immediately fit under (14.1) may sometimes be given an equivalent formulation within that framework, an example of which we would now like to present.

Proposition 14.5. *Let u be a *-partial representation of the group G in a unital C*-algebra. Given two elements g and h in G, the following are equivalent*

(*i*) $u_{gh} = u_g u_h$,

(*ii*) $e_{gh} = e_{gh} e_g$,

where $e_g = u_g u_{g^{-1}}$, as usual.

PROOF. By the C*-identity (11.1.iii), we have

$$\begin{aligned} \|u_{gh} - u_g u_h\|^2 &= \|(u_{gh} - u_g u_h)(u_{gh} - u_g u_h)^*\| \\ &= \|u_{gh}u_{h^{-1}g^{-1}} - u_{gh}u_{h^{-1}}u_{g^{-1}} - u_g u_h u_{h^{-1}g^{-1}} + u_g u_h u_{h^{-1}} u_{g^{-1}}\| \\ &\overset{(9.1.\text{iii})}{=} \|e_{gh} - u_{gh}u_{h^{-1}}u_{g^{-1}}\|. \end{aligned}$$

In addition, we have

$$\begin{aligned} u_{gh}u_{h^{-1}}u_{g^{-1}} &= u_{gh}u_{h^{-1}}u_{g^{-1}}u_g u_{g^{-1}} \\ &= u_{gh}u_{h^{-1}g^{-1}}u_g u_{g^{-1}} \\ &= e_{gh}e_g, \end{aligned}$$

which plugged above gives

$$\|u_{gh} - u_g u_h\|^2 = \|e_{gh} - e_{gh}e_g\|,$$

from where the statement follows immediately. □

As an important application of this idea, one may phrase the fact that a *-partial representation is semi-saturated (see Definition (9.7)), relative to a given length function ℓ on G, by requiring it to satisfy the set of relations

$$e_{gh} - e_{gh}e_g = 0,$$

for all g and h in G such that $\ell(gh) = \ell(g) + \ell(h)$.

One of the main goals in this chapter is to prove that $C^*_{\text{par}}(G, \mathcal{R})$ is isomorphic to a partial crossed product of the form

$$C(\Omega_{\mathcal{R}}) \rtimes G.$$

The partial dynamical system involved in this result will be presented as a restriction of the partial Bernoulli action, introduced in (5.12), to a closed invariant subset (see (2.10)). The first step towards this goal will therefore be to describe the subset $\Omega_{\mathcal{R}} \subseteq \Omega_1$ mentioned above.

Let us consider, for each g in G, the mapping

$$\varepsilon_g : \{0,1\}^G \to \{0,1\}, \tag{14.6}$$

defined by

$$\varepsilon_g(\omega) = \omega_g = [g \in \omega], \quad \forall \omega \in \{0,1\}^G,$$

(see (5.8)). Seeing $\{0,1\}^G$ as a product space, the ε_g are precisely the standard projections, hence continuous functions.

Consider a relation

$$p(e_{g_1}, e_{g_2}, \ldots, e_{g_n}) = 0,$$

of the form (14.1). If we replace each e_g by ε_g in the left-hand-side above we get

$$p(\varepsilon_{g_1}, \varepsilon_{g_2}, \ldots, \varepsilon_{g_n}),$$

which may be interpreted as a complex valued function on $\{0,1\}^G$, namely

$$\omega \in \{0,1\}^G \;\mapsto\; p\big(\varepsilon_{g_1}(\omega), \varepsilon_{g_2}(\omega), \ldots, \varepsilon_{g_n}(\omega)\big) \in \mathbb{C}. \tag{14.7}$$

Definition 14.8. Given a set $\mathcal{R}$ of relations of the form (14.1), we will let $\mathcal{F}_{\mathcal{R}}$ be the set of functions on $\{0,1\}^G$ of the form (14.7), obtained via the above substitution procedure from each relation in $\mathcal{R}$. The *spectrum* of $\mathcal{R}$, denoted by $\Omega_{\mathcal{R}}$, is then defined to be the subset of Ω_1 defined by

$$\Omega_{\mathcal{R}} = \big\{\omega \in \Omega_1 : f(g^{-1}\omega) = 0, \ \forall f \in \mathcal{F}_{\mathcal{R}}, \ \forall g \in \omega\big\}.$$

Proposition 14.9. *For any set $\mathcal{R}$ of relations of the form* (14.1), *one has that $\Omega_{\mathcal{R}}$ is a compact subset of Ω_1, which is moreover invariant under the partial Bernoulli action β defined in* (5.12).

PROOF. In order to prove that $\Omega_{\mathcal{R}}$ is closed in Ω_1, let $\{\omega_i\}_{i \in I}$ be a net in $\Omega_{\mathcal{R}}$ converging to some ω in Ω_1. Given $g \in \omega$, observe that $\varepsilon_g(\omega) = 1$, so ω lies in the open set $\varepsilon_g^{-1}(\{1\})$. We may therefore assume, without loss of generality, that every ω_i lie in $\varepsilon_g^{-1}(\{1\})$. This is to say that $g \in \omega_i$, for every $i \in I$, and since ω_i is in $\Omega_{\mathcal{R}}$, we deduce that

$$f(g^{-1}\omega_i) = 0, \quad \forall i \in I, \quad \forall f \in \mathcal{F}_{\mathcal{R}}.$$

Observing that the correspondence

$$\nu \in \{0,1\}^G \mapsto f(g^{-1}\nu) \in \mathbb{C}$$

is a continuous mapping, we conclude that

$$f(g^{-1}\omega) = \lim_{i \to \infty} f(g^{-1}\omega_i) = 0,$$

proving that $\omega \in \Omega_{\mathcal{R}}$, and hence that $\Omega_{\mathcal{R}}$ is closed. Since Ω_1 is compact, then so is $\Omega_{\mathcal{R}}$.

Recall that the partial Bernoulli action $\beta = \big(\{D_g\}_{g\in G}, \{\beta_g\}_{g\in G}\big)$ is given by

$$D_g = \{\omega \in \Omega_1 \ : \ g \in \omega\}, \quad \text{and} \quad \beta_g(\omega) = g\omega, \quad \forall\, \omega \in D_{g^{-1}}.$$

In order to prove invariance, we must show that $\beta_g(\Omega_{\mathcal{R}} \cap D_{g^{-1}}) \subseteq \Omega_{\mathcal{R}}$, so pick ω in $\Omega_{\mathcal{R}} \cap D_{g^{-1}}$, and let $f \in \mathcal{F}_{\mathcal{R}}$, and $h \in \beta_g(\omega)$ be given. Then $g^{-1}h \in \omega$, and since $\omega \in \Omega_{\mathcal{R}}$, we have that

$$0 = f\big((g^{-1}h)^{-1}\omega\big) = f(h^{-1}g\omega) = f\big(h^{-1}\beta_g(\omega)\big),$$

proving that $\beta_g(\omega)$ is in $\Omega_{\mathcal{R}}$. □

► From now on we will fix an arbitrary set $\mathcal{R}$ of relations of the form (14.1). We may then use (2.10) to restrict the partial Bernoulli action to $\Omega_{\mathcal{R}}$, obtaining a partial action

$$\theta_{\mathcal{R}} = \big(\{D_g^{\mathcal{R}}\}_{g\in G}, \{\theta_g^{\mathcal{R}}\}_{g\in G}\big), \tag{14.10}$$

where

$$D_g^{\mathcal{R}} = \Omega_{\mathcal{R}} \cap D_g = \{\omega \in \Omega_{\mathcal{R}} : g \in \omega\}, \tag{14.11}$$

and

$$\theta_g^{\mathcal{R}}(\omega) = g\omega, \quad \forall\, \omega \in D_{g^{-1}}.$$

It should be noticed that, since each D_g is open in Ω_1, one has that $D_g^{\mathcal{R}}$ is open in $\Omega_{\mathcal{R}}$. This, plus the obvious fact that each $\theta_g^{\mathcal{R}}$ is continuous, means that $\theta_{\mathcal{R}}$ is a topological partial action of G on $\Omega_{\mathcal{R}}$.

Observe that $D_g^{\mathcal{R}}$ is also closed in $\Omega_{\mathcal{R}}$ by (14.11), so (5.7) applies and hence we see that $\theta_{\mathcal{R}}$ admits a Hausdorff globalization. In fact we may describe the globalization of $\theta_{\mathcal{R}}$ without resorting to (5.7) as follows: define

$$X_{\mathcal{R}} = \big\{\omega \in \{0,1\}^G : f(g^{-1}\omega) = 0, \ \forall f \in \mathcal{F}_{\mathcal{R}}, \ \forall g \in \omega\big\},$$

that is, $X_{\mathcal{R}}$ is defined as in (14.8), except that we have replaced "$\omega \in \Omega_1$" by "$\omega \in \{0,1\}^G$". In other words,

$$\Omega_{\mathcal{R}} = X_{\mathcal{R}} \cap \Omega_1.$$

By staring at the above definition of $X_{\mathcal{R}}$, it is clear that $X_{\mathcal{R}}$ is invariant under the global Bernoulli action η described in (5.9).

Notice that $\omega = \varnothing$, namely the empty set, is a member of $X_{\mathcal{R}}$ since the membership condition above is vacuously verified for $\varnothing$. If ω is any element of $X_{\mathcal{R}}$ other than $\varnothing$, choose g in ω, and notice that $1 \in g^{-1}\omega = \eta_{g^{-1}}(\omega)$, so

$$\eta_{g^{-1}}(\omega) \in X_{\mathcal{R}} \cap \Omega_1 = \Omega_{\mathcal{R}},$$

whence

$$\omega = \eta_g\big(\eta_{g^{-1}}(\omega)\big) \in \eta_g(\Omega_{\mathcal{R}}).$$

With this it is easy to see that the orbit of $\Omega_{\mathcal{R}}$ under η is given by

$$\bigcup_{g\in G} \eta_g(\Omega_{\mathcal{R}}) = X_R \setminus \{\varnothing\}.$$

This proves the following:

Proposition 14.12. *The restriction of the global Bernoulli action η to the invariant subset $X_R \setminus \{\varnothing\}$ coincides with the globalization of the partial action $\theta_{\mathcal{R}}$ introduced in* (14.10).

Let us now turn our topological partial action into a C*-algebraic one. For this we invoke (11.6) to obtain a partial action of G on $C(\Omega_{\mathcal{R}})$, which, by abuse of language, we will also denote by $\theta_{\mathcal{R}}$.

As seen in our discussion after the definition of the partial Bernoulli action (5.12), D_g is a compact space for every g in G, and hence so is $D_g^{\mathcal{R}}$. The corresponding ideal of $C(\Omega_{\mathcal{R}})$, namely

$$C_0(D_g^{\mathcal{R}}) = C(D_g^{\mathcal{R}}),$$

is therefore a unital ideal. In particular, we are entitled to use (9.4) in order to get a *-partial representation

$$v : G \to C(\Omega_{\mathcal{R}}) \rtimes G, \tag{14.13}$$

defined by

$$v_g = 1_g \delta_g, \quad \forall g \in G,$$

where 1_g denotes the unit[18] of $C(D_g^{\mathcal{R}})$. Viewed within $C(\Omega_{\mathcal{R}})$, observe that 1_g is the characteristic function of $D_g^{\mathcal{R}}$. By the description of $D_g^{\mathcal{R}}$ given in (14.11), we may then write

$$1_g(\omega) = [g \in \omega], \quad \forall \omega \in \Omega_{\mathcal{R}},$$

which is to say that

$$1_g = \varepsilon_g|_{\Omega_{\mathcal{R}}}. \tag{14.14}$$

Lemma 14.15. *The *-partial representation v defined in* (14.13) *satisfies $\mathcal{R}$.*

PROOF. Given a relation "$p(e_{g_1}, e_{g_2}, \ldots, e_{g_n}) = 0$" in $\mathcal{R}$, we must therefore prove that

$$p(v_{g_1} v_{g_1^{-1}}, v_{g_2} v_{g_2^{-1}}, \ldots, v_{g_n} v_{g_n^{-1}}) = 0. \tag{†}$$

For any g in G, notice that

$$v_g v_{g^{-1}} = (1_g \delta_g)(1_{g^{-1}} \delta_{g^{-1}}) \overset{(8.14)}{=} 1_g \delta_1,$$

so, under the usual identification of $C(\Omega_{\mathcal{R}})$ as a subalgebra of $C(\Omega_{\mathcal{R}}) \rtimes G$ provided by (8.8), we may write

$$v_g v_{g^{-1}} = 1_g = \varepsilon_g|_{\Omega_{\mathcal{R}}}.$$

Therefore we see that the left-hand-side of (†) equals the restriction of

$$p(\varepsilon_{g_1}, \varepsilon_{g_2}, \ldots, \varepsilon_{g_n}),$$

to $\Omega_{\mathcal{R}}$. This is precisely one of the functions in $\mathcal{F}_{\mathcal{R}}$ which, by the very definition of $\Omega_{\mathcal{R}}$ (with $g = 1$) vanishes identically on $\Omega_{\mathcal{R}}$. This verifies (†) and hence completes the proof. □

[18]We could also use the heavier notation $1_g^{\mathcal{R}}$ for the unit of $C(D_g^{\mathcal{R}})$, but this will be unnecessary.

With this we arrive at the main result of this chapter:

Theorem 14.16. *Let G be a group and let $\mathcal{R}$ be a set of relations of the form* (14.1). *Then there exists a *-isomorphism*

$$\varphi : C^*_{par}(G, \mathcal{R}) \to C(\Omega_{\mathcal{R}}) \rtimes G,$$

such that

$$\varphi(u_g) = 1_g \delta_g, \quad \forall\, g \in G,$$

where 1_g denotes the characteristic function of $D^{\mathcal{R}}_g$.

Proof. By (14.15) and (14.4) there exists a *-homomorphism φ satisfying the conditions in the statement, and all we must do is prove φ to be an isomorphism.

We begin by building a covariant representation of $\theta_{\mathcal{R}}$ in $C^*_{\text{par}}(G, \mathcal{R})$. For this, let A be the closed *-subalgebra of $C^*_{\text{par}}(G, \mathcal{R})$ generated by the set $\{e_g : g \in G\}$. Noticing that A is commutative by (13.2.3), we denote by $\hat{A}$ its spectrum.

Consider the mapping

$$h : \gamma \in \hat{A} \mapsto \big(\gamma(e_g)\big)_{g \in G} \in \{0, 1\}^G.$$

Since each e_g is idempotent, we have that $\gamma(e_g) \in \{0, 1\}$, so the above is well defined and the reader will not have any difficulty in proving that h is continuous. We next claim that the range of h is contained in $\Omega_{\mathcal{R}}$. Picking $\gamma \in \hat{A}$, we must then prove that $\omega := h(\gamma) \in \Omega_{\mathcal{R}}$. Evidently

$$[1 \in \omega] = \omega_1 = \gamma(e_1) = \gamma(1) = 1,$$

so $\omega \in \Omega_1$. Let f be a function on $\{0, 1\}^G$ of the form (14.7), associated to a relation in $\mathcal{R}$. Picking any $g \in \omega$, we must therefore prove that

$$f(g^{-1}\omega) = 0.$$

With $p, g_1, \ldots, g_n$, as in (14.7), observe that

$$f(g^{-1}\omega) = p\big(\varepsilon_{g_1}(g^{-1}\omega), \ldots, \varepsilon_{g_n}(g^{-1}\omega)\big). \tag{14.16.1}$$

In order to get a better grasp on the meaning of the above, notice that for any $i = 1, \ldots, n$, one has

$$\varepsilon_{g_i}(g^{-1}\omega) = [g_i \in g^{-1}\omega] = [gg_i \in \omega] = \omega_{gg_i} = \gamma(e_{gg_i}) = \cdots$$

Since $g \in \omega$, we see that $\omega_g = 1$, which is to say that $\gamma(e_g) = 1$, so the above equals

$$\cdots = \gamma(e_g)\gamma(e_{gg_i}) = \gamma(u_g u_{g^{-1}} e_{gg_i}) \overset{(9.8.\text{iii})}{=} \gamma(u_g e_{g_i} u_{g^{-1}}) = \gamma_g(e_{g_i}),$$

where $\gamma_g : A \to \mathbb{C}$ is defined by

$$\gamma_g(b) = \gamma(u_g b u_{g^{-1}}), \quad \forall\, b \in A.$$

It is easy to see that γ_g is a homomorphism, which therefore commutes with polynomials. So by (14.16.1) we have

$$f(g^{-1}\omega) = p\big(\gamma_g(e_{g_1}), \ldots, \gamma_g(e_{g_n})\big) = \gamma_g\big(p(e_{g_1}, \ldots, e_{g_n})\big) = 0.$$

This proves that h indeed maps $\hat{A}$ into $\Omega_{\mathcal{R}}$, so by dualization we obtain a unital *-homomorphism

$$\hat{h} : f \in C(\Omega_{\mathcal{R}}) \mapsto f \circ h \in C(\hat{A}).$$

Observe that $\hat{h}(1_g)$ coincides with the Gelfand transform $\hat{e}_g$ of e_g, because, for every $\gamma \in \hat{A}$, one has

$$\hat{h}(1_g)(\gamma) = 1_g\big(h(\gamma)\big) \overset{(14.14)}{=} \varepsilon_g\big(h(\gamma)\big) = \gamma(e_g) = \hat{e}_g(\gamma).$$

Identifying $C(\hat{A})$ with A via the Gelfand transform, we then have that $\hat{h}$ is a *-homomorphism

$$\hat{h} : C(\Omega_{\mathcal{R}}) \to A,$$

such that $\hat{h}(1_g) = e_g$, for all g in G. We will now prove that the pair $(\hat{h}, u)$ is a covariant representation of $\theta_{\mathcal{R}}$ in $C^*_{\text{par}}(G, \mathcal{R})$, which is to say that

$$u_g \hat{h}(f) u_{g^{-1}} = \hat{h}\big(\theta^{\mathcal{R}}_g(f)\big), \tag{14.16.2}$$

for all $f \in C(D^{\mathcal{R}}_{g^{-1}})$, and all $g \in G$.

Using the Stone-Weierstrass Theorem one may show that the functions of the form $f = 1_{g^{-1}} 1_h$, with h in G, generate $C(D^{\mathcal{R}}_{g^{-1}})$, as a C*-algebra, so it is enough to prove (14.16.2) only for such functions. In this case the right-hand-side of (14.16.2) becomes

$$\hat{h}\big(\theta^{\mathcal{R}}_g(f)\big) = \hat{h}\big(\theta^{\mathcal{R}}_g(1_{g^{-1}} 1_h)\big) \overset{(6.8)}{=} \hat{h}(1_g 1_{gh}) = e_g e_{gh}.$$

The left-hand-side of (14.16.2), on the other hand, is given by

$$\begin{aligned} u_g \hat{h}(f) u_{g^{-1}} &= u_g \hat{h}(1_{g^{-1}} 1_h) u_{g^{-1}} \\ &= u_g e_{g^{-1}} e_h u_{g^{-1}} \\ &= u_g e_h u_{g^{-1}} \\ &\overset{(9.8.\text{iii})}{=} u_g u_{g^{-1}} e_{gh} \\ &= e_g e_{gh}, \end{aligned}$$

so (14.16.2) is seen to hold, as claimed, whence $(\hat{h}, u)$ is indeed a covariant representation. From (13.1) we then obtain a *-homomorphism

$$\hat{h} \times u : C(\Omega_{\mathcal{R}}) \rtimes G \to C^*_{\text{par}}(G, \mathcal{R}),$$

satisfying

$$(\hat{h} \times u)(1_g \delta_g) = \hat{h}(1_g) u_g = e_g u_g = u_g u_{g^{-1}} u_g = u_g,$$

for all g in G. Observing that φ sends u_g to $1_g \delta_g$, and that $C^*_{\text{par}}(G, \mathcal{R})$ is generated by the u_g, one easily proves that $(\hat{h} \times u) \circ \varphi$ is the identity on $C^*_{\text{par}}(G, \mathcal{R})$.

To prove that the composition $\varphi \circ (\hat{h} \times u)$ also coincides with the identity it suffices to prove that $C(\Omega_{\mathcal{R}}) \rtimes G$ is generated by the $1_g \delta_g$.

We have already seen that each $D^{\mathcal{R}}_g$ is generated by the elements $1_g 1_h$, for h ranging in G, so $C(\Omega_{\mathcal{R}}) \rtimes G$ is generated by the elements of the form

$$1_h 1_g \delta_g = (1_h \delta_e)(1_g \delta_g) = (1_h \delta_h)(1_{h^{-1}} \delta_{h^{-1}})(1_g \delta_g).$$

Thus the crossed product is indeed generated by the $1_g \delta_g$, as claimed, so $\varphi \circ (\hat{h} \times u)$ is the identity map, whence φ and $(\hat{h} \times u)$ are each other's inverse, proving that φ is an isomorphism and hence concluding the proof. □

In this and in later chapters we will present several applications of (14.16), describing certain classes of C*-algebras as partial crossed products. In the first such application we will prove a version of (10.9) to the context of C*-algebras, so let us begin by adapting the definition of $\mathbb{K}_{\text{par}}(G)$ given in (10.4) to our context.

Definition 14.17. Given a group G, the *partial group C*-algebra* of G, denoted $C^*_{\rm par}(G)$, is defined to be the algebra $C^*_{\rm par}(G,\mathcal{R})$, where $\mathcal{R}$ is the empty set of relations.

Thus, by (14.4) we see that $C^*_{\rm par}(G)$ is the universal unital C*-algebra for *-partial representations of G (without any further requirements), a fact that may be considered a version of (10.5) to C*-algebras.

Notice that if $\mathcal{R}$ is the empty set of relations, then $\Omega_{\mathcal{R}} = \Omega_1$ so, as an immediate consequence of (14.16) we have:

Corollary 14.18. *For every group G, one has that $C^*_{par}(G)$ is *-isomorphic to the crossed product of $C(\Omega_1)$ by G, relative to the partial Bernoulli action β defined in* (5.12)*, under an isomorphism which sends each canonical generating partial isometry u_g in $C^*_{par}(G)$ to $1_g\delta_g$, where 1_g denotes the characteristic function of D_g.*

It is not hard to see that $C(\Omega_1)$ is the universal unital C*-algebra generated by a set $\mathcal{E} = \{e_g : g \in G\}$, subject to the relations stating that the e_g are commuting self-adjoint idempotents, and that $e_1 = 1$. In other words, $C(\Omega_1)$ is the analogue of the algebra $A_{\rm par}(G)$ introduced in (10.7) within the category of C*-algebras. Therefore (14.18) may be seen as the analogue of (10.9) in the present category.

As our second example, let us consider the universal C*-algebra for semi-saturated partial representations. For this let us suppose that G is a group equipped with a length function ℓ. One may then define a *pseudo-metric* on G by setting

$$d(g,h) = \ell(g^{-1}h), \quad \forall\, g,h \in G.$$

By pseudo-metric we simply mean that d is a nonnegative real valued function satisfying the triangle inequality:

$$d(g,h) \leq d(g,k) + d(k,h), \quad \forall\, g,h,k \in G.$$

Although we will not need it here, one may clearly relate certain properties of ℓ with the remaining axioms for metric spaces.

In Euclidean space we know that the triangle inequality becomes an equality if and only if the three points involved are suitably placed in a straight line. This motivates the following:

Definition 14.19. Let G be a group equipped with a length function ℓ.

(a) Given g and h in G, the *segment* joining g and h is the subset of G given by

$$\overline{gh} = \big\{x \in G : d(g,h) = d(g,x) + d(x,h)\big\}.$$

(b) A subset $\omega \subseteq G$ is said to be *convex* if $\overline{gh} \subseteq \omega$, for all g and h in ω.

The above notion of convexity and the notion of semi-saturated partial representations, as rephrased by (14.5), may be bridged as follows:

Proposition 14.20. *Let G be a group and let $\mathcal{R}_{sat}$ be the set consisting of one relation of the form*

$$e_{gh} - e_{gh}e_g = 0,$$

for each pair of elements g and h in G satisfying $\ell(gh) = \ell(g) + \ell(h)$. Then

$$\Omega_{\mathcal{R}_{sat}} = \big\{\omega \in \Omega_1 : \omega \text{ is convex}\big\}.$$

PROOF. Notice that the corresponding set $\mathcal{F}_{\mathcal{R}_{\text{sat}}}$ of functions on $\{0,1\}^G$ is formed by the functions

$$f(\omega) = [gh \in \omega] - [gh \in \omega][g \in \omega],$$

for all $g, h \in G$, such that $\ell(gh) = \ell(g) + \ell(h)$. For each such function, each $\omega \in \Omega_1$, and each $k \in \omega$, notice that

$$\begin{aligned} f(k^{-1}\omega) &= [gh \in k^{-1}\omega] - [gh \in k^{-1}\omega][g \in k^{-1}\omega] \\ &= [kgh \in \omega] - [kgh \in \omega][kg \in \omega]. \end{aligned}$$

Letting Φ and Ψ denote the formulas "$kgh \in \omega$" and "$kg \in \omega$", respectively, the above may be shortened to

$$f(k^{-1}\omega) = [\Phi] - [\Phi][\Psi].$$

We next claim that the Boolean value of the formula "$\Phi \Rightarrow \Psi$" is given by $1 - f(k^{-1}\omega)$. In fact, using well known properties of Boolean operators we have

$$\begin{aligned} [\Phi \Rightarrow \Psi] &= [\neg\Phi \vee \Psi] \\ &= [\neg\Phi] + [\Psi] - [\neg\Phi][\Psi] \\ &= 1 - [\Phi] + [\Psi] - (1 - [\Phi])[\Psi] \\ &= 1 - [\Phi] + [\Psi] - [\Psi] + [\Phi][\Psi] \\ &= 1 - [\Phi] + [\Phi][\Psi] \\ &= 1 - f(k^{-1}\omega). \end{aligned}$$

Therefore, to say that ω is in $\Omega_{\mathcal{R}_{\text{sat}}}$ is the same as saying that for all g and h in G with $\ell(gh) = \ell(g) + \ell(h)$, one has that the Boolean value of "$\Phi \Rightarrow \Psi$" is equal to 1, obviously meaning that Φ implies Ψ, so

$$(\forall k \in \omega) \quad kgh \in \omega \Rightarrow kg \in \omega.$$

One may easily show that the most general situation in which an element x lies in a segment $\overline{kz}$ is when $x = kg$, and $z = kgh$, where g and h are elements in G such that $\ell(gh) = \ell(g) + \ell(h)$. Thus the above condition for ω to lie in $\Omega_{\mathcal{R}_{\text{sat}}}$ is precisely saying that ω is convex. □

The following is a direct consequence of (14.16):

Corollary 14.21. *Let G be a group equipped with a length function ℓ. Then the set*

$$\Omega_{conv} = \big\{\omega \in \Omega_1 : \omega \text{ is convex}\big\}$$

is a closed subspace of Ω_1, invariant under the partial Bernoulli action. Moreover, denoting by θ the corresponding restricted partial action, one has that $C(\Omega_{conv}) \rtimes_\theta G$ is the universal C^*-algebra for semi-saturated partial representations in the following sense:

(*i*) *The map*

$$g \in G \mapsto 1_g\delta_g \in C(\Omega_{conv}) \rtimes G,$$

where 1_g is the characteristic function of the set $\{\omega \in \Omega_{conv} : g \in \omega\}$, is a semi-saturated *-partial representation.

(*ii*) *Given any semi-saturated* **-partial representation of* G *in a unital* C^**-algebra* B, *there exists a unique* **-homomorphism*

$$\varphi : C(\Omega_{conv}) \rtimes G \to B,$$

such that $\varphi(1_g\delta_g) = u_g$.

Notes and remarks. Most results of this chapter are taken from [**60**].

CHAPTER 15

Hilbert Modules and Morita-Rieffel-Equivalence

One of the main tools to study partial crossed products in the realm of C*-algebras is the theory of Hilbert modules. In this short chapter we will therefore outline some of the main results from that theory which we will need in the sequel. The reader is referred to [**72**], [**18**] and [**78**] for careful treatments of this important subject.

Hilbert modules are also crucial in defining the concept of Morita-Rieffel-equivalence between C*-algebras and between C*-algebraic dynamical systems, which we will also attempt to briefly discuss, while referring the reader to [**98**] for a more extensive treatment and examples.

Definition 15.1. Let A be a C*-algebra. By a *right pre-Hilbert A-module* we mean a complex vector space M, equipped with the structure of a right A-module as well as an *A-valued inner-product*

$$\langle \cdot, \cdot \rangle : M \times M \to A,$$

satisfying

(i) $\langle \xi, \lambda\eta + \eta' \rangle = \lambda\langle \xi, \eta \rangle + \langle \xi, \eta' \rangle$,
(ii) $\langle \xi, \xi \rangle \geq 0$,
(iii) $\langle \xi, \xi \rangle = 0 \Rightarrow \xi = 0$,
(iv) $\langle \xi, \eta a \rangle = \langle \xi, \eta \rangle a$,
(v) $\langle \xi, \eta \rangle = \langle \eta, \xi \rangle^*$,

for every $\xi, \eta, \eta' \in M$, $\lambda \in \mathbb{C}$, and $a \in A$.

Notice that it follows from (15.1.i) and (15.1.v) that $\langle \cdot, \cdot \rangle$ is conjugate-linear in the first variable. Also, from (15.1.iv) and (15.1.v), one has that

$$\langle a\xi, \eta \rangle = a^* \langle \xi, \eta \rangle, \quad \forall\, \xi, \eta, \in M, \quad \forall\, a \in A.$$

It is a well known fact [**88**, Proposition 2.3] that the expression

$$\|\xi\|_2 = \|\langle \xi, \xi \rangle\|^{1/2}, \quad \forall\, \xi \in M,$$

defines a norm on M, and we then say that M is a *right Hilbert A-module* if M is complete relative to this norm. When M is not complete, the Banach space completion of M may be shown to carry the structure of a right Hilbert A-module extending the one of M.

Contrary to the Hilbert space case, a bounded A-linear operator T between two Hilbert modules M and N, does not necessarily admit an adjoint. By an adjoint of T we mean an operator $T^* : N \to M$ satisfying the familiar property

$$\langle T(\xi), \eta \rangle = \langle \xi, T^*(\eta) \rangle, \quad \forall\, \xi \in M, \quad \forall\, \eta \in N.$$

For a counter-example, let A be a C*-algebra and let us view A as a Hilbert module over itself with inner-product defined by

$$\langle a, b\rangle = a^*b, \quad \forall\, a, b \in A.$$

Every ideal $J \trianglelefteq A$ is a sub-Hilbert module of A and the inclusion $\iota : J \to A$ is an isometric A-linear map, hence bounded. However the adjoint of ι may fail to exist under certain circumstances. Suppose, for instance, that A is unital and J is not. If ι^* exists then for every x in J we have

$$x^* = x^*1 = \langle \iota(x), 1\rangle = \langle x, \iota^*(1)\rangle = x^*\iota^*(1).$$

This implies that $\iota^*(1)$ is a unit for J, contradicting our assumptions. Therefore ι^* does not exist.

When studying Hilbert module one therefore usually restricts attention to the *adjointable operators*, meaning the operators which happen to have an adjoint.

The set of all adjointable operators on a Hilbert module M is denoted $\mathscr{L}(M)$. One may prove without much difficulty that $\mathscr{L}(M)$ is a C*-algebra with respect to the composition of operators, the adjoint defined above, and the operator norm.

Given a C*-algebra A, one may likewise define the concept of *left pre-Hilbert A-module*, the only differences relative to Definition (15.1) being that M is now assumed to be a left A-module and axioms (15.1.i) & (15.1.iv) become

(i') $\langle \lambda\xi + \xi', \eta\rangle = \lambda\langle \xi, \eta\rangle + \langle \xi', \eta\rangle$,
(iv') $\langle a\xi, \eta\rangle = a\langle \xi, \eta\rangle$,

for every ξ, ξ', η in M, λ in $\mathbb{C}$, and a in A. The notion of *left Hilbert A-module* is defined in the obvious way.

Lemma 15.2. *Let A be a C*-algebra and let M be a right (resp. left) Hilbert A-module. If $\{v_i\}_i$ is an approximate identity for A, then for every ξ in M, one has that $\xi = \lim_i \xi v_i$ (resp. $\xi = \lim_i v_i\xi$).*

PROOF. We have

$$\begin{aligned}
\|\xi - \xi v_i\|^2 &= \|\langle \xi - \xi v_i, \xi - \xi v_i\rangle\| \\
&= \|\langle \xi, \xi\rangle - \langle \xi, \xi\rangle v_i - v_i^*\langle \xi, \xi\rangle + v_i^*\langle \xi, \xi\rangle v_i\| \\
&\leq \|\langle \xi, \xi\rangle - \langle \xi, \xi\rangle v_i\| + \|v_i\|\|\langle \xi, \xi\rangle v_i - \langle \xi, \xi\rangle\| \overset{i\to\infty}{\longrightarrow} 0.
\end{aligned}$$

A similar argument proves the left-handed version. □

As a consequence of the previous result, we see that $M = [MA]$ (closed linear span), but we can in fact get a slightly more precise result:

Lemma 15.3. *Let A be a C*-algebra and let M be a right (resp. left) Hilbert A-module. Then, for every ξ in M one has that*

$$\xi = \lim_{n\to\infty} \xi\langle \xi, \xi\rangle^{1/n} \quad (\textit{resp. } \xi = \lim_{n\to\infty} \langle \xi, \xi\rangle^{1/n}\xi).$$

Consequently $M = \big[M\langle M, M\rangle\big]$ (resp. $M = \big[\langle M, M\rangle M\big]$).

PROOF. We have

$$\begin{aligned}
\langle \xi - \xi\langle \xi, \xi\rangle^{1/n}, \xi - \xi\langle \xi, \xi\rangle^{1/n}\rangle &= \langle \xi, \xi\rangle - \langle \xi, \xi\rangle\langle \xi, \xi\rangle^{1/n} \\
&\quad - \langle \xi, \xi\rangle^{1/n}\langle \xi, \xi\rangle + \langle \xi, \xi\rangle^{1/n}\langle \xi, \xi\rangle\langle \xi, \xi\rangle^{1/n} \\
&= \langle \xi, \xi\rangle - 2\langle \xi, \xi\rangle^{1+1/n} + \langle \xi, \xi\rangle^{1+2/n} \overset{n\to\infty}{\longrightarrow} 0.
\end{aligned}$$

Therefore $\|\xi - \xi\langle\xi,\xi\rangle^{1/n}\| \overset{n\to\infty}{\longrightarrow} 0$, concluding the proof in the right module case, while a similar argument proves the left-handed version. □

It is interesting to notice that every right Hilbert A-module M is automatically a left Hilbert module over $\mathscr{L}(M)$ as follows: given T in $\mathscr{L}(M)$ and ξ in M, we evidently define the multiplication of T by ξ by

$$T \cdot \xi := T(\xi),$$

thus providing the left module structure. As for the $\mathscr{L}(M)$-valued inner-product, given ξ and η in M, consider the mapping

$$\Omega_{\xi,\eta} : \zeta \in M \mapsto \xi\langle\eta,\zeta\rangle \in M.$$

It is easy to see that $\Omega_{\xi,\eta}$ is an adjointable operator with $\Omega^*_{\xi,\eta} = \Omega_{\xi,\eta}$. We thus define the $\mathscr{L}(M)$-valued inner-product on M by

$$\langle\xi,\eta\rangle_{\mathscr{L}(M)} = \Omega_{\xi,\eta}, \quad \forall\, \xi,\eta \in M.$$

Observe that the two inner-products defined on M satisfy the compatibility condition

$$\langle\xi,\eta\rangle_{\mathscr{L}(M)}\zeta = \xi\langle\eta,\zeta\rangle_B, \quad \forall\, \xi,\eta,\zeta \in M,$$

which motivates our next concept.

Definition 15.4. Given C*-algebras A and B, by a *Hilbert A-B-bimodule* we mean a left Hilbert A-module M, which is also equipped with the structure of a right Hilbert B-module such that, denoting by $\langle\cdot,\cdot\rangle_A$ and $\langle\cdot,\cdot\rangle_B$ the A-valued and B-valued inner-products, respectively, one has that

(i) $(a\xi)b = a(\xi b)$.
(ii) $\langle\xi,\eta\rangle_A\zeta = \xi\langle\eta,\zeta\rangle_B$,

for all $a \in A$, $b \in B$, and $\xi,\eta,\zeta \in M$.

It is possible to show [**18**, Corollary 1.11] that the norms originating from the two inner-products on a Hilbert bimodule agree, meaning that

$$\|\langle\xi,\xi\rangle_A\| = \|\langle\xi,\xi\rangle_B\|, \quad \forall\, \xi \in M.$$

Definition 15.5. Let M be a Hilbert A-B-bimodule.

(a) We say that M is *left (resp. right) full*, if the linear span of the range of $\langle\cdot,\cdot\rangle_A$ (resp. $\langle\cdot,\cdot\rangle_B$) is dense in A (resp. B).
(b) If M is both left and right full, we say that M is an *imprimitivity bimodule.*
(c) If there exists an imprimitivity A-B-bimodule, we say that A and B are *Morita-Rieffel-equivalent.*

A rather common situation in which a Morita-Rieffel-equivalence takes place is as follows: suppose that B is a C*-algebra and A is a closed *-subalgebra of B.

Recall that A is said to be a *hereditary subalgebra* if

$$ABA \subseteq A.$$

This is equivalent to saying that any element b in B, such that $0 \le b \le a$, for some a in A, necessarily satisfies $b \in A$.

On the other hand, A is said to be a *full subalgebra* if the ideal generated by A in B coincides with B, namely

$$[BAB] = B.$$

If A is a hereditary subalgebra of B, the right ideal of B generated by A, namely $M = [AB]$, becomes a Hilbert A-B-bimodule with inner-products defined by

$$\langle x, y\rangle_A = xy^*, \quad \text{and} \quad \langle x, y\rangle_B = x^*y, \quad \forall x, y \in M.$$

Since $A \subseteq M$, it is evident that M is left-full. If we moreover suppose that A is a full subalgebra in the above sense, then M is also right-full, hence an imprimitivity bimodule, whose existence tells us that A and B are Morita-Rieffel-equivalent. Thus:

Proposition 15.6. *If A is a full, hereditary, closed *-subalgebra of a C^*-algebra B, then A and B are Morita-Rieffel-equivalent.*

Morita-Rieffel-equivalent C*-algebras share numerous interesting properties relating to representation theory and K-theory [**98**], [**17**]. One of the most striking results in this field states that if A and B are separable[19] C*-algebras, then they are Morita-Rieffel-equivalent if and only if they are *stably isomorphic*, meaning that

$$\mathcal{K} \otimes A \simeq \mathcal{K} \otimes B,$$

where $\mathcal{K}$ is the algebra of compact operators on a separable infinite dimensional Hilbert space (see [**17**] for more details).

The concept of Morita-Rieffel-equivalence has an important counterpart for C*-algebraic partial dynamical systems:

Definition 15.7. Let G be a group and suppose that for each $k = 1, 2$ we are given a C*-algebraic partial dynamical system

$$\theta^k = \big(A^k,\ G,\ \{A^k_g\}_{g\in G},\ \{\theta^k_g\}_{g\in G}\big).$$

We will say that θ^1 and θ^2 are *Morita-Rieffel-equivalent* if there exists a Hilbert A^1-A^2-bimodule M, and a (set-theoretical) partial action

$$\gamma = \big(\{M_g\}_{g\in G},\ \{\gamma_g\}_{g\in G}\big)$$

of G on M, such that for all $k = 1, 2$, and all g in G, one has that

(i) M_g is a norm closed, sub-A^1-A^2-bimodule of M,
(ii) $A^k_g = [\langle M_g, M_g\rangle_{A^k}]$ (closed linear span),
(iii) γ_g is a complex linear map,
(iv) $\big\langle \gamma_g(\xi), \gamma_g(\eta)\big\rangle_{A^k} = \theta^k_g\big(\langle \xi, \eta\rangle_{A^k}\big)$, for all $\xi, \eta \in M_{g^{-1}}$.

In this case we say that

$$\gamma = \big(M,\ G,\ \{M_g\}_{g\in G},\ \{\gamma_g\}_{g\in G}\big)$$

is an *imprimitivity system* for θ^1 and θ^2.

Speaking of (15.7.iv), observe that $\langle \xi, \eta\rangle_{A^k}$ lies in the domain of θ^k_g by (15.7.ii).

Using (15.7.iv), and the fact that C*-algebra automorphisms are necessarily isometric, one sees that the γ_g must be isometric as well.

Notice that each M_g may be seen as a full Hilbert A^1_g-A^2_g-bimodule, so necessarily A^1_g and A^2_g are Morita-Rieffel-equivalent. This applies in particular to $g = 1$, so A^1 and A^2 must be Morita-Rieffel-equivalent as well.

---

[19] In fact this result is known to hold under less stringent conditions, namely if A and B possess strictly positive elements. However most of the applications we have in mind are to the separable case, so we will not bother to deal with non-separable algebras here, unless the separability condition is irrelevant.

Proposition 15.8. *If θ^1 and θ^2 are Morita-Rieffel-equivalent partial dynamical systems, and γ is an imprimitivity system for θ^1 and θ^2, as in (15.7), then, given ξ in $M_{g^{-1}}$, one has that*

$$\gamma_g(a\xi) = \theta^1_g(a)\gamma_g(\xi), \quad \forall\, a \in D^1_{g^{-1}},$$

and

$$\gamma_g(\xi b) = \gamma_g(\xi)\,\theta^2_g(b), \quad \forall\, b \in D^2_{g^{-1}}.$$

PROOF. Focusing on the first assertion, let $\eta \in M_{g^{-1}}$. Then by (15.7.iv) we have

$$\begin{aligned} \langle \gamma_g(a\xi), \gamma_g(\eta)\rangle_{A^1} &= \theta^1_g\big(\langle a\xi, \eta\rangle_{A^1}\big) \\ &= \theta^1_g(a)\theta^1_g\big(\langle \xi, \eta\rangle_{A^1}\big) \\ &= \theta^1_g(a)\langle \gamma_g(\xi), \gamma_g(\eta)\rangle_{A^1} \\ &= \big\langle \theta^1_g(a)\gamma_g(\xi), \gamma_g(\eta)\big\rangle_{A^1}. \end{aligned}$$

Since η is arbitrary, and since $\gamma_g(\eta)$ can take on any value in M_g, we conclude that

$$\gamma_g(a\xi) = \theta^1_g(a)\gamma_g(\xi).$$

The second assertion is proved similarly. □

In view of the above result, it is not reasonable to expect the γ_g to be bi-module maps (hence not adjointable either) relative to any of the available Hilbert module structures. However, if we see each M_g as a *ternary C*-ring* [**107**] under the ternary operation

$$\{\xi, \eta, \zeta\}_g := \xi\langle \eta, \zeta\rangle_{A^2},$$

it easily follows from (15.8) that γ_g is an isomorphism of ternary C*-rings.

One of our long term goals, unfortunately not to be achieved too soon, will be proving that Morita-Rieffel-equivalent partial actions lead to Morita-Rieffel-equivalent crossed products. For this it is important to introduce the concept of linking algebra.

► So, suppose for the time being that A and B are C*-algebras, and that M is a Hilbert A-B-bimodule.

We first define the *adjoint Hilbert bimodule* as follows. Let M^* be any set admitting a bijective function

$$\xi \in M \mapsto \xi^* \in M^*.$$

We then define the structures of vector space, left B-module, and right A-module on M^* by

$$\begin{aligned} \xi^* + \lambda\eta^* &= (\xi + \bar{\lambda}\eta)^*, \\ b\xi^* &= (\xi b^*)^*, \\ \xi^* a &= (a^*\xi)^*, \end{aligned}$$

for all ξ and η in M, λ in $\mathbb{C}$, a in A and b in B. We further define A- and B-valued inner-products on M^* by

$$\langle \xi^*, \eta^*\rangle_A = \langle \xi, \eta\rangle_A, \quad \text{and} \quad \langle \xi^*, \eta^*\rangle_B = \langle \xi, \eta\rangle_B,$$

for all ξ and η in M.

The reader might be annoyed by the fact that, unlike module structures, the inner-products defined on M^* have not changed relative to the original ones. However this turns out to be the only sensible choice, and one will quickly be convinced of this by checking the appropriate axioms, which the reader is urged to do. Once this is done M^* becomes a Hilbert B-A-bimodule.

We then write

$$L = \begin{pmatrix} A & M \\ M^* & B \end{pmatrix},$$

simply meaning the cartesian product $A \times M \times M^* \times B$, denoted in a slightly unusual way.

We make L a complex-vector space in the obvious way, and a *-algebra by introducing multiplication and adjoint operations as follows:

$$\begin{pmatrix} a_1 & \xi_1 \\ \eta_1^* & b_1 \end{pmatrix} \begin{pmatrix} a_2 & \xi_2 \\ \eta_2^* & b_2 \end{pmatrix} = \begin{pmatrix} a_1a_2 + \langle \xi_1, \eta_2 \rangle_A & a_1\xi_2 + \xi_1 b_2 \\ \eta_1^* a_2 + b_1\eta_2^* & \langle \eta_1, \xi_2 \rangle_B + b_1 b_2 \end{pmatrix},$$

and

$$\begin{pmatrix} a & \xi \\ \eta^* & b \end{pmatrix}^{\star} = \begin{pmatrix} a^* & \eta \\ \xi^* & b^* \end{pmatrix}.$$

In order to give L a norm, we define a representation π_B of L as adjointable operators on the right Hilbert B-module $M \oplus B$ (where B is seen as a right Hilbert B-module in the obvious way) by

$$\pi_B \begin{pmatrix} a_1 & \xi_1 \\ \eta_1^* & b_1 \end{pmatrix} \begin{pmatrix} \xi_2 \\ b_2 \end{pmatrix} = \begin{pmatrix} a_1\xi_2 + \xi_1 b_2 \\ \langle \eta_1, \xi_2 \rangle_B + b_1 b_2 \end{pmatrix},$$

and another representation π_A on the right Hilbert A-module $A \oplus M^*$, by

$$\pi_A \begin{pmatrix} a_1 & \xi_1 \\ \eta_1^* & b_1 \end{pmatrix} \begin{pmatrix} a_2 \\ \eta_2^* \end{pmatrix} = \begin{pmatrix} a_1a_2 + \langle \xi_1, \eta_2 \rangle_A \\ \eta_1^* a_2 + b_1\eta_2^* \end{pmatrix}.$$

We then define a norm on L by

$$\|c\| = \max\big\{\|\pi_A(c)\|, \|\pi_B(c)\|\big\}, \quad \forall c \in L,$$

and one may then prove that L becomes a C*-algebra with the above structure [**18**, Proposition 2.3].

Definition 15.9. Given C*-algebras A and B, and a Hilbert A-B-bimodule M, the *linking algebra* of M is the C*-algebra L described above.

The linking algebra relative to the imprimitivity bimodule implementing a Morita-Rieffel-equivalence between partial actions also carries a partial action, as we will now show:

Proposition 15.10. *Let G be a group and suppose we are given C^*-algebraic partial dynamical systems*

$$\alpha = \big(A, G, \{A_g\}_{g\in G}, \{\alpha_g\}_{g\in G}\big), \quad \text{and} \quad \beta = \big(B, G, \{B_g\}_{g\in G}, \{\beta_g\}_{g\in G}\big).$$

Suppose, moreover, that α and β are Morita-Rieffel-equivalent partial actions, and that

$$\gamma = \big(M,\ G,\ \{M_g\}_{g\in G},\ \{\gamma_g\}_{g\in G}\big)$$

is an imprimitivity system for α and β. Letting L be the linking algebra of M, for *each g in G, we have that*

(*i*) *the subset L_g of L defined by*

$$L_g = \begin{pmatrix} A_g & M_g \\ M_g^* & B_g \end{pmatrix}$$

is a closed two-sided ideal,

(*ii*) *the mapping*

$$\lambda_g : \begin{pmatrix} a & \xi \\ \eta^* & b \end{pmatrix} \in L_{g^{-1}} \longmapsto \begin{pmatrix} \alpha_g(a) & \gamma_g(\xi) \\ \gamma_g(\eta)^* & \beta_g(b) \end{pmatrix} \in L_g.$$

is a **-isomorphism,*

(*iii*) *the pair*

$$\lambda = \big(\{L_g\}_{g\in G},\ \{\lambda_g\}_{g\in G}\big),$$

is a C^*-algebraic partial action of G on L.

PROOF. We first claim that if $\xi \in M$, and $\eta \in M_g$, then

$$\langle \xi, \eta\rangle_A \in A_g, \quad \text{and} \quad \langle \xi, \eta\rangle_B \in B_g. \tag{15.10.1}$$

In order to see this, let $\{v_i\}_i$ be an approximate identity for B_g. Viewing M_g as a right Hilbert B_g-module, we get from (15.2) that $\eta = \lim_i \eta v_i$, so

$$\langle \xi, \eta\rangle_B = \langle \xi, \lim_i \eta v_i\rangle_B = \lim_i \langle \xi, \eta\rangle_B v_i \in B_g.$$

Using a similar reasoning one shows that $\langle \xi, \eta\rangle_A \in A_g$.

Of course, if it is ξ, rather than η, which belongs to M_g, then (15.10.1) still hold (by taking adjoints).

We will need another fact of a similar nature, that is, if a is in A_g, b is in B_g, and ξ is in M, then

$$a\xi \in M_g, \quad \text{and} \quad \xi b \in M_g. \tag{15.10.2}$$

To prove it, notice that by (15.7.ii), we may assume that $b = \langle \eta, \zeta\rangle_B$, for some η and ζ in M_g. In this case we have

$$\xi b = \xi \langle \eta, \zeta\rangle_B \overset{(15.4.\text{ii})}{=} \langle \xi, \eta\rangle_A \zeta \in M_g.$$

The proof that $a\xi \in M_g$ is similar.

We therefore have that, in any one of the expressions:

$$ab,\ a\xi,\ \xi b,\ \langle \xi, \eta\rangle_A \text{ and } \langle \xi, \eta\rangle_B,$$

where a is in A, b is in B, and ξ and η are in M, if one of the two terms involved belongs to a set named with a subscript "g", then so does the whole expression.

Keeping this principle in mind, and staring at the definition of the multiplication operation in L for a while, one then sees that L_g is in fact a two-sided ideal of L.

In [**18**, Proposition 2.3] it is proven that the norm topology of L coincides with the product topology, when L is viewed as $A \times M \times M^* \times B$. Thus, since A_g is closed in A, B_g is closed in B and M_g is closed in M, we see that L_g is closed in L.

The proof that each γ_g is a *-isomorphism is an easy consequence of (15.7.iv) and (15.8), and the corresponding property for α_g and β_g.

Finally, the last point is easily verified by observing that λ is the direct sum of four partial actions. □

This result will later be used to prove that Morita-Rieffel-equivalent partial actions lead to Morita-Rieffel-equivalent crossed products. The strategy for doing so will be to form the crossed product of L by G, and then make sense of the expression

$$L \rtimes G = \begin{pmatrix} A \rtimes G & M \rtimes G \\ (M \rtimes G)^* & B \rtimes G \end{pmatrix},$$

so $M \rtimes G$ will be seen to be an imprimitivity bimodule implementing a Morita-Rieffel-equivalence between $A \rtimes G$ and $B \rtimes G$. The biggest hurdle we will face, and the reason we need to further develop our theory, is showing $A \rtimes G$ and $B \rtimes G$ to be isomorphic to subalgebras of $L \rtimes G$. Even though we may easily find *-homomorphisms

$$A \rtimes G \to L \rtimes G, \quad \text{and} \quad B \rtimes G \to L \rtimes G,$$

proving these to be injective requires some extra work. As soon as we have the appropriate tools, we will return to this point and we will prove the following:

Theorem 15.11. *If*

$$\alpha = \big(A, G, \{A_g\}_{g\in G}, \{\alpha_g\}_{g\in G}\big), \quad \textit{and} \quad \beta = \big(B, G, \{B_g\}_{g\in G}, \{\beta_g\}_{g\in G}\big)$$

are Morita-Rieffel-equivalent C-algebraic partial dynamical systems, then $A \rtimes G$ and $B \rtimes G$ are Morita-Rieffel-equivalent C*-algebras.*

Notes and remarks. Inspired by Kaplansky's C*-modules, Hilbert modules were introduced by Paschke in [**88**]. The notion of Morita-Rieffel-equivalence, sometimes also referred to as *strong Morita equivalence*, was introduced by Rieffel in [**98**], adapting the purely algebraic concept of Morita equivalence [**83**] to operator algebras.

Morita-Rieffel-equivalence of actions of groups on C*-algebras were considered independently in [**31**] and [**26**] and, for the case of partial actions, in [**1**] and [**2**], where a version of Theorem (15.11) for reduced crossed products appeared. When we are ready to prove (15.11), we will take care of both the reduced and full versions.

Part 2

Fell Bundles

CHAPTER 16

Fell Bundles

If one wishes to understand the structure of a given algebra, a traditional method is to try to decompose it in the direct sum of ideals. However, when working with simple algebras, such a decomposition is evidently unavailable.

An alternative method is to try to find a grading (see (8.10)) of our algebra, in which case we will have decomposed it in its *homogeneous spaces*. These are of course subspaces of the given algebra but they can also be seen as separate entities. In other words, starting from a graded algebra we may see the collection of its homogeneous spaces as the parts we are left with after *disassembling* the algebra along its grading. The study of the separate pieces might then hopefully shed some light on the structure of our algebra.

From now on we will employ the concept of Fell bundles, introduced by J. M. G. Fell under the name of *C*-algebraic bundles*, in order to deal with disassembled C*-algebras[20]. However it is crucial to note that, in the category of C*-algebras, the concept of a grading, to be defined shortly, only requires the direct sum of the homogeneous spaces to be *dense*, so the process of passing from a graded C*-algebra to its corresponding Fell bundle, that is, the collection formed by its homogeneous spaces, involves a significant loss of information: there is no straightforward process to reassemble a graded C*-algebra from its parts! In other words, there are examples of non-isomorphic graded C*-algebras whose associated Fell bundles are indistinguishable.

While this can be considered a weakness of the method, it often helps to organize one's tasks in two broad groups. In order to understand the structure of a graded C*-algebra one should therefore attempt to separately understand:

(1) the structure of its associated Fell bundle, and
(2) the way in which the various parts are pieced together.

Breaking up assignments along these lines frequently makes a lot of sense because the two tasks often require very different tool sets.

In this and the forthcoming chapters we plan to provide both sets of tools for dealing with graded C*-algebras: (1) understanding the structure of Fell bundles will be done by proving that, under suitable hypothesis, every Fell bundle arises from a partial dynamical system, while (2) understanding how to piece together the homogeneous spaces to reassemble the algebra will be done via the theory of cross-sectional algebras and amenability.

[20]It is important to stress that, over topological groups, Fell bundles carry in its topology a lot more information than we will be using here. However, since all of our groups are discrete, this extra topological data will not be relevant for us here.

Definition 16.1. A *Fell bundle* (also known as a *C^*-algebraic bundle*) over a group G is a collection

$$\mathscr{B} = \{B_g\}_{g\in G}$$

of Banach spaces, each of which is called a *fiber*. In addition, the *total space* of $\mathscr{B}$, namely the disjoint union of all the B_g's, which we also denote by $\mathscr{B}$, by abuse of language, is equipped with a multiplication operation and an involution

$$\cdot : \mathscr{B} \times \mathscr{B} \to \mathscr{B}, \qquad * : \mathscr{B} \to \mathscr{B},$$

satisfying the following properties for all g and h in G, and all b and c in $\mathscr{B}$:

(a) $B_g B_h \subseteq B_{gh}$,
(b) multiplication is bi-linear from $B_g \times B_h$ to B_{gh},
(c) multiplication on $\mathscr{B}$ is associative,
(d) $\|bc\| \leq \|b\|\|c\|$,
(e) $(B_g)^* \subseteq B_{g^{-1}}$,
(f) involution is conjugate-linear from B_g to $B_{g^{-1}}$,
(g) $(bc)^* = c^*b^*$,
(h) $b^{**} = b$,
(i) $\|b^*\| = \|b\|$,
(j) $\|b^*b\| = \|b\|^2$,
(k) $b^*b \geq 0$ in B_1.

Axioms (a–d) above define what is known as a *Banach algebraic bundle*. Adding (e–i) gives the definition of a *Banach *-algebraic bundle*.

Observe that axioms (a–j) imply that B_1 is a C*-algebra with the restricted operations. We will often refer to B_1 as the *unit fiber algebra*.

With respect to axiom (k), notice that the reference to positivity there is to be taken with respect to the standard order relation in B_1, seen as a C*-algebra. Should one prefer to avoid any reference to this order relation, an alternative formulation of (k) is to require that for each b in $\mathscr{B}$, there exists some a in B_1, such that $b^*b = a^*a$.

As already hinted upon, a concept which is closely related to Fell bundles is the notion of graded C*-algebras, which is not equivalent, and hence should be distinguished from its purely algebraic counterpart (8.10).

Definition 16.2. Let B be a C*-algebra and G be a group. We say that a linearly independent collection $\{B_g\}_{g\in G}$ of closed subspaces of B is a *C^*-grading* for B, if $\bigoplus_{g\in G} B_g$ is dense in B, and for every g and h in G, one has that

(i) $B_g B_h \subseteq B_{gh}$,
(ii) $B_g^* \subseteq B_{g^{-1}}$.

In this case we say that B is a *G-graded C^*-algebra* and each B_g is called a *grading subspace*.

The reason why the above is not a special case of (8.10) is that here the direct sum of the B_g's is only required to be dense in B.

16.3. If one is given a G-graded C*-algebra B, then the collection of all grading subspaces will clearly form a Fell bundle with the norm, the multiplication operation and the adjoint operation borrowed from B.

We will see that, conversely, every Fell bundle may be obtained from a G-graded C*-algebra B, as above, although B is not uniquely determined since there might be many ways to complete the direct sum of the B_g's.

One of the reasons why one is interested in studying Fell bundles rather than graded C*-algebras is to avoid getting distracted by the problems related to the different completions mentioned above.

16.4. An important example of Fell bundles is given by the *group bundle*

$$\mathscr{B} = \mathbb{C} \times G,$$

where G is any group. Each fiber B_g is defined to be $\mathbb{C} \times \{g\}$, with usual linear and norm structure, and with operations

$$(\lambda, g)(\mu, h) = (\lambda\mu, gh), \quad \text{and} \quad (\lambda, g)^* = (\bar{\lambda}, g^{-1}),$$

for all $\lambda, \mu \in \mathbb{C}$, and $g, h \in G$.

16.5. One of the main reasons we are interested in Fell bundles is because they may be built from C*-algebraic partial dynamical systems. In order to describe this construction, let us fix a C*-algebraic partial action

$$\theta = \big(\{D_g\}_{g\in G},\ \{\theta_g\}_{g\in G}\big)$$

of a group G on a C*-algebra A. We will begin the construction of our Fell bundle by defining its total space to be

$$\mathscr{B} = \{(b, g) \in A \times G : b \in D_g\}.$$

Inspired by the notation introduced in (8.3.1), we will write $b\delta_g$ to refer to (b, g), whenever $b \in D_g$. We may then identify the fibers of our bundle as

$$B_g = \{b\delta_g : b \in D_g\}.$$

The linear structure and the norm on each B_g is borrowed from D_g, while the multiplication operation is defined exactly as in (8.3.2), namely

$$(a\delta_g)(b\delta_h) = \theta_g\big(\theta_{g^{-1}}(a)b\big)\delta_{gh}, \quad \forall\, a \in D_g, \quad \forall\, b \in D_h.$$

We finally borrow the definition of the involution from (8.9), namely

$$(a\delta_g)^* = \theta_{g^{-1}}(a^*)\delta_{g^{-1}}, \quad \forall\, g \in G, \quad \forall\, a \in D_g.$$

Proposition 16.6. *Given a C*-algebraic partial action*

$$\theta = \big(\{D_g\}_{g\in G},\ \{\theta_g\}_{g\in G}\big)$$

of a group G on a C-algebra A, one has that $\mathscr{B}$, with the above operations, is a Fell bundle over G, henceforth called the semi-direct product bundle relative to θ.*

PROOF. The task of checking the axioms in (16.1) is mostly routine. We restrict ourselves to a few comments. With respect to associativity, we observe that it follows by the same argument used in (8.7). Axioms (d), (i) and (j), namely the axioms referring to the norm structure, all follow easily from the fact that partial automorphisms of C*-algebras are isometric. Axioms (g) and (h) may be proved following the ideas used in the proof of (8.9), while (k) easily follows from (8.14.e) and the fact that partial automorphisms preserve positivity. □

16.7. Another important class of examples of Fell bundles is given in terms of partial representations. In order to describe these, let

$$u : G \to A$$

be a *-partial representation of a given group G in a unital C*-algebra A. For each g in G, consider the closed linear subspace $B_g^u \subseteq A$, spanned by the elements of the form

$$u_{h_1}u_{h_2}\cdots u_{h_n},$$

where $n \geq 1$ is any integer, the h_i are in G, and

$$h_1\cdots h_n = g.$$

It is elementary to check that

$$(B_g^u)^* = B_{g^{-1}}^u, \quad \text{and} \quad B_g^u B_h^u \subseteq B_{gh}^u,$$

for all g and h in G, so the collection

$$\mathscr{B}^u = \{B_g^u\}_{g\in G}$$

is seen to be a Fell bundle with the operations borrowed from A.

Proposition 16.8. *Let u be a *-partial representation of the group G in the unital C^*-algebra A. For each g in G, let $e_g = u_g u_{g^{-1}}$, as usual. Then, regarding the Fell bundle $\mathscr{B}^u$ defined above, we have:*

(*i*) *the unit fiber algebra B_1^u coincides with the C^*-algebra generated by the set $\{e_g : g \in G\}$,*
(*ii*) *B_1^u is abelian,*
(*iii*) *for each g in G, one has that $u_g B_1^u = B_g^u = B_1^u u_g$.*

Proof. Given $h_1, \ldots, h_n$ in G, we claim that

$$u_{h_1}u_{h_2}\cdots u_{h_n} = e_{p_1}e_{p_2}\cdots e_{p_n}u_{p_n}, \tag{16.8.1}$$

where

$$p_k = h_1\cdots h_k, \quad \forall 1 \leq k \leq n.$$

If $n = 1$, we have

$$u_{h_1} = u_{h_1}u_{h_1^{-1}}u_{h_1} = e_{h_1}u_{h_1} = e_{p_1}u_{p_1},$$

proving the claim in this case. Assuming that $n \geq 2$, we have

$$\begin{aligned} u_{h_1}u_{h_2}u_{h_3}\cdots u_{h_n} &= u_{h_1}u_{h_1^{-1}}u_{h_1}u_{h_2}u_{h_3}\cdots u_{h_n} \\ &\overset{(9.1.\text{iii})}{=} u_{h_1}u_{h_1^{-1}}u_{h_1h_2}u_{h_3}\cdots u_{h_n} \\ &= e_{h_1}u_{h_1h_2}u_{h_3}\cdots u_{h_n}, \end{aligned}$$

and the claim follows by induction.

If we moreover assume that $h_1\cdots h_n = 1$, so that $u_{h_1}u_{h_2}\cdots u_{h_n}$ is an arbitrary generator of B_1^u, we deduce from the claim that

$$u_{h_1}u_{h_2}\cdots u_{h_n} = e_{h_1}e_{h_1h_2}\cdots e_{h_1h_2\cdots h_n},$$

proving (i), and then (ii) follows from (9.8.iv).

Let us next prove (iii). Observing that u_g lies in B_g^u, it is immediate that $u_g B_1^u \subseteq B_g^u$. On the other hand, notice that if $h_1\cdots h_n = g$, and b is defined by

$b = u_{h_1} u_{h_2} \cdots u_{h_n}$, then

$$\begin{aligned} be_{g^{-1}} &= bu_{g^{-1}}u_g \\ &\overset{(16.8.1)}{=} e_{p_1}e_{p_2}\cdots e_{p_n}u_{p_n}u_{g^{-1}}u_g \\ &\overset{(9.8.\text{i})}{=} e_{p_1}e_{p_2}\cdots e_{p_n}u_{p_n} \\ &= b. \end{aligned}$$

Since B^u_g is the closed linear span of the set of elements b, as above, we then see that

$$be_{g^{-1}} = b, \quad \forall\, b \in B^u_g.$$

For each such b, we then have that

$$b = be_{g^{-1}} = bu_{g^{-1}}u_g \in B^u_1 u_g,$$

because $bu_{g^{-1}}$ is clearly in B^u_1. This proves that $B^u_g = B^u_1 u_g$, and the remaining statement in (iii) follows by taking adjoints. □

It is interesting to notice that the Fell bundle associated to a given *-partial representation, as discussed above, may also be described as a semi-direct product bundle. In fact, one may adapt (10.1) to the context of C*-algebras, obtaining a partial action of G on B^u_1, whose associated semi-direct product bundle is isomorphic to $\mathscr{B}^u$. We leave the verification of these statements as an easy exercise.

▶ From now on we fix an arbitrary Fell bundle $\mathscr{B} = \{B_g\}_{g\in G}$.

Lemma 16.9. *If $\{v_i\}_i$ is an approximate identity for B_1, then, for every b in any B_g, one has that*

$$b = \lim_i bv_i = \lim_i v_i b.$$

PROOF. Noticing that B_g is a Hilbert B_1-B_1-bimodule in a natural way, the result follows from (15.2). □

Using square brackets to denote closed linear span, notice that for any two fibers B_g and B_h of $\mathscr{B}$, we have that $[B_gB_h]$ is a closed linear subspace of B_{gh}. Questions related to whether or not $[B_gB_h]$ coincide with B_{gh} will have a special significance for us so let us introduce two concepts which are based on this. Recall from (4.9) and (9.7) that the term *semi-saturated* has already been defined both in the context of partial actions and of partial representation. We will now extend it to the context of Fell bundles.

Definition 16.10. Let $\mathscr{B} = \{B_g\}_{g\in G}$ be a Fell bundle.

(a) We say that $\mathscr{B}$ is *saturated* if $[B_gB_h] = B_{gh}$, for every g and h in G.
(b) If G is moreover equipped with a length function ℓ, we say that $\mathscr{B}$ is *semi-saturated* (with respect to the given length function ℓ) if, for all g and h in G satisfying $\ell(gh) = \ell(g) + \ell(h)$, one has that $[B_gB_h] = B_{gh}$.

As already seen, the unit fiber B_1 of a Fell bundle is always a C*-algebra, and it is easy to see that for each g in G, one has that $[B_gB_{g^{-1}}]$ is an ideal of B_1.

Lemma 16.11. *Let $\mathscr{B} = \{B_g\}_{g\in G}$ be a Fell bundle, and let $g \in G$. Given an approximate identity $\{u_i\}_i$ for $[B_gB_{g^{-1}}]$, and another approximate identity $\{v_i\}_i$ for $[B_{g^{-1}}B_g]$, one has that*

$$b = \lim_i u_i b = \lim_i bv_i, \quad \forall\, b \in B_g.$$

PROOF. This is an immediate consequence of (15.2), once we notice that B_g is a Hilbert $[B_g B_{g^{-1}}]$-$[B_{g^{-1}} B_g]$-bimodule. □

Lemma 16.12. *For every g in G, one has that $[B_g B_{g^{-1}} B_g] = B_g$.*

PROOF. Given b in B_g, choose an approximate identity $\{u_i\}_i$ for $[B_g B_{g^{-1}}]$. Then

$$b = \lim_i u_i b \in [B_g B_{g^{-1}} B_g],$$

proving that $B_g \subseteq [B_g B_{g^{-1}} B_g]$. The reverse inclusion is obvious. □

We may use this to give a simpler characterization of saturated Fell bundles.

Proposition 16.13. *A necessary and sufficient condition for a Fell bundle $\mathscr{B} = \{B_g\}_{g\in G}$ to be saturated is that*

$$[B_g B_{g^{-1}}] = B_1, \quad \forall\, g \in G.$$

PROOF. That the condition is necessary is obvious. Conversely, notice that for all g and h in G, we have

$$\begin{aligned} B_{gh} &\overset{(16.12)}{=} [B_{gh} B_{(gh)^{-1}} B_{gh}] \\ &\subseteq [B_1 B_{gh}] \\ &= [B_g B_{g^{-1}} B_{gh}] \\ &\subseteq [B_g B_h] \\ &\subseteq B_{gh}. \end{aligned}$$

Therefore equality holds throughout, proving that $\mathscr{B}$ is saturated. □

We may easily characterize saturatedness and semi-saturatedness for semi-direct product bundles:

Proposition 16.14. *Let $\theta = \big(\{D_g\}_{g\in G},\ \{\theta_g\}_{g\in G}\big)$ be a C^*-algebraic partial action of the group G on the C^*-algebra A, and let $\mathscr{B}$ be the associated semi-direct product bundle.*

(i) Then $\mathscr{B}$ is saturated if and only if θ is a global action.

(ii) If G is moreover equipped with a length function, then $\mathscr{B}$ is a semi-saturated Fell bundle (according to (16.10.b)*) if and only if θ is a semi-saturated partial action (according to* (4.9)*).*

PROOF. Given g in G, we have

$$\begin{aligned} [B_g B_{g^{-1}}] &= [(D_g \delta_g)(D_{g^{-1}} \delta_{g^{-1}})] \\ &\overset{(8.14)}{=} [D_g \theta_g(D_{g^{-1}}) \delta_1] \\ &= D_g \delta_1. \end{aligned} \tag{16.14.1}$$

So, to say that $[B_g B_{g^{-1}}] = B_1$ is equivalent to saying that $D_g = A$. This proves (i).

Focusing on (ii), pick g and h in G, with $\ell(gh) = \ell(g) + \ell(h)$. Reasoning as in (2.2), it is easy to see that the range of $\theta_g \circ \theta_h$ is given by $D_g \cap D_{gh}$. Since $\theta_g \circ \theta_h$ is a restriction of the bijective mapping θ_{gh}, it is clear that these two maps coincide if and only if they have the same range, that is,

$$\theta_g \circ \theta_h = \theta_{gh} \ \Leftrightarrow\ D_g \cap D_{gh} = D_{gh} \ \Leftrightarrow\ D_{gh} \subseteq D_g.$$

Assuming that $\mathscr{B}$ is semi-saturated, we have that $B_{gh} = [B_g B_h]$, so

$$D_{gh}\delta_1 \overset{(16.14.1)}{=} [B_{gh}B_{h^{-1}g^{-1}}] = [B_g B_h B_{h^{-1}g^{-1}}] \subseteq [B_g B_{g^{-1}}] \overset{(16.14.1)}{=} D_g\delta_1,$$

so $D_{gh} \subseteq D_g$, whence $\theta_g \circ \theta_h = \theta_{gh}$, as seen above, proving θ to be semi-saturated.

Conversely, assuming that θ is semi-saturated, we have

$$\begin{aligned} B_{gh} &\overset{(16.12)}{=} [B_{gh}B_{h^{-1}g^{-1}}B_{gh}] \\ &\overset{(16.14.1)}{=} \big[(D_{gh}\delta_1)B_{gh}\big] \\ &\subseteq \big[(D_g\delta_1)B_{gh}\big] \\ &\overset{(16.14.1)}{=} [B_g B_{g^{-1}}B_{gh}] \\ &\subseteq [B_g B_h], \end{aligned}$$

which says that $\mathscr{B}$ is semi-saturated. □

Our next result relates saturatedness for partial representations and for Fell bundles.

Proposition 16.15. *Given a *-partial representation u of a group G in a nonzero unital C^*-algebra A, consider its associated Fell bundle $\mathscr{B}^u$, as described in* (16.7). *Then:*

(*i*) *$\mathscr{B}^u$ is a saturated Fell bundle if and only if u is a unitary group representation. In this case $\mathscr{B}^u$ is isomorphic to the group bundle $\mathbb{C} \times G$.*

(*ii*) *$\mathscr{B}^u$ is a semi-saturated Fell bundle if and only if u is a semi-saturated partial representation.*

PROOF. Assuming that $\mathscr{B}^u$ is saturated, let g be in G. Then

$$B_1^u = [B_g^u B_{g^{-1}}^u] \overset{(16.8.\text{iii})}{=} [B_1^u u_g u_{g^{-1}} B_1^u] = [e_g B_1^u].$$

This is to say that the ideal generated by the idempotent element e_g coincides with the unital algebra B_1^u, but this may only happen if $e_g = 1$. Since g is arbitrary, this easily implies that u is a unitary group representation.

Conversely, supposing that u is a unitary group representation, notice that whenever $h_1 \cdots h_n = g$, one has that $u_{h_1}u_{h_2}\cdots u_{h_n} = u_g$, which is to say that $B_g^u = \mathbb{C}u_g$. It is then easy to see that $\mathscr{B}^u$ is isomorphic to the group bundle $\mathbb{C} \times G$, whence a saturated Fell bundle.

With respect to (ii), pick g and h in G, with $\ell(gh) = \ell(g) + \ell(h)$. Then, assuming u to be semi-saturated, we have that $u_g u_h = u_{gh}$, so

$$[B_g^u B_h^u] = [B_1^u u_g u_h B_1^u] = [B_1^u u_{gh} B_1^u] = [B_1^u B_{gh}^u] \overset{(16.9)}{=} B_{gh}^u,$$

so $\mathscr{B}^u$ is seen to be semi-saturated.

In order to prove the converse, given any g in G, we claim that

$$e_g b = b, \quad \forall g \in B_g^u.$$

To see this, notice that, if $b \in B_g^u$, we may write $b = u_g a$, for some a in B_1^u, by (16.8.iii), so

$$e_g b = e_g u_g a = u_g u_{g^{-1}} u_g a = u_g a = b,$$

proving the claim. If $\mathscr{B}^u$ is semi-saturated, and still under the assumption that $\ell(gh) = \ell(g) + \ell(h)$, we then immediately see that

$$e_g b = b, \quad \forall g \in B_{gh}^u,$$

because $B^u_{gh} = [B^u_g B^u_h]$. In particular, since u_{gh} is in B^u_{gh}, we deduce that

$$e_g e_{gh} = e_g u_{gh} u_{gh^{-1}} = u_{gh} u_{gh^{-1}} = e_{gh}.$$

Using (14.5) we then conclude that $u_{gh} = u_g u_h$, thus verifying that u is semi-saturated. □

► Let us now return to the above situation in which $\mathscr{B} = \{B_g\}_{g\in G}$ denotes an arbitrary, fixed Fell bundle.

Every element c of B_1 defines *left and right multiplication operators*

$$L_c : b \in B_g \mapsto cb \in B_g, \quad \text{and} \quad R_c : b \in B_g \mapsto bc \in B_g.$$

We will next show that the above may be extended to multipliers of B_1.

Proposition 16.16. *Given any m in the multiplier algebra $\mathcal{M}(B_1)$, there is a unique pair of bounded linear maps*

$$L_m, R_m : B_g \to B_g,$$

such that, for all a in B_1, and all b in B_g, one has that

(i) $L_m(ba) = L_m(b)a$,
(ii) $R_m(ab) = aR_m(b)$,
(iii) $aL_m(b) = (am)b$,
(iv) $R_m(b)a = b(ma)$.

Moreover, if $\{u_i\}_i$ is any approximate identity for B_1, then

$$L_m(b) = \lim_i (u_i m)b, \quad \textit{and} \quad R_m(b) = \lim_i b(mu_i), \quad \forall\, b \in B_g.$$

PROOF. Let $\{v_i\}_i$ be an approximate identity for B_1. Addressing uniqueness, notice that if b is in B_g, we have that

$$L_m(b) \overset{(16.9)}{=} \lim_i v_i L_m(b) \overset{\text{(iii)}}{=} \lim_i (v_i m)b.$$

One similarly proves that R_m is unique. As for existence, we claim that the limits

$$\lim_i (v_i m)b, \quad \text{and} \quad \lim_i b(mv_i)$$

exist. Checking the Cauchy condition relative to the first limit we have, for all i and j, that

$$\begin{aligned}
\|(v_i m)b - (v_j m)b\|^2 &= \big\|\big((v_i m)b - (v_j m)b\big)\big(b^*(m^* v_i) - b^*(m^* v_j)\big)\big\| \\
&= \|(v_i m)bb^*(m^* v_i) - (v_i m)bb^*(m^* v_j) - (v_j m)bb^*(m^* v_i) \\
&\qquad + (v_j m)bb^*(m^* v_j)\| \\
&\leq \|v_i m(bb^*)m^* v_i - v_i m(bb^*)m^* v_j\| \\
&+ \|v_j m(bb^*)m^* v_i - v_j m(bb^*)m^* v_j\| \\
&\leq \|v_i\|\|m(bb^*)m^* v_i - m(bb^*)m^* v_j\| \\
&\qquad + \|v_j\|\|m(bb^*)m^* v_i - m(bb^*)m^* v_j\| \\
&\overset{i,j\to\infty}{\longrightarrow} 0.
\end{aligned}$$

This proves the existence of the first limit, while the second limit is shown to exist by a similar argument. We may then define L_m and R_m as in the last sentence of the statement, and the proofs of (i–iv) are now mostly routine, so we restrict ourselves to proving (iii), leaving the proofs of the other points to the reader. Given a in B_1, and b in B_g, the left-hand-side of (iii) equals

$$aL_m(b) = a\lim_i(v_i m)b = \lim_i(av_i m)b = (am)b.$$

□

The following result gives a concrete way to describe the *fiber-wise multipliers* given by (16.16) in the case of semi-direct product bundles:

Proposition 16.17. *Let $\mathscr{B}$ be the semi-direct product bundle associated to a given C^*-algebraic partial dynamical system*

$$\theta = \big(A,\ G,\ \{D_g\}_{g\in G},\ \{\theta_g\}_{g\in G}\big).$$

Then for every multiplier $m \in \mathcal{M}(A)$, and every a in any D_g, one has that

$$L_m(a\delta_g) = (ma)\delta_g, \quad \textit{and} \quad R_m(a\delta_g) = \theta_g\big(\theta_{g^{-1}}(a)m\big)\delta_g.$$

Proof. Picking an approximate identity $\{v_i\}_i$ for A, we have that $\{v_i\delta_1\}_i$ is an approximate identity for $B_1 = A\delta_1$, so for any a in D_g, we have that

$$L_m(a\delta_g) = \lim_i(v_i m\delta_1)a\delta_g = \lim_i(v_i ma)\delta_g = (ma)\delta_g,$$

proving the first identity. As for the second one, we have

$$\begin{aligned} R_m(a\delta_g) &= \lim_i(a\delta_g)(mv_i\delta_1) \\ &= \lim_i \theta_g\big(\theta_{g^{-1}}(a)mv_i\big)\delta_g \\ &= \theta_g\big(\theta_{g^{-1}}(a)m\big)\delta_g. \end{aligned}$$

□

The reader is urged to compare the above formula for $R_m(a\delta_g)$ with the multiplication $(a\delta_g)(m\delta_1)$, should m be an element of A.

We will now present the first process of assembling a C*-algebra from a given Fell bundle. This will produce the biggest possible outcome.

► Recall that $\mathscr{B} = \{B_g\}_{g\in G}$ denotes a Fell bundle which has been fixed near the beginning of this chapter.

Definition 16.18. By a *section* of $\mathscr{B}$ we shall mean any function y from G to the total space of $\mathscr{B}$, such that $y_g \in B_g$, for every g in G. We will moreover denote by $C_c(\mathscr{B})$ the collection of all finitely supported sections. Given two sections y and z in $C_c(\mathscr{B})$, we define their *convolution product* by

$$(y \star z)_g = \sum_{h\in G} y_h z_{h^{-1}g}, \quad \forall\, g \in G.$$

We also define an *adjoint operation* by

$$y^*_g = (y_{g^{-1}})^*, \quad \forall\, g \in G.$$

With respect to the convolution product above, notice that for every h in G we have

$$y_h z_{h^{-1}g} \in B_h B_{h^{-1}g} \subseteq B_g,$$

so all of the summands lie in the same vector space, hence the sum is well defined as an element of B_g.

We leave it for the reader to prove the following easy fact:

Proposition 16.19. *With the above operations one has that $C_c(\mathscr{B})$ is an associative *-algebra.*

As an example, notice that for the semi-direct product bundle $\mathscr{B}$ described in (16.6), we have that $C_c(\mathscr{B})$ is isomorphic to $A \rtimes_{\text{alg}} G$.

Definition 16.20. A *-representation of a Fell bundle $\mathscr{B} = \{B_g\}_{g\in G}$ in a *-algebra C is a collection $\pi = \{\pi_g\}_{g\in G}$ of linear maps

$$\pi_g : B_g \to C,$$

such that

(i) $\pi_g(b)\pi_h(c) = \pi_{gh}(bc)$,
(ii) $\pi_g(b)^* = \pi_{g^{-1}}(b^*)$,

for all $g, h \in G$, and all $b \in B_g$, and $c \in B_h$.

Given a Fell bundle $\mathscr{B}$, for each g in G we will denote by j_g the natural inclusion of B_g in $C_c(\mathscr{B})$, namely the map

$$j_g : B_g \to C_c(\mathscr{B})$$

defined by

$$j_g(b)|_h = \begin{cases} b, & \text{if } h = g, \\ 0, & \text{otherwise.} \end{cases} \tag{16.21}$$

Proposition 16.22. *The collection of maps $j = \{j_g\}_{g\in G}$ is a *-representation of $\mathscr{B}$ in $C_c(\mathscr{B})$.*

PROOF. Left to the reader. □

We would now like to construct a C*-algebra from $C_c(\mathscr{B})$ in the same way we built the C*-algebraic partial crossed product from its algebraic counterpart. For this we need to obtain a bound for C*-seminorms on $C_c(\mathscr{B})$, much like we did in (11.9).

Proposition 16.23. *Let p be a C*-seminorm on $C_c(\mathscr{B})$. Then, for every y in $C_c(\mathscr{B})$, one has that*

$$p(y) \leq \sum_{g\in G} \|y_g\|.$$

PROOF. Using (16.22) we have for every $b \in B_g$, that

$$p\big(j_g(b)\big)^2 = p\big(j_g(b)^* j_g(b)\big) = p\big(j_{g^{-1}}(b^*) j_g(b)\big) = p\big(j_1(b^*b)\big).$$

Observe that the composition $p \circ j_1$ is clearly a C*-seminorm on B_1. So, as already mentioned in (11.8), we have

$$p\big(j_1(b^*b)\big) \leq \|b^*b\| = \|b\|^2.$$

The conclusion is thus that

$$p\big(j_g(b)\big) \leq \|b\|.$$

For a general y in $C_c(\mathscr{B})$, we may write $y = \sum_{g\in G} j_g(y_g)$, so

$$p(y) \leq \sum_{g\in G} p\big(j_g(y_g)\big) \leq \sum_{g\in G} \|y_g\|.$$

This concludes the proof. □

Given y in $C_c(\mathscr{B})$, define

$$\|y\|_{\max} = \sup_p \ p(y), \tag{16.24}$$

where p ranges in the collection of all C*-seminorms on $C_c(\mathscr{B})$. By the above result we see that $\|\cdot\|_{\max}$ is finite for every y, and hence it is a well defined C*-seminorm on $C_c(\mathscr{B})$.

Definition 16.25. The *cross sectional C^*-algebra* of $\mathscr{B}$, denoted $C^*(\mathscr{B})$, is the C*-algebra obtained by taking the quotient of $C_c(\mathscr{B})$ by the ideal consisting of the elements of norm zero and then completing it.

We will soon see that $\|\cdot\|_{\max}$ is in fact a norm, and hence that $C_c(\mathscr{B})$ may be embedded as a dense subalgebra of $C^*(\mathscr{B})$. Meanwhile we will call the canonical mapping arising from the completion process by

$$\kappa : C_c(\mathscr{B}) \to C^*(\mathscr{B}).$$

Since κ is a *-homomorphism, and since j is a *-representation, the following is evident:

Proposition 16.26. *For each g in G, let us denote by $\hat{\jmath}_g$ the composition*

$$B_g \xrightarrow{j_g} C_c(\mathscr{B}) \xrightarrow{\kappa} C^*(\mathscr{B}).$$

Then $\hat{\jmath} = \{\hat{\jmath}_g\}_{g\in G}$ is a *-representation of $\mathscr{B}$ in $C^*(\mathscr{B})$, henceforth called the universal representation.

We have already seen that B_1 is always a C*-algebra and it is easy to see that $\hat{\jmath}_1$ is necessarily a *-homomorphism. In fact we have:

Proposition 16.27. *For every Fell bundle $\mathscr{B}$, the map*

$$\hat{\jmath}_1 : B_1 \to C^*(\mathscr{B})$$

is a non-degenerate *-homomorphism.

Proof. Follows immediately from (16.9). □

The result stated below is simply the recognition that the C*-algebraic partial crossed product defined in (11.11) is a special case of the cross sectional C*-algebra.

Proposition 16.28. *Let*

$$\theta = \big(\{D_g\}_{g\in G},\ \{\theta_g\}_{g\in G}\big)$$

be a C^*-algebraic partial action of a group G on a C^*-algebra A, and let $\mathscr{B}$ be the corresponding semi-direct product bundle. Then $A \rtimes G$ is naturally isomorphic to $C^*(\mathscr{B})$.

In the case of the group bundle $\mathbb{C}\times G$, the cross sectional C*-algebra is denoted $C^*(G)$ and is called the *group C^*-algebra* of G.

The cross sectional C*-algebra of a Fell bundle possesses the following universal property with respect to bundle representations:

Proposition 16.29. *Let $\pi = \{\pi_g\}_{g\in G}$ be a *-representation of the Fell bundle $\mathscr{B}$ in a C^*-algebra C. Then there exists a unique *-homomorphism*

$$\varphi : C^*(\mathscr{B}) \to C,$$

such that

$$\varphi\big(\hat{\jmath}_g(b)\big) = \pi_g(b), \quad \forall\, g \in G, \quad \forall\, b \in B_g.$$

We will say that φ is the integrated form of π.

PROOF. Given π, define $\varphi_0 : C_c(\mathscr{B}) \to C$, by

$$\varphi_0(y) = \sum_{g\in G} \pi_g(y_g), \quad \forall\, y \in C_c(\mathscr{B}).$$

It is a routine matter to prove that φ_0 is a *-homomorphism, whence the formula

$$p(y) = \|\varphi_0(y)\|, \quad \forall\, y \in C_c(\mathscr{B}),$$

defines a C*-seminorm on $C_c(\mathscr{B})$, which is therefore dominated by $\|\cdot\|_{\max}$. This says that φ_0 is continuous and so it extends to the completion, providing the required map. □

Notes and remarks. Fell bundles were introduced in the late '60s by J. M. G. Fell in [**62**], under the name *C*-algebraic bundles*, setting the stage for a far reaching generalization of Harmonic Analysis. See [**63**] for a very extensive treatment of this and other important concepts.

Fell bundles over topological groups are defined in a different fashion since the topology of the group must also be taken into account. However, since we have chosen to work only with discrete groups, we do not need to worry about any extra topological conditions.

CHAPTER 17

Reduced Cross-Sectional Algebras

The construction of the cross-sectional C*-algebra for a Fell bundle given in (16.25) shares with (11.11), as well as many other universal constructions, a norm defined via a supremum (see (16.24) and (11.10)). While we have so far verified that these maximal norms are finite, we have not worried about them being zero! In other words, we are facing the potentially tragic situation in which the cross-sectional algebra of nontrivial Fell bundles may turn out to be the zero algebra! By (16.28) this would also affect C*-algebraic partial crossed products.

In order to rule out this undesirable situation, given a Fell bundle $\mathscr{B}$, we must provide nontrivial C*-seminorms on $C_c(\mathscr{B})$, and one of the best ways to do so is through representation theory. That is, given any representation ρ of $C_c(\mathscr{B})$ on a Hilbert space, we may define a seminorm by $p(y) = \|\rho(y)\|$, which will be nontrivial as long as ρ is nontrivial.

However, instead of representing $C_c(\mathscr{B})$ on a Hilbert space, our representation will be on a Hilbert module over the unit fiber algebra B_1, which we now set out to construct. As a byproduct we will describe our second process of assembling a C*-algebra from a given Fell bundle.

► From now on we fix an arbitrary Fell bundle $\mathscr{B} = \{B_g\}_{g\in G}$.

By (16.22) we have that j_1 is a *-homomorphism through which we may view B_1 as a subalgebra of $C_c(\mathscr{B})$. This makes $C_c(\mathscr{B})$ a right B_1-module in a standard way and we will now introduce a B_1-valued inner-product on $C_c(\mathscr{B})$ as follows

$$\langle y, z\rangle = \sum_{g\in G} (y_g)^* z_g, \quad \forall\, y, z \in C_c(\mathscr{B}).$$

The easy verification that this is indeed an inner-product is left to the reader. Once this is done we have that $C_c(\mathscr{B})$ is a right pre-Hilbert B_1-module.

Definition 17.1. We shall denote by $\ell^2(\mathscr{B})$ the right Hilbert B_1-module obtained by completing $C_c(\mathscr{B})$ under the norm $\|\cdot\|_2$ arising from the inner-product defined above.

Seeing $C_c(\mathscr{B})$ as a dense subspace of $\ell^2(\mathscr{B})$, we may view j_g as a map from B_g to $\ell^2(\mathscr{B})$, and it is interesting to remark that this is an isometry, since, for every b in any B_g, one has

$$\|j_g(b)\|_2^2 = \|\langle j_g(b), j_g(b)\rangle\| = \|b^*b\| = \|b\|^2.$$

Our next step will be the construction of a representation of $\mathscr{B}$ within $\mathscr{L}\big(\ell^2(\mathscr{B})\big)$. The following technical fact will be useful in that direction:

Lemma 17.2. *Given $g, h \in G$, $b \in B_g$, and $c \in B_h$, one has that*

$$c^*b^*bc \leq \|b\|^2 c^*c.$$

PROOF. It is well known that a positive element h in a C*-algebra A satisfies $h \leq \|h\|$, in the sense that $v^*(\|h\| - h)v \geq 0$, for every v in A. Here, we either temporarily work in the unitization of A, or we simply write $v^*(\|h\|-h)v$ to actually mean $\|h\|v^*v - v^*hv$.

Picking an approximate identity $\{v_i\}_i$ for B_1, we then have for all i that,

$$v_i^*(\|b\|^2 - b^*b)v_i \geq 0.$$

So, there exists some a_i in B_1 such that

$$v_i^*(\|b\|^2 - b^*b)v_i = a_i^*a_i.$$

Applying (16.1.k) for $a_ic \in B_h$, and using (16.9), we get

$$\begin{aligned} 0 \leq (a_ic)^*a_ic &= c^*a_i^*a_ic \\ &= c^*v_i^*(\|b\|^2 - b^*b)v_ic \\ &\overset{i\to\infty}{\longrightarrow} c^*(\|b\|^2 - b^*b)c \\ &= \|b\|^2c^*c - c^*b^*bc, \end{aligned}$$

concluding the proof. □

Proposition 17.3. *Given b in any B_g, the operator*

$$\lambda_g(b): \ y \ \in \ C_c(\mathscr{B}) \ \longmapsto \ j_g(b) \star y \ \in \ C_c(\mathscr{B})$$

is continuous relative to the Hilbert module norm on $C_c(\mathscr{B})$, and hence extends to a bounded operator on $\ell^2(\mathscr{B})$, still denoted by $\lambda_g(b)$ by abuse of language, such that $\|\lambda_g(b)\| = \|b\|$, and which moreover satisfies

$$\lambda_g(b)\big(j_h(c)\big) = j_{gh}(bc), \quad \forall\, h \in G, \quad \forall\, c \in B_h.$$

PROOF. Notice that for every y in $C_c(\mathscr{B})$ and every h in G, the formula for the convolution product gives

$$\lambda_g(b)y|_h = by_{g^{-1}h}.$$

Therefore, if c is in B_h, then for every k in G, we have that

$$\lambda_g(b)\big(j_h(c)\big)|_k = bj_h(c)_{g^{-1}k} = \delta_{h,g^{-1}k}bc = \delta_{gh,k}bc = j_{gh}(bc)_k,$$

proving the last assertion in the statement. Addressing the boundedness of $\lambda_g(b)$, given y in $C_c(\mathscr{B})$, notice that

$$\begin{aligned} \langle\lambda_g(b)y, \lambda_g(b)y\rangle &= \sum_{h\in G} (y_{g^{-1}h})^*b^*by_{g^{-1}h} \\ &= \sum_{h\in G} (y_h)^*b^*by_h \\ &\overset{(17.2)}{\leq} \|b\|^2 \sum_{h\in G} (y_h)^*y_h \\ &= \|b\|^2\langle y, y\rangle. \end{aligned}$$

This implies that $\|\lambda_g(b)y\|_2 \leq \|b\|\|y\|_2$, from where we see that $\lambda_g(b)$ is bounded and $\|\lambda_g(b)\| \leq \|b\|$.

In order to prove the reverse inequality let ξ be the element of $\ell^2(\mathscr{B})$ obtaining by mapping b^* into $\ell^2(\mathscr{B})$ via $j_{g^{-1}}$. Recalling that $j_{g^{-1}}$ is an isometry, we have that $\|\xi\|_2 = \|b\|$. We then have

$$\lambda_g(b)\xi = \lambda_g(b)j_{g^{-1}}(b^*) = j_g(b) \star j_{g^{-1}}(b^*) = j_1(bb^*),$$

so

$$\|b\|^2 = \|b^*b\| = \|j_1(bb^*)\|_2 = \|\lambda_g(b)\xi\|_2 \leq \|\lambda_g(b)\|\|\xi\|_2 = \|\lambda_g(b)\|\|b\|.$$

This proves that $\|\lambda_g(b)\| \geq \|b\|$, completing the proof. □

We may now finally introduce the first nontrivial representation of our bundle.

Proposition 17.4. *The collection of maps*

$$\lambda = \{\lambda_g\}_{g\in G}$$

introduced above is a representation of $\mathscr{B}$ in $\mathscr{L}\left(\ell^2(\mathscr{B})\right)$, henceforth called the regular representation of $\mathscr{B}$.

PROOF. For b in any B_g, and y, z in $C_c(\mathscr{B})$, we have

$$\langle \lambda_g(b)y, z\rangle = \sum_{h\in G} (y_{g^{-1}h})^* b^* z_h = \sum_{h\in G} (y_h)^* b^* z_{gh} = \langle y, \lambda_{g^{-1}}(b^*)z\rangle.$$

Since $C_c(\mathscr{B})$ is dense in $\ell^2(\mathscr{B})$, we conclude from the above that

$$\langle \lambda_g(b)\xi, \eta\rangle = \langle \xi, \lambda_{g^{-1}}(b^*)\eta\rangle, \quad \forall\, \xi, \eta \in \ell^2(\mathscr{B}),$$

and hence that $\lambda_g(b)$ is an adjointable operator, with $\lambda_g(b)^* = \lambda_{g^{-1}}(b^*)$. This also proves (16.20.ii). Given b_g in B_g, b_h in B_h, and y in $C_c(\mathscr{B})$, we have

$$\begin{aligned}\lambda_g(b_g)\big(\lambda_h(b_h)y\big) &= j_g(b_g) \star j_h(b_h) \star y \\ &\overset{(16.22)}{=} j_{gh}(b_g b_h) \star y \\ &= \lambda_{gh}(b_g b_h)y.\end{aligned}$$

Again because $C_c(\mathscr{B})$ is dense in $\ell^2(\mathscr{B})$, we get $\lambda_g(b_g)\lambda_h(b_h) = \lambda_{gh}(b_g b_h)$, proving (16.20.i) and concluding the proof. □

By (16.29), the regular representation of $\mathscr{B}$ gives rise to a *-homomorphism

$$\Lambda : C^*(\mathscr{B}) \to \mathscr{L}\left(\ell^2(\mathscr{B})\right), \tag{17.5}$$

satisfying $\Lambda \circ \hat{\jmath}_g = \lambda_g$, for any g in G.

Definition 17.6. The *-homomorphism Λ defined above will be called the *regular representation of $C^*(\mathscr{B})$*, and its range, which we will denote by $C^*_{\text{red}}(\mathscr{B})$, will be called the *reduced cross sectional C^*-algebra* of $\mathscr{B}$.

We thus see that $C^*_{\text{red}}(\mathscr{B})$ is isomorphic to the quotient of $C^*(\mathscr{B})$ by the null space of Λ.

$$B_g \xrightarrow{\;j_g\;} C_c(\mathscr{B}) \xrightarrow{\;\kappa\;} C^*(\mathscr{B}) \xrightarrow{\;\Lambda\;} C^*_{\text{red}}(\mathscr{B}). \tag{17.7}$$

In the special case of the group bundle $\mathbb{C} \times G$, the reduced cross sectional C*-algebra is denoted $C^*_{\text{red}}(G)$, and it is called the *reduced group C^*-algebra* of G.

The consequences of the existence of the regular representation are numerous. We begin with a technical remark which should be seen as a generalization of Fourier coefficients to Fell bundles.

Lemma 17.8. *For each g in G, there exists a contractive*[21] *linear mapping*

$$E_g : C^*_{red}(\mathscr{B}) \to B_g,$$

such that, for every h in G, and every b in B_h, one has that

$$E_g\big(\lambda_h(b)\big) = \begin{cases} b, & \text{if } g = h, \\ 0, & \text{if } g \neq h. \end{cases}$$

For any given z in $C^*_{red}(\mathscr{B})$ we will refer to $E_g(z)$ as the gth Fourier coefficient of z.

PROOF. The map

$$P_g : y \in C_c(\mathscr{B}) \to y_g \in B_g,$$

is easily seen to be continuous for the Hilbert module norm, and hence it extends to a continuous mapping, also denoted P_g, from $\ell^2(\mathscr{B})$ to B_g.

Letting $\{v_i\}_i$ be an approximate identity for B_1, we then claim that the limit

$$\lim_i P_g\Big(z\big(j_1(v_i)\big)\Big) \tag{17.8.1}$$

exists for every z in $C^*_{\text{red}}(\mathscr{B})$. In order to prove our claim, let us first check it for $z = \lambda_h(b)$, where $h \in G$, and $b \in B_h$. In this case, for every i, we have

$$\begin{aligned} P_g\Big(z\big(j_1(v_i)\big)\Big) &= P_g\Big(\lambda_h(b)\big(j_1(v_i)\big)\Big) \\ &= P_g\big(j_h(b) \star j_1(v_i)\big) \\ &= P_g\big(j_h(bv_i)\big) \\ &= \delta_{g,h} bv_i, \end{aligned}$$

where $\delta_{g,h}$ is the Kronecker symbol. Thus the limit in (17.8.1) exists and coincides with $\delta_{g,h}b$, by (16.9). It follows that the limit also exists for every z in

$$\sum_{h\in G} \lambda_h(B_h) = \Lambda\big(\kappa(C_c(\mathscr{B}))\big),$$

which is clearly dense in $C^*_{\text{red}}(\mathscr{B})$. Since the operators involved in (17.8.1) are uniformly bounded, we conclude that the above limit indeed exists for all z in $C^*_{\text{red}}(\mathscr{B})$. We may thus define

$$E_g(z) = \lim_i P_g\Big(z\big(j_1(v_i)\big)\Big), \quad \forall\, z \in C^*_{\text{red}}(\mathscr{B}),$$

observing that the calculation above gives $E_g\big(\lambda_h(b)\big) = \delta_{g,h}b$, for all b in B_h.

Finally, notice that since $\|P_g\| = 1$, and $\|v_i\| \leq 1$, then

$$\|E_g(z)\| = \lim_i \Big\| P_g\Big(z\big(j_1(v_i)\big)\Big)\Big\| \leq \|z\|.$$

□

The more technical aspects having been taken care of, we may now reap the benefits:

Proposition 17.9. *Let $\mathscr{B}$ be a Fell bundle. Then:*

(*i*) *the map $\kappa : C_c(\mathscr{B}) \to C^*(\mathscr{B})$ is injective,*
(*ii*) *the map $\Lambda \circ \kappa : C_c(\mathscr{B}) \to C^*_{red}(\mathscr{B})$ is injective,*
(*iii*) *there exists a C^*-seminorm on $C_c(\mathscr{B})$ that is a norm,*

[21] A linear map T is said to be *contractive* if $\|T(x)\| \leq \|x\|$, for all x in its domain.

(*iv*) *for every* g *in* G, *the map* $\hat{\jmath}_g : B_g \to C^*(\mathscr{B})$ *is isometric,*
(*v*) *for every* g *in* G, *the map* $\lambda_g : B_g \to C^*_{red}(\mathscr{B})$ *is isometric,*
(*vi*) $C^*(\mathscr{B})$ *is a graded* C^**-algebra, with grading subspaces* $\hat{\jmath}_g(B_g)$,
(*vii*) $C^*_{red}(\mathscr{B})$ *is a graded* C^**-algebra, with grading subspaces* $\lambda_g(B_g)$.

PROOF. (ii) Given $y \in C_c(\mathscr{B})$ such that $\Lambda\big(\kappa(y)\big) = 0$, write $y = \sum_{h\in G} j_h(y_h)$. Then, for every g in G, we have

$$0 = E_g\big(\Lambda(\kappa(y))\big) = \sum_{h\in G} E_g\Big(\Lambda\big(\kappa(j_h(y_h))\big)\Big) \overset{(17.7)}{=} \sum_{h\in G} E_g\big(\lambda_h(y_h)\big) \overset{(17.8)}{=} y_g,$$

hence $y = 0$.

(i) Follows immediately from (ii).

(iii) In view of (i), it is enough to take $p(y) := \|y\|_{\max} = \|\kappa(y)\|$.

(iv) Given b in any B_g, on the one hand we have that

$$\|\hat{\jmath}_g(b)\| = \|\kappa\big(j_g(b)\big)\| = \|j_g(b)\|_{\max} \overset{(16.23)}{\leq} \|b\|,$$

and on the other

$$\|\hat{\jmath}_g(b)\| \geq \|\Lambda\big(\hat{\jmath}_g(b)\big)\| \overset{(17.7)}{=} \|\lambda_g(b)\| \overset{(17.3)}{=} \|b\|.$$

(v) This is just (17.3). It is included here for comparison purposes only.

(vii) By (v) we have that each $\lambda_g(B_g)$ is a closed subspace of $C^*_{\rm red}(\mathscr{B})$. To show that they are independent, suppose that

$$\sum_{g\in G} z_g = 0,$$

with $z_g \in \lambda_g(B_g)$, the sum having only finitely many nonzero terms. For each g, write $z_g = \lambda_g(y_g)$, with $y_g \in B_g$. We may then view y as an element of $C_c(\mathscr{B})$ and we have

$$\Lambda\big(\kappa(y)\big) = \sum_{g\in G} \Lambda\big(\kappa\big(j_g(y_g)\big)\big) = \sum_{g\in G} \lambda_g(y_g) = \sum_{g\in G} z_g = 0,$$

whence $y = 0$ by (ii), and consequently $z_g = \lambda_g(y_g) = 0$, for all g in G. This shows that the $\lambda_g(B_g)$ form an independent collection of subspaces. It is then easy to see that

$$\bigoplus_{g\in G} \lambda_g(B_g) = \Lambda\big(\kappa(C_c(\mathscr{B}))\big), \tag{17.9.1}$$

which is clearly dense in $C^*_{\rm red}(\mathscr{B})$. In order to check the remaining conditions (16.2.i)–(16.2.ii) , observe that λ is a representation by (17.4), so for all $g, h \in G$, we have that

$$\lambda_g(B_g)\lambda_h(B_g) = \lambda_{gh}(B_{gh}),$$

and

$$\lambda_g(B_g)^* = \lambda_{g^{-1}}(B_g^*) = \lambda_{g^{-1}}(B_{g^{-1}}).$$

Point (vi) may be proved in a similar fashion and so it is left for the reader. □

Recall from (16.28) that, given a C*-algebraic partial action of a group G on a C*-algebra A, the crossed product $A \rtimes G$ is isomorphic to the cross sectional C*-algebra of the semi-direct product bundle. Inspired by this we may give the following:

Definition 17.10. The *reduced crossed product* of a C*-algebra A by a partial action θ of a group G, denoted $A \rtimes_{\text{red}} G$, is the reduced cross sectional algebra of the corresponding semi-direct product bundle.

Applying (17.9) to semi-direct product bundles we obtain:

Corollary 17.11. *Let θ be a C*-algebraic partial action of a group G on a C*-algebra A. Then:*

(*i*) *the map $\kappa : A \rtimes_{alg} G \to A \rtimes G$ is injective,*
(*ii*) *the map $\Lambda \circ \kappa : A \rtimes_{alg} G \to A \rtimes_{red} G$ is injective,*
(*iii*) *there exists a C*-seminorm on $A \rtimes_{alg} G$ that is a norm,*
(*iv*) *for every g in G, and all $a \in D_g$, one has that $\|a\delta_g\| = \|a\|$,*
(*v*) *for every g in G, and all $a \in D_g$, one has that $\|\Lambda(a\delta_g)\| = \|a\|$,*
(*vi*) *$A \rtimes G$ is a graded C*-algebra, with grading subspaces $D_g\delta_g$,*
(*vii*) *$A \rtimes_{red} G$ is a graded C*-algebra, with grading subspaces $\Lambda(D_g\delta_g)$.*

We have already briefly indicated that the E_g of (17.8) are to be thought of as Fourier coefficients. In fact, when the bundle in question is the group bundle over $\mathbb{Z}$, namely $\mathbb{C} \times \mathbb{Z}$, then the reduced cross sectional algebra is isomorphic to the algebra of all continuous functions on the unit circle. Moreover, for any such function f, one has that $E_n(f)$ coincides with the Fourier coefficient $\hat{f}(n)$.

We will now present a few results about Fell bundles which are motivated by Fourier analysis and which emphasize further similarities between these theories. Our first such result should be compared with the well known fact that the matrix of a multiplication operator on $L^2(S^1)$, relative to the standard basis, is a Laurent matrix, that is, a doubly infinite matrix with constant diagonals.

Proposition 17.12. *Given $z \in C^*_{red}(\mathscr{B})$, one has that*

$$\langle j_g(b), z j_h(c)\rangle = b^* E_{gh^{-1}}(z)c,$$

for all $g, h \in G$, $b \in B_g$, and $c \in B_h$.

PROOF. It is clearly enough to prove the result for $z = \sum_{k\in G} \lambda_k(y_k)$, with each $y_k \in B_k$, the sum having only finitely many nonzero terms. In this case we have

$$\begin{aligned}\langle j_g(b), z j_h(c)\rangle &= \Big\langle j_g(b), \sum_{k\in G} \lambda_k(y_k) j_h(c)\Big\rangle \\ &= \Big\langle j_g(b), \sum_{k\in G} j_k(y_k) \star j_h(c)\Big\rangle \\ &= \Big\langle j_g(b), \sum_{k\in G} j_{kh}(y_k c)\Big\rangle \\ &= b^* y_{gh^{-1}} c \\ &\overset{(17.8)}{=} b^* E_{gh^{-1}}(z)c.\end{aligned}$$

□

Another well known result in Classical Harmonic Analysis states that, if all of the Fourier coefficients of a continuous function f vanish, then $f = 0$. Our next result should be considered as a generalization of this fact.

Proposition 17.13. *Given $z \in C^*_{red}(\mathscr{B})$, the following are equivalent:*

(i) $E_1(z^*z) = 0$,
(ii) $E_g(z) = 0$, *for every g in G,*
(iii) $z = 0$.

PROOF. (i) $\Rightarrow$ (iii). Given b in any B_g, we have

$$\langle zj_g(b), zj_g(b)\rangle = \langle j_g(b), z^*zj_g(b)\rangle \overset{(17.12)}{=} b^*E_1(z^*z)b = 0,$$

whence $zj_g(b) = 0$. Since the $j_g(b)$ span a dense subspace of $\ell^2(\mathscr{B})$, we have $z = 0$.

(ii) $\Rightarrow$ (iii). Given b in any B_g, and c in any B_h, we have

$$\big\langle j_g(b), zj_h(c)\big\rangle = b^*E_{gh^{-1}}(z)c = 0.$$

Since the $j_g(b_g)$ span a dense subspace of $\ell^2(\mathscr{B})$, we see that $zj_h(c) = 0$. As above, this gives $z = 0$.

(i) $\Leftarrow$ (iii) $\Rightarrow$ (ii). Obvious. □

A very important fact in Classical Harmonic Analysis is Parseval's identity, which asserts that, for a continuous function f on the unit circle $\mathbb{T}$ (in fact also for more general functions), one has that

$$\int_{\mathbb{T}} |f(z)|^2\,dz = \sum_{n\in\mathbb{Z}} |\hat{f}(n)|^2.$$

We will now present a generalization of this to Fell bundles by seeing the right-hand side above in terms of our version of Fourier coefficients, while the left-hand-side is interpreted based on the fact that, in the classical case, the integral of a function coincides with its zeroth Fourier coefficient.

We begin with a preparatory result, the second part of which is a generalization of yet another important fact in Harmonic Analysis.

Lemma 17.14. *(a) Let $y \in C_c(\mathscr{B})$, and let $z = \sum_{g\in G} \lambda_g(y_g)$. Then*

$$\sum_{g\in G} E_g(z)^*E_g(z) = E_1(z^*z).$$

(b) (Bessel's inequality) For every $z \in C^*_{red}(\mathscr{B})$, and for every finite subset $K \subseteq G$, one has that

$$\sum_{g\in K} E_g(z)^*E_g(z) \le E_1(z^*z).$$

PROOF. Initially observe that if z is as in (a), then by (17.8) one has that $E_g(z) = y_g$, so the sum in (a) is actually a finite sum. We then have

$$\begin{aligned} E_1(z^*z) &= E_1\Big(\sum_{g,h\in G} \lambda_g(y_g)^*\lambda_h(y_h)\Big) \\ &\overset{(16.20)}{=} E_1\Big(\sum_{g,h\in G} \lambda_{g^{-1}h}(y_g^* y_h)\Big) \\ &\overset{(17.8)}{=} \sum_{g\in G} y_g^* y_g \\ &= \sum_{g\in G} E_g(z)^* E_g(z), \end{aligned}$$

proving (a). In order to prove (b), assume first that z is as in (a). Then it is clear that for every finite set $K \subseteq G$, one has that

$$E_1(z^*z) \overset{(a)}{=} \sum_{g\in G} E_g(z)^*E_g(z) \geq \sum_{g\in K} E_g(z)^*E_g(z).$$

Since the set of elements z considered above is dense in $C^*_{\text{red}}(\mathscr{B})$, and since both the left- and the right-hand expressions above represent continuous functions of the variable z, the result follows by taking limits. □

We now wish to show that the conclusion in (17.14.a) in fact holds for every z in $C^*_{\text{red}}(\mathscr{B})$. Evidently the sum might now have infinitely many nonzero terms, so we need to discuss the kind of convergence to be expected. We will see that unconditional, rather than absolute convergence is the appropriate notion to be used here.

Recall that a series $\sum_{i\in I} x_i$ in a Banach space is said to be *unconditionally convergent* with sum s if, for every $\varepsilon > 0$, there exists a finite set $F_0 \subseteq I$, such that for every finite set $F \subseteq I$, with $F_0 \subseteq F$, one has that

$$\Big\| s - \sum_{i\in F} x_i \Big\| < \varepsilon.$$

This is equivalent to saying that s is the limit of the net $\big\{\sum_{i\in F} x_i\big\}_F$, indexed by the directed set formed by all finite subsets $F \subseteq I$, ordered by inclusion.

Proposition 17.15. *(Parseval's identity) For every z in $C^*_{\text{red}}(\mathscr{B})$, one has that*

$$\sum_{g\in G} E_g(z)^*E_g(z) = E_1(z^*z),$$

where the series converges unconditionally.

PROOF. Given a finite subset $K \subseteq G$, write $K = \{g_1, g_2, \ldots, g_n\}$, and let us consider the mapping

$$E_K : z \in C^*_{\text{red}}(\mathscr{B}) \mapsto \begin{pmatrix} E_{g_1}(z) & 0 & \ldots & 0 \\ E_{g_2}(z) & 0 & \ldots & 0 \\ \vdots & \vdots & \ddots & \vdots \\ E_{g_n}(z) & 0 & \ldots & 0 \end{pmatrix} \in M_n\big(C^*_{\text{red}}(\mathscr{B})\big),$$

where, as usual, we are identifying each B_g as a subspace of $C^*_{\text{red}}(\mathscr{B})$ via the regular representation. It is evident that E_K is a linear mapping, and we claim that it is contractive. In fact, given z in $C^*_{\text{red}}(\mathscr{B})$, we have that

$$\begin{aligned}\|E_K(z)\|^2 &= \|E_K(z)^*E_K(z)\| \\ &= \Big\| \sum_{g\in K} E_g(z)^*E_g(z) \Big\| \\ &\overset{(17.14.b)}{\leq} \|E_1(z^*z)\| \\ &\overset{(17.8)}{\leq} \|z^*z\| \\ &= \|z\|^2,\end{aligned}$$

proving our claim. Given $z, z_0 \in C^*_{\text{red}}(\mathscr{B})$, we then have that

$$\|E_K(z)^*E_K(z) - E_K(z_0)^*E_K(z_0)\| \leq \|z\|\|z - z_0\| + \|z - z_0\|\|z_0\|,$$

from where one sees that the collection of (non-linear) functions

$$\Big\{E_K(\cdot)^*E_K(\cdot) : K \subseteq G,\ |K| < \infty\Big\},$$

is equicontinuous. Observe that each function in the above set takes values in the $(1, 1)$ upper left corner of our matrix algebra. Moreover, for every z in $C^*_{\text{red}}(\mathscr{B})$, we have that

$$\big(E_K(z)^*E_K(z)\big)_{1,1} = \sum_{g\in K} E_g(z)^*E_g(z).$$

By (17.14.a) we see that, as $K\uparrow G$, the expression above converges to $E_1(z^*z)$ on the dense set formed by the finite sums, as in (17.14.a). Being equicontinuous, it therefore converges to $E_1(z^*z)$, for every z in $C^*_{\text{red}}(\mathscr{B})$. □

Notes and remarks. The construction of the reduced cross sectional algebra of a Fell bundle was introduced in [**49**] in analogy with the reduced crossed products defined for C*-dynamical systems, and McClanahan's reduced partial crossed products [**81**, Section 3]. Our approach involves representations on Hilbert modules, rather than Hilbert spaces, allowing us to focus on a single representation at the expense of a bit more abstraction.

CHAPTER 18

Fell's Absorption Principle

The classical version of Fell's absorption principle[22] [**19**, Theorem 2.5.5] states that the tensor product of any unitary group representation by the left-regular representation is weakly-contained in the left-regular representation. The main goal of this chapter is to prove a version of this powerful principle for Fell bundles.

In order to set up the context let us first introduce our notation for the left regular representation of the group G: we will denote by $\{e_g\}_{g\in G}$ the canonical basis of $\ell^2(G)$, and by λ^G the regular representation of G on $\ell^2(G)$, so that

$$\lambda^G_g(e_h) = e_{gh}, \quad \forall\, g, h \in G.$$

Our slightly unusual choice of notation "λ^G", as opposed to "λ", is due to a potential conflict between this and our previous notation for the regular representation of a Fell bundle introduced in (17.4).

Given a Fell bundle

$$\mathscr{B} = \{B_g\}_{g\in G}$$

and a representation $\pi = \{\pi_g\}_{g\in G}$ of $\mathscr{B}$ on a Hilbert space H, let us consider another representation of $\mathscr{B}$, this time on $H \otimes \ell^2(G)$, by putting

$$(\pi_g \otimes \lambda^G)(b) = \pi_g(b) \otimes \lambda^G_g, \quad \forall\, b \in B_g.$$

It is an easy exercise to check that

$$\pi \otimes \lambda^G := \{\pi_g \otimes \lambda^G_g\}_{g\in G} \tag{18.1}$$

is indeed a *-representation of $\mathscr{B}$. The integrated form of $\pi \otimes \lambda^G$, according to (16.29), is then a *-representation

$$\varphi : C^*(\mathscr{B}) \to \mathscr{L}\big(H \otimes \ell^2(G)\big), \tag{18.2}$$

satisfying

$$\varphi\big(\hat{\jmath}_g(b)\big) = \pi_g(b) \otimes \lambda^G_g, \quad \forall\, g \in G, \quad \forall\, b \in B_g. \tag{18.3}$$

Proposition 18.4. *(Fell's absorption principle for Fell bundles). Let π be a representation of the Fell bundle $\mathscr{B}$ on a Hilbert space H. Also let φ be the integrated form of the representation $\pi \otimes \lambda^G$ described above. Then φ vanishes on the kernel of the regular representation Λ, and hence factors through $C^*_{red}(\mathscr{B})$, providing a representation ψ of the latter,*

$$\begin{array}{ccc} C^*(\mathscr{B}) & \xrightarrow{\hspace{3em}} & \mathscr{L}\big(H \otimes \ell^2(G)\big) \\ & {}_{\Lambda}\searrow \quad \nearrow_{\psi} & \\ & C^*_{red}(\mathscr{B}) & \end{array}$$

[22] Apparently Fell has not attempted to generalize this for Fell bundles.

such that $\psi \circ \Lambda = \varphi$. *In addition, if* π_1 *is faithful, then* ψ *is also faithful.*

PROOF. Consider the linear mapping F from $\mathscr{L}\big(H \otimes \ell^2(G)\big)$ into itself sending each bounded operator to its diagonal relative to the decomposition

$$H \otimes \ell^2(G) = \bigoplus_{g \in G} H \otimes e_g.$$

To be precise,

$$F(T) = \sum_{g \in G} P_g T P_g, \quad \forall\, T \in \mathscr{L}\big(H \otimes \ell^2(G)\big),$$

where each P_g is the orthogonal projection onto $H \otimes e_g$, and the sum is interpreted in the strong operator topology. It is a well known fact that F is a *conditional expectation*[23] onto the algebra of diagonal operators, which is moreover faithful. We next claim that

$$F\big(\varphi(y)\big) = \pi_1\Big(E_1\big(\Lambda(y)\big)\Big) \otimes 1, \quad \forall\, y \in C^*(\mathscr{B}). \tag{18.4.1}$$

Since both sides represent continuous linear maps on $C^*(\mathscr{B})$, it is enough to check it for $y = \hat{\jmath}_g(b)$, where $b \in B_g$. In this case we have

$$F\big(\varphi(y)\big) = F\big(\varphi(\hat{\jmath}_g(b))\big) \overset{(18.3)}{=} F\big(\pi_g(b) \otimes \lambda^G_g\big) = \delta_{g,1}\pi_1(b) \otimes 1.$$

In the last step above we have used the fact that

$$\big(\pi_g(b) \otimes \lambda^G_g\big)(H \otimes e_h) \subseteq H \otimes e_{gh}, \quad \forall\, h \in G,$$

so that $\pi_g(b) \otimes \lambda^G_g$ has zero diagonal entries when $g \neq 1$, while it is a diagonal operator when $g = 1$. Focusing now on the right-hand-side of (18.4.1), we have

$$\pi_1\Big(E_1\big(\Lambda(y)\big)\Big) = \pi_1\Big(E_1\big(\Lambda(\hat{\jmath}_g(b))\big)\Big) \overset{(17.7)}{=} \pi_1\Big(E_1\big(\lambda_g(b)\big)\Big) \overset{(17.8)}{=} \delta_{g,1}\pi_1(b),$$

hence proving (18.4.1).

Given y in $C^*(\mathscr{B})$ such that $\Lambda(y) = 0$, we then have that $\Lambda(y^*y) = 0$, so (18.4.1) gives

$$0 = F\big(\varphi(y^*y)\big) = F\big(\varphi(y)^*\varphi(y)\big),$$

which implies that $\varphi(y) = 0$, since F is faithful. This proves that φ vanishes on the kernel of Λ.

Assuming that π_1 is faithful, we will prove that the kernel of Λ in fact coincides with the kernel of φ, from where the last sentence of the statement will follow. We have already seen above that $\mathrm{Ker}(\Lambda) \subseteq \mathrm{Ker}(\varphi)$, so we need only prove the reverse inclusion. Assuming that $\varphi(y) = 0$, we deduce from (18.4.1) that $\pi_1\Big(E_1\big(\Lambda(y^*y)\big)\Big) = 0$, and since π_1 is faithful, also that

$$0 = E_1\big(\Lambda(y^*y)\big) = E_1\big(\Lambda(y)^*\Lambda(y)\big).$$

From (17.13) we then get $\Lambda(y) = 0$, so $\mathrm{Ker}(\varphi) \subseteq \mathrm{Ker}(\Lambda)$. □

Recall from [**19**, Section 3.3] that if A and B are C*-algebras, then the algebraic tensor product $A \odot B$ may have many different C*-norms, the smaller of which, usually denoted by $\|\cdot\|_{\min}$, is called the *minimal* or *spatial tensor norm*, while the biggest one, usually denoted by $\|\cdot\|_{\max}$, is called the *maximal* norm.

---

[23] A *conditional expectation* is a linear mapping F from a C*-algebra B onto a closed *-subalgebra A which is positive, idempotent, contractive and an A-bimodule map. We moreover say that F is *faithful* if $F(b^*b) = 0 \Rightarrow b = 0$.

Accordingly, we denote by $A \otimes_{\min} B$ and $A \otimes_{\max} B$, the completions of $A \odot B$ under $\|\cdot\|_{\min}$ and $\|\cdot\|_{\max}$, respectively. These are often referred to as the *minimal* and *maximal* tensor products of A and B.

If A and B are faithfully represented on Hilbert spaces H and K, respectively, then the natural representation of $A \odot B$ on $H \otimes K$ is isometric for the minimal norm, and this is precisely the reason why this norm is also called the spatial norm. See [**19**] for a comprehensive study of tensor products of C*-algebras.

In our next result we use the minimal tensor product to give a slightly different, but sometimes more useful way to state Fell's absorption principle:

Corollary 18.5. *Let $\mathscr{B} = \{B_g\}_{g\in G}$ be a Fell bundle and let $\pi = \{\pi_g\}_{g\in G}$ be a *-representation of $\mathscr{B}$ in a C^*-algebra A. Then there is a *-homomorphism*

$$\psi : C^*_{red}(\mathscr{B}) \to A \underset{min}{\otimes} C^*_{red}(G),$$

such that

$$\psi\big(\lambda_g(b)\big) = \pi_g(b) \otimes \lambda^G_g, \quad \forall g \in G, \quad \forall b \in B_g.$$

Moreover, if π_1 is faithful, then so is ψ.

PROOF. Supposing, without loss of generality, that A is an algebra of operators on a Hilbert space H, we may view π as a representation of $\mathscr{B}$ on H. Letting φ and ψ be as in (18.4), we have for all b in any B_g, that

$$\psi\big(\lambda_g(b)\big) = \psi\big(\Lambda(\hat{\jmath}_g(b))\big) = \varphi\big(\hat{\jmath}_g(b)\big) = \pi_g(b) \otimes \lambda^G_g,$$

as required. Observing that $C^*_{\rm red}(G)$ is precisely the subalgebra of operators on $\ell^2(G)$ generated by the range of the left-regular representation λ^G, we see from the above computation that $\psi\big(\lambda_g(b)\big)$ lies in the spatial tensor product of A by $C^*_{\rm red}(G)$, which, by [**19**, Theorem 3.3.11], is isomorphic to $A \otimes_{\min} C^*_{\rm red}(G)$. We may then view ψ as a map from $C^*_{\rm red}(\mathscr{B})$ to $A \otimes_{\min} C^*_{\rm red}(G)$, as needed. Finally, assuming that π_1 is faithful, (18.4) implies that ψ is also faithful. □

Fell's absorption principle may be used to produce a variety of maps simultaneously involving the full and the reduced cross-sectional C*-algebras, such as the following:

Proposition 18.6. *Given a Fell bundle $\mathscr{B} = \{B_g\}_{g\in G}$, there exists an injective *-homomorphism*

$$\tau : C^*_{red}(\mathscr{B}) \to C^*(\mathscr{B}) \underset{min}{\otimes} C^*_{red}(G),$$

such that

$$\tau\big(\lambda_g(b)\big) = \hat{\jmath}_g(b) \otimes \lambda^G_g, \quad \forall g \in G, \quad \forall b \in B_g.$$

PROOF. Apply (18.5) to $\pi = \hat{\jmath}$. □

A similar result is as follows:

Proposition 18.7. *Given a Fell bundle $\mathscr{B} = \{B_g\}_{g\in G}$, there exists an injective *-homomorphism*

$$\sigma : C^*_{red}(\mathscr{B}) \to C^*_{red}(\mathscr{B}) \underset{min}{\otimes} C^*_{red}(G),$$

such that

$$\sigma\big(\lambda_g(b)\big) = \lambda_g(b) \otimes \lambda^G_g, \quad \forall g \in G, \quad \forall b \in B_g.$$

PROOF. Apply (18.5) to $\pi = \lambda$. □

A very interesting question arises when one replaces the minimal tensor product by the maximal one in (18.7). To be precise, given a Fell bundle $\mathscr{B} = \{B_g\}_{g\in G}$, for each g in G, consider the mapping

$$\mu_g : B_g \to C^*_{\text{red}}(\mathscr{B}) \underset{\max}{\otimes} C^*_{\text{red}}(G),$$

given by

$$\mu_g(b) = \lambda_g(b) \otimes \lambda^G_g, \quad \forall\, b \in B_g.$$

It is easy to see that $\mu = \{\mu_g\}_{g\in G}$ is a *-representation of $\mathscr{B}$, the integrated form of which is then a *-homomorphism

$$\mathscr{S} : C^*(\mathscr{B}) \to C^*_{\text{red}}(\mathscr{B}) \underset{\max}{\otimes} C^*_{\text{red}}(G), \tag{18.8}$$

satisfying

$$\mathscr{S}\big(\hat{\jmath}_g(b)\big) = \lambda_g(b) \otimes \lambda^G_g, \quad \forall\, b \in B_g.$$

Observe that the right-hand side of the above expression is identical to the expression defining σ in (18.7), although there we used the minimal tensor product, while here we are using the maximal one. In any case, since the above formula for $\mathscr{S}$ involves not one, but two regular representations, one is left wondering whether or not it factors through the reduced cross-sectional algebra. The answer is a bit surprising: we will prove below that $\mathscr{S}$ is injective!

Proposition 18.9. *Given any Fell bundle $\mathscr{B} = \{B_g\}_{g\in G}$, there exists an injective *-homomorphism*

$$\mathscr{S} : C^*(\mathscr{B}) \to C^*_{red}(\mathscr{B}) \underset{max}{\otimes} C^*_{red}(G),$$

such that

$$\mathscr{S}\big(\hat{\jmath}_g(b)\big) = \lambda_g(b) \otimes \lambda^G_g, \quad \forall\, b \in B_g.$$

PROOF. The proof will of course consist in verifying that the map $\mathscr{S}$ introduced in (18.8) is injective. Choose a faithful representation ρ of $C^*(\mathscr{B})$ on a Hilbert space H, and apply (18.5) to $\rho \circ \hat{\jmath}$, obtaining a faithful representation

$$\psi : C^*_{\text{red}}(\mathscr{B}) \to \mathscr{L}(H) \underset{\min}{\otimes} C^*_{\text{red}}(G) \subseteq \mathscr{L}\left(H \otimes \ell^2(G)\right),$$

such that

$$\psi(\lambda_g(b)) = \rho(\hat{\jmath}_g(b)) \otimes \lambda^G_g, \quad \forall\, b \in B_g. \tag{18.9.1}$$

Let r^G be the *right regular representation* of G on $\ell^2(G)$, given on the standard orthonormal basis $\{e_g\}_{g\in G}$ of $\ell^2(G)$ by

$$r^G_g(e_h) = e_{hg^{-1}}, \quad \forall\, g, h \in G.$$

We claim that there exist a *-representation

$$\tilde{r^G} : C^*_{\text{red}}(G) \to \mathscr{L}\left(H \otimes \ell^2(G)\right),$$

such that

$$\tilde{r^G}(\lambda^G_g) = 1 \otimes r^G_g, \quad \forall\, g \in G.$$

To see this, consider the unitary operator U on $H \otimes \ell^2(G)$ given by

$$U(\xi \otimes e_g) = \xi \otimes e_{g^{-1}}, \quad \forall\, \xi \in H, \quad \forall\, g \in G.$$

Then, for every $g, h \in G$, and $\xi \in H$, we have

$$\begin{aligned} U^*(1 \otimes \lambda_g^G)U(\xi \otimes e_h) &= U^*(id \otimes \lambda_g^G)(\xi \otimes e_{h^{-1}}) \\ &= U^*(\xi \otimes e_{gh^{-1}}) \\ &= \xi \otimes e_{hg^{-1}} \\ &= (1 \otimes r_g^G)(\xi \otimes e_h), \end{aligned}$$

so $U^*(1 \otimes \lambda_g^G)U = 1 \otimes r_g^G$, for all g in G, and the desired representation may then be defined by

$$\tilde{r}^G(x) = U^*(1 \otimes x)U, \quad \forall x \in C^*_{\rm red}(G).$$

It is easy to see that the range of ψ commutes with the range of $\tilde{r}^G$, so by the universal property of the maximal norm [**19**, Theorem 3.3.7], we get a representation

$$\psi \times \tilde{r}^G : C^*_{\rm red}(\mathscr{B}) \underset{\max}{\otimes} C^*_{\rm red}(G) \to \mathscr{L}\big(H \otimes \ell^2(G)\big),$$

such that

$$(\psi \times \tilde{r}^G)(x \otimes y) = \psi(x)\tilde{r}^G(y), \quad \forall x \in C^*_{\rm red}(\mathscr{B}), \quad \forall y \in C^*_{\rm red}(G).$$

Consider the map φ defined to be the composition

$$\varphi : C^*(\mathscr{B}) \xrightarrow{\mathscr{S}} C^*_{\rm red}(\mathscr{B}) \underset{\max}{\otimes} C^*_{\rm red}(G) \xrightarrow{\psi\times\tilde{r}^G} \mathscr{L}\big(H \otimes \ell^2(G)\big),$$

and observe that, for b in any B_g, we have that

$$\begin{aligned} \varphi\big(\hat{\jmath}_g(b)\big) &= (\psi \times \tilde{r}^G)\Big(\mathscr{S}\big(\hat{\jmath}_g(b)\big)\Big) \\ &= (\psi \times \tilde{r}^G)\big(\lambda_g(b) \otimes \lambda_g^G\big) \\ &\overset{(18.9.1)}{=} \Big(\rho\big(\hat{\jmath}_g(b)\big) \otimes \lambda_g^G\Big)(1 \otimes r_g^G) \\ &= \rho\big(\hat{\jmath}_g(b)\big) \otimes \lambda_g^G r_g^G. \end{aligned}$$

Notice that the above operator, when applied to a vector of the form $\xi \otimes e_1$, with $\xi \in H$, produces

$$\varphi\big(\hat{\jmath}_g(b)\big)\big|_{\xi\otimes e_1} = \rho\big(\hat{\jmath}_g(b)\big)\xi \otimes e_1.$$

Since the $\hat{\jmath}_g(b)$ span a dense subset of $C^*(\mathscr{B})$, we conclude that

$$\varphi(y)\big|_{\xi\otimes e_1} = \rho(y)\xi \otimes e_1, \quad \forall y \in C^*(\mathscr{B}), \quad \forall \xi \in H.$$

Therefore, assuming that $\mathscr{S}(y) = 0$, for some y in $C^*(\mathscr{B})$, we also have $\varphi(y) = 0$, whence $\rho(y)\xi = 0$, for all ξ in H, so $\rho(y) = 0$. Since ρ was supposed to be injective on $C^*(\mathscr{B})$, we deduce that $y = 0$. This proves that $\mathscr{S}$ is injective. □

Notes and remarks. Proposition (18.9) first appeared in [**7**, Theorem 6.2]. It is based on an idea verbally suggested to me by Eberhard Kirchberg at the CRM in Barcelona, in 2011.

CHAPTER 19

Graded C*-Algebras

We have seen in (17.9.vi) & (17.9.vii) that a Fell bundle $\mathscr{B}$ gives rise to two graded C*-algebras, namely

$$C^*(\mathscr{B}), \quad \text{and} \quad C^*_{\text{red}}(\mathscr{B}).$$

It is not hard to see that the Fell bundles obtained by disassembling these two algebras, as in (16.3), are both isomorphic to the original Fell bundle $\mathscr{B}$. Since there are situations in which $C^*(\mathscr{B})$ and $C^*_{\text{red}}(\mathscr{B})$ are not isomorphic[24], we see that the correspondence between Fell bundles and graded C*-algebras is not a perfect one. The goal of this chapter is thus to study this rather delicate relationship.

Given a graded C*-algebra B with associated Fell bundle $\mathscr{B} = \{B_g\}_{g\in G}$, by (16.29) we see that there is a unique C*-algebra epimorphism

$$\varphi : C^*(\mathscr{B}) \to B,$$

which is the identity on each B_g (identified both with a subspace of $C^*(\mathscr{B})$ and of B in the natural way). This says that $C^*(\mathscr{B})$ is, in a sense, the biggest C*-algebra whose associated Fell bundle is $\mathscr{B}$. Our next result will show that the reduced cross sectional algebra is on the other extreme of the range. It is also a very economical way to show a C*-algebra to be graded.

Theorem 19.1. *Let B be a C*-algebra and assume that for each g in a group G, there is associated a closed linear subspace $B_g \subseteq B$ such that, for all g and h in G, one has*

(*i*) $B_gB_h \subseteq B_{gh}$,

(*ii*) $B_g^* = B_{g^{-1}}$,

(*iii*) $\sum_{g\in G} B_g$ *is dense in B. Assume, in addition, that there is a bounded linear map*

$$F : B \to B_1,$$

such that F is the identity map on B_1, and that F vanishes on each B_g, for $g \neq 1$. Then

(*a*) *The B_g are independent and hence $\{B_g\}_{g\in G}$ is a grading for B.*

(*b*) *F is a conditional expectation onto B_1.*

(*c*) *If $\mathscr{B}$ denotes the associated Fell bundle, then there exists a C*-algebra epimorphism*

$$\psi : B \to C^*_{red}(\mathscr{B}),$$

such that $\psi(b) = \lambda_g(b)$, for each g in G, and each b in B_g.

[24]An example is the group bundle over a non-amenable group.

PROOF. If $x = \sum_{g\in G} b_g$ is a finite sum[25] with $b_g \in B_g$, then

$$x^*x = \sum_{g,h\in G} b_g^*b_h = \sum_{k\in G}\Big(\sum_{g\in G} b_g^*b_{gk}\Big).$$

Observing that $b_g^*b_{gk}$ lies in B_k, we see that

$$F(x^*x) = \sum_{g\in G} b_g^*b_g.$$

Therefore, if $x = 0$, then each $b_g = 0$, which shows the independence of the B_g's. This also shows that F is positive.

Given a in B_1, it is easy to see that

$$F(ax) = aF(x), \quad \text{and} \quad F(xa) = F(x)a, \quad \forall x \in B,$$

by first checking on finite sums, as above. So, apart from contractivity, (b) is proven.

Define a right pre-Hilbert B_1-module structure on B via the B_1-valued inner product

$$\langle x, y\rangle = F(x^*y), \quad \forall x, y \in B.$$

For $b, x \in B$, using the positivity of F, we have that

$$\langle bx, bx\rangle = F(x^*b^*bx) \le \|b\|^2F(x^*x) = \|b\|^2\langle x, x\rangle.$$

So, the left multiplication operators

$$L_b : x \in B \mapsto bx \in B$$

are bounded and hence extend to the completion X of B (after moding out vectors of norm zero). It is then easy to show that

$$L : b \in B \mapsto L_b \in \mathscr{L}(X)$$

is a C*-algebra homomorphism. Let

$$x = \sum_{g\in G} b_g, \quad \text{and} \quad y = \sum_{g\in G} c_g$$

be finite sums with $b_g, c_g \in B_g$, and regard both x and y as elements of X. We then have

$$\langle x, y\rangle = F\Big(\sum_{g,h\in G} b_g^*c_h\Big) = \sum_{g\in G} b_g^*c_g = \Big\langle \sum_{g\in G} j_g(b_g), \sum_{g\in G} j_g(c_g)\Big\rangle,$$

where the last inner product is that of $\ell^2(\mathscr{B})$. This is the key ingredient in showing that the formula

$$U\Big(\sum_{g\in G} b_g\Big) = \sum_{g\in G} j_g(b_g)$$

may be used to define an isometry of Hilbert B_1-modules

$$U : X \to \ell^2(\mathscr{B}).$$

Here it is important to remark that the continuity of F ensures that the set of finite sums $\sum_{g\in G} b_g$ is not only dense in B but also in X. For b in B_g and c in B_h we have

$$UL_b(c) = U(bc) = j_{gh}(bc) = j_g(b) \star j_h(c) = \lambda_g(b)j_h(c) = \lambda_g(b)U(c).$$

Since the finite sums $\sum_{h\in G} c_h$ are dense in X, as observed above, we conclude that

$$UL_bU^* = \lambda_g(b).$$

[25]By *finite sum* we actually mean that the set of indices g, for which $b_g \neq 0$, is finite.

This implies that, for all b in any B_g, the operator UL_bU^* belongs to $C^*_{\text{red}}(\mathscr{B})$, and hence the same holds for an arbitrary b in B. This defines a map

$$\psi : b \in B \mapsto UL_bU^* \in C^*_{\text{red}}(\mathscr{B}),$$

which satisfies the requirements in (c).

The only remaining task, the proof of the contractivity of F, now follows easily because $F = E_1 \circ \psi$, where E_1 is given by (17.8). □

The map ψ, above, should be thought of as a generalized regular representation of B.

From now on we will be mostly interested in graded algebras possessing a conditional expectation, so we make the following:

Definition 19.2. A C*-grading $\{B_g\}_{g\in G}$ on the C*-algebra B is said to be a *topological grading* if there exists a (necessarily unique) conditional expectation of B onto B_1, vanishing on all B_g, for $g \neq 1$, as in (19.1).

Proposition 19.3. *If $\mathscr{B}$ is a Fell bundle, then both $C^*(\mathscr{B})$ and $C^*_{red}(\mathscr{B})$ are topologically graded C*-algebras.*

PROOF. With respect to $C^*_{\text{red}}(\mathscr{B})$, notice that the mapping $E_1 : C^*_{\text{red}}(\mathscr{B}) \to B_1$ given by (17.8) is a conditional expectation by (19.1), clearly satisfying all of the required conditions.

The statement for $C^*(\mathscr{B})$ follows immediately by taking the conditional expectation

$$C^*(\mathscr{B}) \xrightarrow{\Lambda} C^*_{\text{red}}(\mathscr{B}) \xrightarrow{E_1} B_1.$$

□

Not all graded C*-algebras are topologically graded, as the following example shows: suppose that X is a closed subset of the unit circle $\mathbb{T}$, and let z be the standard generator of the algebra $C(X)$, namely the inclusion function $X \hookrightarrow \mathbb{C}$. Letting

$$B_n = \mathbb{C}z^n, \quad \forall\, n \in \mathbb{Z},$$

it is easy to see that $B_nB_m \subseteq B_{n+m}$, and $B_n^* = B_{-n}$. In order to decide whether or not the B_n are linearly independent subspaces of $C(X)$, suppose that a finite sum $\sum_{n\in\mathbb{Z}} b_n = 0$, where each $b_n \in B_n$. Write $b_n = a_nz^n$, for some scalar a_n, so that

$$\sum_{n\in\mathbb{Z}} b_n = \sum_{n\in\mathbb{Z}} a_nz^n =: p,$$

which is to say that p is a trigonometric polynomial. The condition that we need, namely

$$p = 0 \quad \Rightarrow \quad (\forall n \in \mathbb{Z},\ a_n = 0)$$

does not always hold, as X could be contained in the set of roots of p, but if we assume that X is infinite, the above implication is clearly true. In this case the B_n are linearly independent and so $\{B_n\}_{n\in\mathbb{Z}}$ is a grading for $C(X)$.

We claim that when X is a proper infinite subset of $\mathbb{T}$, we do not have a topological grading. To see this, assume the contrary, which is to say that there exists a *continuous* conditional expectation $F : C(X) \to \mathbb{C}$, such that

$$F\Big(\sum_{n\in\mathbb{Z}} a_nz^n\Big) = a_0,$$

for every trigonometric polynomial $\sum_{n\in\mathbb{Z}} a_n z^n$. Equivalently

$$F(p|_X) = \frac{1}{2\pi}\int_0^{2\pi} p(e^{it})\,dt, \tag{19.4}$$

for every trigonometric polynomial p, but since these are dense in $C(\mathbb{T})$, the above integral formula holds for every p in $C(\mathbb{T})$.

Since we are assuming that X is not the whole of $\mathbb{T}$, we may use Tietze's extension Theorem to obtain a nonzero, positive $p \in C(\mathbb{T})$, vanishing on X. Plugging p into (19.4) will then bring about a contradiction, thus proving that our grading does not admit a conditional expectation.

Recalling our discussion about $C^*(\mathscr{B})$ being the biggest graded algebra for a given Fell bundle, we now see that $C^*_{\text{red}}(\mathscr{B})$ is the smallest such, at least among topologically graded algebras.

Summarizing we obtain the following important consequence of (19.1.c) and the remark preceding the statement of (19.1):

Theorem 19.5. *Let B be a topologically graded C^*-algebra with grading $\mathscr{B} = \{B_g\}_{g\in G}$. Then there exists a commutative diagram of surjective *-homomorphism:*

$$\begin{array}{ccc} C^*(\mathscr{B}) & \xrightarrow{\quad\Lambda\quad} & C^*_{red}(\mathscr{B}) \\ & {\scriptstyle\varphi}\searrow \quad \nearrow{\scriptstyle\psi} & \\ & B & \end{array}$$

Another important consequence of (19.1) is the existence of *Fourier coefficients* in topologically graded algebras as follows:

Corollary 19.6. *Let B be a topologically graded C^*-algebra with grading $\{B_g\}_{g\in G}$. Then, for every g in G, there exists a contractive linear map*

$$F_g : B \to B_g,$$

such that, for all finite sums $x = \sum_{g\in G} b_g$, with $b_g \in B_g$, one has $F_g(x) = b_g$. Moreover, given $h \in G$, and $b \in B_h$, one has that

$$F_g(bx) = bF_{h^{-1}g}(x), \quad \text{and} \quad F_g(xb) = F_{gh^{-1}}(x)b,$$

for any x in B.

PROOF. For the existence it is enough to define $F_g = E_g \circ \psi$, where E_g is as in (17.8), and ψ is given by (19.1.c). In order to prove the last two identities, one easily checks them for finite sums of the form $x = \sum_{k\in G} b_k$, with $b_k \in B_k$, and the result then follows by density of the set formed by such elements. □

A measure of the size of a graded C*-algebra relative to the corresponding reduced cross sectional algebra is evidently the null space of the regular representation. The next result gives a characterization of this null space.

Proposition 19.7. *Given a topologically graded C^*-algebra B with conditional expectation F, let ψ be its regular representation, as in* (19.1.c). *Then*

$$Ker(\psi) = \{b \in B : F(b^*b) = 0\}.$$

PROOF. Observing that $F = E_1 \circ \psi$, we have

$$F(b^*b) = E(\psi(b)^*\psi(b)),$$

from where we see that $F(b^*b) = 0$, if and only if $\psi(b) = 0$, by (17.13). □

This can be employed to give a useful characterization of $C^*_{\text{red}}(\mathscr{B})$ among graded algebras:

Proposition 19.8. *Suppose we are given a topologically graded C^*-algebra B with grading $\mathscr{B} = \{B_g\}_{g\in G}$, and conditional expectation F. If F is faithful, in the sense that*

$$F(b^*b) = 0 \;\Rightarrow\; b = 0, \quad \forall\, b \in B,$$

then B is canonically isomorphic to $C^*_{red}(\mathscr{B})$.

PROOF. The regular representation ψ of (19.1.c) will be an isomorphism by (19.7). □

Notes and remarks. Theorems (19.1) and (19.5) were first proved in [**49**].

CHAPTER 20

Amenability for Fell Bundles

Theorem (19.5) tells us that the topologically graded C*-algebras whose associated Fell bundles coincide with a given Fell bundle $\mathscr{B}$ are to be found among the quotients of $C^*(\mathscr{B})$ by ideals contained in the kernel of the regular representation Λ.

It is therefore crucial to understand the kernel of Λ and, in particular, to determine conditions under which Λ is injective. In the case of the group bundle over G, it is well known that the injectivity of Λ is equivalent to the amenability of G.

In this chapter we will extend to Fell bundles some of the techniques pertaining to the theory of amenable groups, including the important notion of amenability of group actions introduced by Anantharaman-Delaroche [**6**]. We begin with some terminology.

Definition 20.1. A Fell bundle $\mathscr{B}$ is said to be *amenable* if the regular representation

$$\Lambda : C^*(\mathscr{B}) \to C^*_{\text{red}}(\mathscr{B})$$

is injective.

Amenability, when applied to the context of groups, may be characterized by a large number of equivalent conditions [**65**], such as the equality of the full and reduced group C*-algebras [**89**, Theorem 7.3.9]. Many of these conditions may be rephrased for Fell bundles although they are not always known to be equivalent. One therefore has many possibly inequivalent choices when defining the term *amenable* in the context of Fell bundles. Definition (20.1) is just one among many alternatives.

An immediate consequence of (19.5) is:

Proposition 20.2. *Let $\mathscr{B}$ be an amenable Fell bundle. Then all topologically graded C*-algebras whose associated Fell bundles coincide with $\mathscr{B}$ are isomorphic to each other.*

The next technical result is intended to pave the way for the introduction of sufficient conditions for the amenability of a given Fell bundle. The idea is to produce *wrong way maps* which will later be used to prove the injectivity of Λ.

Lemma 20.3. *Let $\mathscr{B} = \{B_g\}_{g\in G}$ be a Fell bundle. Given a finitely supported function $a : G \to B_1$, the formula*

$$V(z) = \sum_{g,h\in\mathscr{G}} \jmath_g\big(a(gh)^* E_g(z) a(h)\big), \quad \forall\, z \in C^*_{red}(\mathscr{B}), \tag{20.3.1}$$

gives a well defined completely positive linear map from $C^*_{red}(\mathscr{B})$ to $C^*(\mathscr{B})$, such that

$$\|V\| \leq \Big\| \sum_{g\in\mathscr{G}} a(g)^* a(g) \Big\|.$$

Identifying each B_g as a subspace of $C^*_{red}(\mathscr{B})$ or $C^*(\mathscr{B})$, as appropriate, one moreover has

$$V(b) = \sum_{h\in\mathscr{G}} a(gh)^* b a(h), \quad \forall\, b \in B_g.$$

PROOF. Let

$$\rho : C^*(\mathscr{B}) \to \mathscr{L}(H)$$

be any faithful *-representation of $C^*(\mathscr{B})$ on a Hilbert space H. For each g in G, let $\pi_g = \rho \circ \hat{j}_g$, so that $\pi = \{\pi_g\}_{g\in G}$ is a *-representation of $\mathscr{B}$ on H. Consider the representation $\pi \otimes \lambda^G$ of $\mathscr{B}$ on $H \otimes \ell^2(G)$ described in (18.1), and let

$$\varphi : C^*(\mathscr{B}) \to \mathscr{L}\left(H \otimes \ell^2(G)\right)$$

be the integrated form of $\pi \otimes \lambda^G$. By (18.4), we have that φ vanishes on $\mathrm{Ker}(\Lambda)$, so it factors through $C^*_{\mathrm{red}}(\mathscr{B})$, giving a *-homomorphism ψ such that the diagram

$$\begin{array}{ccc} C^*(\mathscr{B}) & \xrightarrow{\quad\varphi\quad} & \mathscr{L}\left(H \otimes \ell^2(G)\right) \\ & {}_{\Lambda}\searrow \quad \nearrow_{\psi} & \\ & C^*_{\mathrm{red}}(\mathscr{B}) & \end{array}$$

commutes. We next consider the operator

$$A : H \to H \otimes \ell^2(\mathscr{G})$$

given by

$$A(\xi) = \sum_{g\in\mathscr{G}} \alpha(g)\xi \otimes e_g, \quad \forall\, \xi \in H,$$

where $\alpha(g) = \pi_1\big(a(g)\big)$. In order to compute the norm of A, let $\xi \in H$. Then

$$\begin{aligned} \|A(\xi)\|^2 &= \sum_{g\in G} \|\alpha(g)\xi\|^2 \\ &= \sum_{g\in G} \langle \alpha(g)^*\alpha(g)\xi, \xi\rangle \\ &= \Big\langle \sum_{g\in G} \alpha(g)^*\alpha(g)\xi, \xi\Big\rangle \\ &\leq \Big\| \sum_{g\in G} \alpha(g)^*\alpha(g)\Big\| \, \|\xi\|^2, \end{aligned}$$

from where we deduce that

$$\|A\| \leq \Big\| \sum_{g\in G} \alpha(g)^*\alpha(g)\Big\|^{1/2} \leq \Big\| \sum_{g\in G} a(g)^* a(g)\Big\|^{1/2}. \tag{23.3.2}$$

Considering the completely positive map

$$W : T \in \mathscr{L}\left(H \otimes \ell^2(G)\right) \longmapsto A^*TA \in \mathscr{L}(H),$$

observe that for b in any B_g, and every ξ in H, one has

$$\begin{aligned} W\big(\pi_g(b) \otimes \lambda^G_g\big)\xi &= A^*\big(\pi_g(b) \otimes \lambda^G_g\big)\Big(\sum_{h\in G} \alpha(h)\xi \otimes e_h\Big) \\ &= A^*\Big(\sum_{h\in G} \pi_g(b)\alpha(h)\xi \otimes e_{gh}\Big) = \sum_{h\in G} \alpha(gh)^*\pi_g(b)\alpha(h)\xi, \end{aligned}$$

where we have used the fact that $A^*(\eta\otimes e_k)=\alpha(k)^*\eta$, which the reader will have no difficulty in proving. Summarizing, we have that

$$W\big(\pi_g(b)\otimes\lambda_g^G\big)=\sum_{h\in G}\alpha(gh)^*\pi_g(b)\alpha(h).$$

If we then define V to be the composition

$$V\ :\ C^*_{\text{red}}(\mathscr{B})\xrightarrow{\psi}\mathscr{L}\left(H\otimes\ell^2(G)\right)\xrightarrow{W}\mathscr{L}(H),$$

we have for b in any B_g that

$$\begin{aligned}V\big(\lambda_g(b)\big)&=W\big(\psi(\lambda_g(b))\big)\\&=W\big(\varphi(\hat{\jmath}_g(b))\big)\\&\overset{(16.29)}{=}W\big(\pi_g(b)\otimes\lambda_g^G\big)\\&=\sum_{h\in G}\alpha(gh)^*\pi_g(b)\alpha(h)\\&=\rho\Big(\sum_{h\in G}\hat{\jmath}_g\big(a(gh)^*ba(h)\big)\Big).\end{aligned}$$

Recalling that $C^*_{\text{red}}(\mathscr{B})$ is generated by the $\lambda_g(B_g)$, we see that the range of V is contained in the range of ρ, which is isomorphic to $C^*(\mathscr{B})$, since we have chosen ρ to be faithful. Identifying $C^*(\mathscr{B})$ with its image under ρ, we may then view V as a map

$$V:C^*_{\text{red}}(\mathscr{B})\to C^*(\mathscr{B}),$$

such that

$$V\big(\lambda_g(b)\big)=\sum_{h\in G}\hat{\jmath}_g\big(a(gh)^*ba(h)\big),\quad\forall\,b\in B_g.\tag{20.3.3}$$

Observe that, with the appropriate identifications

$$B_g\simeq\lambda_g(B_g)\subseteq C^*_{\text{red}}(\mathscr{B}),\quad\text{and}\quad B_g\simeq\hat{\jmath}_g(B_g)\subseteq C^*(\mathscr{B}),$$

we have therefore proven the last identity in the statement. For a finite sum

$$z=\sum_{g\in G}\lambda_g(b_g),$$

with each b_g in B_g, we then have by (20.3.3) that

$$\begin{aligned}V(z)&=\sum_{g\in G}V\big(\lambda_g(b_g)\big)\\&=\sum_{g,h\in G}\hat{\jmath}_g\big(a(gh)^*b_ga(h)\big)\\&=\sum_{g,h\in G}\hat{\jmath}_g\big(a(gh)^*E_g(z)a(h)\big),\end{aligned}$$

proving (20.3.1) for all z of the above form. Observe that if (g,h) is a pair of group elements for which the corresponding term in the last sum above is nonzero, then both h and gh must lie in the finite support of a, which we henceforth denote by A. For each such pair we then have

$$g=(gh)h^{-1}\in AA^{-1},$$

so

$$(g,h)\in AA^{-1}\times A.$$

Consequently the sum in (20.3.1) is finite, hence representing a continuous function of z. Since the set of z's considered above is dense, (20.3.1) is proven.

In order to estimate the norm of V we have

$$\|V\| \leq \|W\| \leq \|A\|^2 \overset{(23.3.2)}{\leq} \Big\| \sum_{g\in G} a(g)^* a(g) \Big\|,$$

as desired. □

This puts us in position to describe an important concept.

Definition 20.4. Let $\mathscr{B} = \{B_g\}_{g\in G}$ be a Fell bundle. By a *Cesaro net* for $\mathscr{B}$ we mean a net $\{a_i\}_{i\in I}$ of finitely supported functions

$$a_i : \mathscr{G} \to B_1,$$

such that

(i) $\sup_{i\in I} \Big\| \sum_{g\in G} a_i(g)^* a_i(g) \Big\| < \infty$,

(ii) $\lim_{i\to\infty} \sum_{h\in G} a_i(gh)^* b a_i(h) = b$, for all b in each B_g.

If a Cesaro net exists, we will say that $\mathscr{B}$ has the *approximation property.*

Sometimes one might have difficulty in checking (20.4.ii) for every single b in B_g, so the following might come in handy:

Proposition 20.5. *Let $\mathscr{B} = \{B_g\}_{g\in G}$ be a Fell bundle and let $\{a_i\}_{i\in I}$ be a Cesaro net for $\mathscr{B}$, except that* (20.4.ii) *is only known to hold for b in a total*[26] *subset of each B_g. Then* (20.4.ii) *does hold for all b in B_g. Consequently $\{a_i\}_{i\in I}$ is a Cesaro net and $\mathscr{B}$ has the approximation property.*

Proof. For each i in I, and each g in G, consider the map

$$T_i \ : \ b \in B_g \ \mapsto \ \sum_{h\in\mathscr{G}} a_i(gh)^* b a_i(h) \ \in \ B_g.$$

By hypothesis we have that the T_i converge to the identity map of B_g, pointwise on a dense set. To get the conclusion it is then enough to establish that the T_i are uniformly bounded.

Letting V_i be the map provided by (20.3) in terms of a_i, notice that T may be seen as a restriction of V_i, once B_g is identified with $\lambda_g(B_g) \subseteq C^*_{\text{red}}(\mathscr{B})$, and with $\hat{\jmath}_g(B_g) \subseteq C^*(\mathscr{B})$. Since these are isometric copies of B_g by (17.9.iv)–(17.9.v), we deduce that $\|T_i\| \leq \|V_i\|$, from where the desired uniform boundedness follows, hence concluding the proof. □

The relationship between the approximation property and the amenability of Fell bundles is the main result of this chapter:

Theorem 20.6. *If a Fell bundle $\mathscr{B}$ has the approximation property, then it is amenable.*

Proof. Given a Cesaro net $\{a_i\}_{i\in I}$ as in (20.4), let us consider, for each i in I, the map

$$V_i : C^*_{\text{red}}(\mathscr{B}) \to C^*(\mathscr{B})$$

provided by (20.3), relative to a_i. Define a map Φ_i from $C^*(\mathscr{B})$ to itself to be the composition $\Phi_i = V_i \circ \Lambda$, and observe that, by hypothesis we have

$$\lim_{i\to\infty} \Phi_i(b) = b,$$

[26] A *total set* in a normed space is a set spanning a dense subspace.

for every b in each B_g. Because the b's span a dense subspace of $C^*(\mathscr{B})$, and because the Φ_i's are uniformly bounded, we conclude that $\lim_{i\to\infty}\Phi_i(y) = y$, for every y in $C^*(\mathscr{B})$.

Now, if y is in the kernel of the regular representation, that is, if y is in $C^*(\mathscr{B})$ and $\Lambda(y) = 0$, then

$$y = \lim_{i\to\infty}\Phi_i(y) = \lim_{i\to\infty}V_i(\Lambda(y)) = 0,$$

which proves that Λ is injective. □

A somewhat trivial example of this situation is that of bundles over amenable groups.

Theorem 20.7. *Let $\mathscr{B}$ be a Fell bundle over the amenable group $\mathscr{G}$. Then $\mathscr{B}$ satisfies the approximation property and hence is amenable.*

PROOF. According to [**89**, 7.3.8], there exists a net $\{f_i\}_{i\in I}$ in the unit sphere of $\ell^2(G)$ such that

$$\lim_{i\to\infty}\sum_{h\in\mathscr{G}}\overline{f_i(gh)}f_i(h) = 1, \quad \forall\, g\in G.$$

By truncating the f_i to larger and larger subsets of G we may assume them to be finitely supported. Letting $\{u_j\}_{j\in J}$ be an approximate identity for B_1, one checks that the doubly indexed net

$$\{a_{i,j}\}_{(i,j)\in I\times J}$$

defined by $a_{i,j}(g) = f_i(g)u_j$ is a Cesaro net for $\mathscr{B}$. □

There is another proof that $\mathscr{B}$ is amenable when G is amenable based on Fell's absorption principle, as follows: if G is amenable then the trivial representation is weakly contained in the left-regular representation, which means that there is a *-homomorphism

$$\varepsilon : C^*_{\mathrm{red}}(G)\to\mathbb{C}$$

such that $\varepsilon(\lambda^G_g) = 1$, for all g in G. Taking into account that $C^*_{\mathrm{red}}(G)$ is nuclear by [**19**, Theorem 2.6.8], and using the map τ given by (18.6), one has that the map φ defined by the composition

$$\varphi\ :\ C^*_{\mathrm{red}}(\mathscr{B}) \xrightarrow{\tau} C^*(\mathscr{B}) \underset{\min}{\otimes} C^*_{\mathrm{red}}(G) \simeq C^*(\mathscr{B}) \underset{\max}{\otimes} C^*_{\mathrm{red}}(G) \xrightarrow{id\otimes\varepsilon} C^*(\mathscr{B})$$

satisfies

$$\varphi\big(\lambda_g(b)\big) = \hat{\jmath}_g(b), \quad \forall\, g\in G, \quad \forall\, b\in B_g,$$

whence $\varphi\circ\Lambda$ is the identity of $C^*(\mathscr{B})$, and hence Λ is injective.

One of the major questions we are usually confronted with in Classical Harmonic Analysis is whether or not a function on the unit circle may be reconstructed from its Fourier coefficients. The same question may be stated for any topologically graded algebra:

Question 20.8. Let B be a topologically graded C*-algebra and let F_g be the *Fourier coefficient operators* given by (19.6). Given an element x in B, can we reconstruct x if we are given the value of $F_g(x)$ for every g in G?

There are two possible interpretations of this question, depending on what one means by *to reconstruct.* On the one hand, (17.13) implies that if x and y are two elements of $C^*_{\mathrm{red}}(\mathscr{B})$ having the same Fourier coefficients, then $x = y$. This means that there is at most one x in $C^*_{\mathrm{red}}(\mathscr{B})$ with a given *Fourier transform*, but it gives us no algorithm to actually produce such an x. Of course, if x is a finite sum as in (19.6), then we see that

$$x = \sum_{g\in G} F_g(x),$$

the sum having only finitely many nonzero terms. Any attempt at generalizing this beyond the easy case of finite sums will necessarily require a careful analysis of the convergence of the above series. The fact that, over the group bundle $\mathbb{C} \times \mathbb{Z}$, this reduces to the usual question of convergence of Fourier series, is a stark warning that question (20.8) does not have a straightforward answer.

In the classical case of functions on the unit circle, the lack of convergence of Fourier series is circumvented by Cesaro sums, an analogue of which we shall now discuss.

Definition 20.9. Let B be a topologically graded C*-algebra with grading $\{B_g\}_{g\in G}$, and let F_g be the Fourier coefficient operators given by (19.6). A bounded linear map $S : B \to B$, is said to be a *summation process* if,

(i) $S \circ F_g = 0$, for all but finitely many $g \in G$,
(ii) $S(x) = \sum_{g\in G} S(F_g(x))$, for all x in B.

For example, choosing a finite subset $K \subseteq G$, one obtains a summation process by defining

$$S(x) = \sum_{g\in K} F_g(x), \quad \forall x \in B.$$

The following result answers question (20.8) in the affirmative for Fell bundles possessing the approximation property:

Proposition 20.10. *Suppose that we are given a topologically graded C^*-algebra B whose grading $\mathscr{B} = \{B_g\}_{g\in G}$ admits a Cesaro net $\{a_i\}_{i\in I}$ (and hence satisfies the approximation property). Then the maps $S_i : B \to B$, given by*

$$S_i(x) = \sum_{g,h\in G} a_i(gh)^* F_g(x) a_i(h), \quad \forall x \in B,$$

form a bounded net $\{S_i\}_{i\in I}$ of completely positive summation processes converging pointwise to the identity map of B.

PROOF. Since $\mathscr{B}$ is amenable we may invoke (20.2) to assume without loss of generality that $B = C^*_{\text{red}}(\mathscr{B})$. By (20.3) we have that the S_i indeed form a well defined uniformly bounded net of completely positive linear maps. Moreover, as in the proof of (20.6), one checks that the S_i converge pointwise to the identity map.

We will now prove that the S_i are summation processes. With respect to (20.9.i), suppose that $S_i \circ F_g \neq 0$. Then there is some $b \in B_g$, such that

$$0 \neq S_i(b) = \sum_{h\in G} a_i(gh)^* b a_i(h).$$

In particular there must be at least one h for which both $a_i(gh)$ and $a_i(h)$ are nonzero. If A is the support of a_i, then this implies that $gh, h \in A$, whence

$$g = (gh)h^{-1} \in AA^{-1}.$$

Thus $S_i \circ F_g = 0$, for all g not belonging to the finite set AA^{-1}, proving (20.9.i).

If x is a finite sum $x = \sum_{g\in G} b_g$, with each b_g in B_g, then clearly

$$x = \sum_{g\in G} F_g(x).$$

This implies (20.9.ii) for such an x and hence also for all x in B, by continuity of both sides of (20.9.ii), now that we know that the sum involved is finite. □

Given a C*-algebraic partial dynamical system

$$\theta = \big(A,\ G,\ \{D_g\}_{g\in G},\ \{\theta_g\}_{g\in G}\big)$$

we may consider the associated semi-direct product bundle $\mathscr{B}$, and we could try to determine conditions on θ which would say that $\mathscr{B}$ satisfies the approximation property. Unraveling the definition of the approximation property for the special case of semi-direct product bundles, the condition is the existence of a net $\{a_i\}_{i\in I}$ of finitely supported functions

$$a_i : G \to A,$$

which is bounded in the sense that

$$\sup_{i\in I} \Big\| \sum_{g\in G} a_i(g)^* a_i(g) \Big\| < \infty,$$

and such that

$$\lim_{i\to\infty} \sum_{h\in G} a_i(gh)^* \theta_g\big(c a_i(h)\big) = \theta_g(c), \quad \forall c \in D_{g^{-1}}.$$

One arrives at this by simply plugging $b = \theta_g(c)\delta_g$ in (20.4.ii), where c is in $D_{g^{-1}}$.

In [**109**] Zimmer introduced a notion of amenability of global group actions on measure spaces (see also [**108**]), which was later adapted by Anantharaman-Delaroche [**6**] to actions on C*-algebras. A slightly different version of this concept, taken from [**19**, Definition 4.3.1], is as follows:

Definition 20.11. Let β be a global action of a discrete group G on a unital C*-algebra A. Then β is said to be an *amenable action* if there exists a sequence $\{T_i\}_{i\in\mathbb{N}}$ of finitely supported functions

$$T_i : G \to \mathcal{Z}(A),$$

where $\mathcal{Z}(A)$ stands for the center of A, such that

(i) $T_i(g) \geq 0$, for all $i \in \mathbb{N}$, and all $g \in G$,
(ii) $\sum_{g\in G} T_i(g)^2 = 1$,
(iii) $\lim_{i\to\infty} \|T_i - T_i^g\|_2 = 0$, for all g in G, where

$$T_i^g(h) = \beta_g\big(T_i(g^{-1}h)\big), \quad \forall h \in G,$$

and $\|T\|_2$ is defined for all finitely supported functions $T : G \to A$ by

$$\|T\|_2 = \Big\| \sum_{g\in G} T(g)^* T(g) \Big\|^{\frac{1}{2}}.$$

We would now like to relate the above notion with the approximation property for Fell bundles.

Proposition 20.12. *Let β be a global amenable action of a group G on a unital C*-algebra A. Then the corresponding semi-direct product bundle has the approximation property.*

PROOF. Let $\mathscr{B} = \{B_g\}_{g\in G}$ be the semi-direct product bundle for β, so that each $B_g = A\delta_g$. As usual we will identify B_1 with A in the obvious way.

Choosing $\{T_i\}_{i\in\mathbb{N}}$ as in (20.11), let us consider each T_i as a B_1-valued function, and let us prove that $\{T_i\}_{i\in\mathbb{N}}$ satisfies the conditions of (20.4).

Observing that (20.4.i) follows immediately from (20.11.ii), we concentrate our efforts on proving (20.4.ii). For this, choose any element b in any B_g, so that necessarily $b = c\delta_g$, for some c in A. Then

$$\begin{aligned}\sum_{h\in G} T_i(gh)^* b T_i(h) &= \sum_{h\in G} T_i(gh) c\delta_g T_i(h) \\ &= \sum_{h\in G} T_i(gh) c\beta_g\big(T_i(h)\big)\delta_g \\ &= \sum_{k\in G} T_i(k)\beta_g\big(T_i(g^{-1}k)\big) \\ &= \sum_{k\in G} T_i(k) T_i^g(k)\; b \overset{i\to\infty}{\longrightarrow} b,\end{aligned}$$

where the last step follows from [**19**, Lemma 4.3.2]. This verifies (20.4.ii), so the proof is concluded. □

We would now like to point out a couple of results from [**51**], whose proofs are perhaps a bit too long to be included here, so we shall restrict ourselves to giving their statements. The following is taken from [**51**, Theorems 4.1 & 6.3]:

Theorem 20.13. *Let $\mathscr{G}$ be any set and let $\mathbb{F}$ be the free group on $\mathscr{G}$, equipped with the usual length function.*

(*i*) *Given a semi-saturated *-partial representation u of $\mathbb{F}$ in a C^*-algebra A, which is also orthogonal, in the sense that*

$$u_g^* u_h = 0,$$

for any two distinct $g, h \in \mathscr{G}$, then the associated Fell bundle $\mathscr{B}^u$, introduced in (16.7)*, satisfies the approximation property and hence is amenable.*

(*ii*) *Given a semi-saturated Fell bundle $\mathscr{B} = \{B_g\}_{g\in G}$ over $\mathbb{F}$, which is also orthogonal, in the sense that*

$$B_g^* B_h = \{0\},$$

for any two distinct $g, h \in \mathscr{G}$, then $\mathscr{B}$ is amenable.

Notes and remarks. The definition of the approximation property given in (20.4) is inspired by the equivalent conditions of [**6**, Théorème 3.3]. The question of whether the converse of Theorem (20.6) holds is a most delicate one. In the special case when $\mathscr{B}$ is the semi-direct product bundle relative to a global action of an exact group on an abelian C*-algebra, a converse of Theorem (20.6) has been proven in [**80**]. Another open question of interest is whether (20.12) admits a converse.

CHAPTER 21

Functoriality for Fell Bundles

So far we have been studying individual Fell bundles, but we may also see them as objects of a category. To do so we will introduce a notion of *morphism* between Fell bundles as well as the notion of a *Fell sub-bundle.*

One of the most disturbing questions we will face is whether or not the cross sectional algebra of a Fell sub-bundle is a subalgebra of the cross sectional algebra of the ambient Fell bundle. Strange as it might seem, the natural map between these algebras is not always injective! Attempting to dodge this anomaly we will see that under special hypothesis (heredity or the existence of a conditional expectation) sub-bundles lead to bona fide subalgebras.

Definition 21.1. Given Fell bundles $\mathscr{A} = \{A_g\}_{g\in G}$ and $\mathscr{B} = \{B_g\}_{g\in G}$, a *morphism* from $\mathscr{A}$ to $\mathscr{B}$ is a collection $\varphi = \{\varphi_g\}_{g\in G}$ of linear maps

$$\varphi_g : A_g \to B_g,$$

such that

(i) $\varphi_g(a)\varphi_h(b) = \varphi_{gh}(ab)$, and
(ii) $\varphi_g(a)^* = \varphi_{g^{-1}}(a^*)$,

for all g and h in G, and for all a in A_g and b in A_h.

Observe that if φ is as above then φ_1 is evidently a *-homomorphism from A_1 to B_1, hence φ_1 is necessarily continuous. Also, given a in any A_g, we have that

$$\begin{aligned}\|\varphi_g(a)\|^2 &= \|\varphi_g(a)^*\varphi_g(a)\|\\ &= \|\varphi_{g^{-1}}(a^*)\varphi_g(a)\|\\ &= \|\varphi_1(a^*a)\| \leq \|a^*a\|\\ &= \|a\|^2,\end{aligned}$$

so indeed all of the φ_g are necessarily continuous.

If every φ_g is bijective, it is easy to see that

$$\varphi^{-1} := \{\varphi_g^{-1}\}_{g\in G}$$

is also a morphism. In this case we say that φ is an *isomorphism.* If an isomorphism exists between the Fell bundles $\mathscr{A}$ and $\mathscr{B}$, we say that $\mathscr{A}$ and $\mathscr{B}$ are *isomorphic Fell bundles.*

Morphisms between Fell bundles lead to *-homomorphisms between the corresponding cross sectional C*-algebras as we will now show.

Proposition 21.2. *Let $\varphi = \{\varphi_g\}_{g\in G}$ be a morphism from the Fell bundle $\mathscr{A} = \{A_g\}_{g\in G}$ to the Fell bundle $\mathscr{B} = \{B_g\}_{g\in G}$. Then there exists a unique *-homomorphism*

$$\varphi : C^*(\mathscr{A}) \to C^*(\mathscr{B}),$$

(denoted by φ by abuse of language) such that

$$\varphi\big(\hat{\jmath}^A_g(a)\big) = \hat{\jmath}^B_g\big(\varphi_g(a)\big),$$

for all a in any A_g, where we denote by $\hat{\jmath}^A$ and $\hat{\jmath}^B$ the universal representations of $\mathscr{A}$ and $\mathscr{B}$, respectively.

Proof. For each g in G, consider the composition

$$\pi_g : A_g \xrightarrow{\varphi_g} B_g \xrightarrow{\hat{\jmath}^B_g} C^*(\mathscr{B}).$$

It is then routine to check that $\pi = \{\pi_g\}_{g\in G}$ is a *-representation of $\mathscr{A}$ in $C^*(\mathscr{B})$, so the result follows from (16.29). □

As above, we will now show that morphisms between Fell bundles lead to *-homomorphisms between the corresponding reduced cross sectional C*-algebras:

Proposition 21.3. *Let $\varphi = \{\varphi_g\}_{g\in G}$ be a morphism from the Fell bundle $\mathscr{A} = \{A_g\}_{g\in G}$ to the Fell bundle $\mathscr{B} = \{B_g\}_{g\in G}$. Then there exists a unique *-homomorphism*

$$\varphi_{red} : C^*_{red}(\mathscr{A}) \to C^*_{red}(\mathscr{B}),$$

such that

$$\varphi_{red}\big(\lambda^A_g(a)\big) = \lambda^B_g\big(\varphi_g(a)\big),$$

for all a in any A_g, where we denote by λ^A and λ^B the regular representations of $\mathscr{A}$ and $\mathscr{B}$, respectively. Moreover, if φ_1 is injective, then so is φ_{red}.

Proof. Consider the diagram

$$\begin{array}{ccc}
C^*(\mathscr{A}) & \xrightarrow{\ \varphi\ } & C^*(\mathscr{B}) \\
\downarrow{\scriptstyle \Lambda'} & \searrow^{\psi} & \downarrow{\scriptstyle \Lambda} \\
C^*_{\mathrm{red}}(\mathscr{A}) & & C^*_{\mathrm{red}}(\mathscr{B}) \\
\downarrow{\scriptstyle E_1'} & & \downarrow{\scriptstyle E_1'} \\
A_1 & \xrightarrow{\ \varphi_1\ } & B_1
\end{array}$$

where φ is given by (21.2), Λ' and Λ are the regular representations of $\mathscr{A}$ and $\mathscr{B}$, respectively, and E_1' and E_1 are the conditional expectations given by (17.8) for $\mathscr{A}$ and $\mathscr{B}$, respectively. By checking on $C_c(\mathscr{A})$, it is then easy to see that the outside of the diagram is commutative.

Defining $\psi = \Lambda \circ \varphi$, we then claim that the null space of ψ contains the null space of Λ'. To see this, notice that for any $y \in C^*(\mathscr{A})$, we have that

$$\begin{aligned}
\Lambda'(a) = 0 &\overset{(17.13)}{\iff} E_1'\big(\Lambda'(a^*a)\big) = 0 \\
&\implies \varphi_1\Big(E_1'\big(\Lambda'(a^*a)\big)\Big) = 0 \\
&\iff E_1\big(\psi(a^*a)\big) = 0 \\
&\overset{(17.13)}{\iff} \psi(a) = 0.
\end{aligned}$$

This shows our claim, so we see that ψ factors through $C^*_{\text{red}}(\mathscr{A})$, producing the map φ_{red} mentioned in the statement and one now easily checks that it satisfies all of the required conditions.

In case φ_1 is injective, the only non-reversible arrow "$\Longrightarrow$" above may be replaced by a reversible one, from where it follows that the null space of ψ in fact coincides with the null space of Λ', which is to say that the factored map, namely φ_{red}, is injective. □

Morphisms between Fell bundles have properties somewhat similar to morphisms between C*-algebras, as we shall now discuss.

Proposition 21.4. *Let $\varphi = \{\varphi_g\}_{g\in G}$ be a morphism from the Fell bundle $\mathscr{A} = \{A_g\}_{g\in G}$ to the Fell bundle $\mathscr{B} = \{B_g\}_{g\in G}$.*
(a) If φ_1 is injective, then all of the φ_g are isometric.
(b) If the range of each φ_g is dense in B_g, then all of the φ_g are in fact surjective.

PROOF. Since A_1 and B_1 are C*-algebras, under the hypothesis in (a), we have that φ_1 is in fact isometric. Given a in any A_g, we then have that

$$\|\varphi_g(a)\|^2 = \|\varphi_g(a)^*\varphi_g(a)\| = \|\varphi_1(a^*a)\| = \|a^*a\| = \|a\|^2.$$

This proves that φ_g is isometric.

Supposing now that all of the φ_g have dense range, it is easy to see that the *-homomorphism $\varphi_{\text{red}} : C^*_{\text{red}}(\mathscr{A}) \to C^*_{\text{red}}(\mathscr{B})$, given by (21.3), has dense range. Since φ_{red} is a *-homomorphism between C*-algebras, we deduce that φ_{red} is surjective. Thus, given b in any B_g, there exists some y in $C^*_{\text{red}}(\mathscr{A})$ such that

$$\varphi_{\text{red}}(y) = \lambda^B_g(b).$$

Observe that φ_{red} satisfies

$$\varphi_g\big(E^A_g(x)\big) = E^B_g\big(\varphi_{\text{red}}(x)\big), \quad \forall\, g \in G, \quad \forall\, x \in C^*_{\text{red}}(\mathscr{A}),$$

where E^A_g and E^B_g are given by (17.8). This may be easily checked by first assuming that $x = \lambda_h(a)$, for a in any A_h. Thus

$$\varphi_g\big(E^A_g(y)\big) = E^B_g\big(\varphi_{\text{red}}(y)\big) = E^B_g\big(\lambda^B_g(b)\big) = b,$$

proving that φ_g is surjective. □

Definition 21.5. Let $\mathscr{B} = \{B_g\}_{g\in G}$ be a Fell bundle. By a *Fell sub-bundle* of $\mathscr{B}$ we mean a collection $\mathscr{A} = \{A_g\}_{g\in G}$ of closed subspaces $A_g \subseteq B_g$, such that

(i) $A_g A_h \subseteq A_{gh}$, and
(ii) $A_g^* \subseteq A_{g^{-1}}$,

for all g and h in G.

It is evident that a Fell sub-bundle is itself a Fell bundle with the restricted operations, and moreover the inclusion from $\mathscr{A}$ into $\mathscr{B}$ is a morphism. Thus, given a Fell sub-bundle $\mathscr{A}$ of a Fell bundle $\mathscr{B}$, from (21.2) and (21.3) we get *-homomorphisms

$$C^*(\mathscr{A}) \xrightarrow{\iota} C^*(\mathscr{B}), \quad \text{and} \quad C^*_{\text{red}}(\mathscr{A}) \xrightarrow{\iota_{\text{red}}} C^*_{\text{red}}(\mathscr{B}), \tag{21.6}$$

the second of which is injective by the last sentence of (21.3).

This of course raises the question as to whether ι is also injective but, surprisingly, this is not always the case. We would now like to discuss a rather anomalous phenomenon leading up to the failure of injectivity for ι.

Recall that a C*-algebra C is said to be *exact* if, whenever

$$0 \to J \to A \to B \to 0$$

is an exact sequence of C*-algebras, the corresponding sequence

$$0 \to J \underset{\min}{\otimes} C \to A \underset{\min}{\otimes} C \to B \underset{\min}{\otimes} C \to 0$$

is also exact. Likewise, a group G is said to be *exact* if $C^*_{\text{red}}(G)$ is an exact C*-algebra. See [**19**, Chapters 2 and 5] for more details.

All amenable groups are exact, but there are many non-amenable exact groups, such as the free group on two or more generators [**19**, Proposition 5.1.8]. The first examples of non-exact groups were found by Gromov in [**66**], shattering a previously held belief that all groups were exact.

Proposition 21.7. *Let G be an exact group. Then the following are equivalent:*

(*i*) *for every Fell bundle $\mathscr{B}$ over G, and every Fell sub-bundle $\mathscr{A}$ of $\mathscr{B}$, the map ι of* (21.6) *is injective,*

(*ii*) *G is amenable.*

PROOF. By [**19**, Theorem 5.1.7] there exists a compact space X and a global amenable action β of G on X. Letting $\mathscr{B}$ be the semi-direct product bundle for the corresponding action of G on $C(X)$, we have that $\mathscr{B}$ satisfies the approximation property by (20.12), and hence $C^*(\mathscr{B})$ is naturally isomorphic to $C^*_{\text{red}}(\mathscr{B})$ via the regular representation Λ, by (20.6).

Since β is a global action, and $C(X)$ is a unital algebra, $1\delta_g$ is an element of B_g, for each g in G, and we may then consider the Fell sub-bundle $\mathscr{A} = \{A_g\}_{g\in G}$ of $\mathscr{B}$ given by

$$A_g = \mathbb{C}\delta_g, \quad \forall\, g \in G.$$

By (i) we then have that the map

$$\iota : C^*(\mathscr{A}) \to C^*(\mathscr{B})$$

given by (21.6) is injective. Therefore the composition

$$C^*(\mathscr{A}) \xrightarrow{\iota} C^*(\mathscr{B}) \xrightarrow{\Lambda} C^*_{\text{red}}(\mathscr{B})$$

is also injective. Letting $\tau = E_1 \circ \Lambda \circ \iota$, where E_1 is given by (17.8), it is easy to see that τ takes values in the subspace $\mathbb{C}\delta_1 \subseteq B_1$, so we may view τ as a complex linear functional on $C^*(\mathscr{A})$, which is faithful in the sense that

$$\tau(a^*a) = 0 \;\Rightarrow\; a = 0, \quad \forall\, a \in C^*(\mathscr{A}),$$

because E_1 is a faithful conditional expectation by (17.13).

Recall that $C^*(\mathscr{A})$ is a topologically graded algebra by (19.3), and it is easy to see that τ is its canonical conditional expectation. Therefore it follows from (19.7) that the regular representation

$$\Lambda^{\mathscr{A}} : C^*(\mathscr{A}) \to C^*_{\text{red}}(\mathscr{A})$$

is injective. Since the full (resp. reduced) cross sectional C*-algebra of $\mathscr{A}$ is nothing but the full (resp. reduced) group C*-algebra of G, we conclude from [**19**, Theorem 2.6.8] that G is amenable.

The converse implication is easily proven by observing that for both $\mathscr{B}$ and $\mathscr{A}$, the reduced and full cross sectional algebras coincide by (20.7), so the result follows from the last sentence of (21.3). □

21.8. Given any non-amenable exact group G, such as the free group on two generators, the result above implies that there is a Fell bundle $\mathscr{B}$ over G, and a Fell sub-bundle $\mathscr{A}$ of $\mathscr{B}$, for which the canonical mapping ι of (21.6) is non-injective.

The reader might have noticed that the last sentence of (21.3), relating to the injectivity of φ_{red}, has no counterpart in (21.2). In fact, as the above analysis shows, it is impossible to add to (21.2) a sentence similar to the last one of (21.3).

From our perspective, one of the most important examples of Fell sub-bundles is as follows:

Proposition 21.9. *Let $\beta = \big(B, G, \{B_g\}_{g\in G}, \{\beta_g\}_{g\in G}\big)$ be a C^*-algebraic partial dynamical system, and let A be a β-invariant, closed *-subalgebra of B. Also let $\alpha = \big(\{A_g\}_{g\in G}, \{\alpha_g\}_{g\in G}\big)$ be the restriction of β to A, as defined in* (2.10). *Letting $\mathscr{A}$ and $\mathscr{B}$ be the semi-direct product bundles relative to α and β, respectively, one has that $\mathscr{A}$ is a Fell sub-bundle of $\mathscr{B}$ (see below for the case when A is not invariant).*

PROOF. Recall that A_g is defined by

$$A_g = A \cap B_g,$$

so evidently $A_g\delta_g$ is a closed subspace of $B_g\delta_g$. We leave it for the reader to verify that the multiplication and adjoint operations of $\mathscr{A}$ are precisely the restrictions of the corresponding operations on $\mathscr{B}$, thus proving that $\mathscr{A}$ is indeed a Fell sub-bundle of $\mathscr{B}$, as desired. □

Definition 21.10. Let $\mathscr{B} = \{B_g\}_{g\in G}$ be a Fell bundle and let us be given a Fell sub-bundle $\mathscr{A} = \{A_g\}_{g\in G}$ of $\mathscr{B}$.

(a) We say that $\mathscr{A}$ is a *hereditary Fell sub-bundle* of $\mathscr{B}$ if

$$A_g B_h A_k \subseteq A_{ghk}, \quad \forall\, g, h, k \in G.$$

(b) We say that $\mathscr{A}$ is an *ideal* of $\mathscr{B}$ if

$$A_g B_h \subseteq A_{gh}, \quad \text{and} \quad B_g A_h \subseteq A_{gh}, \quad \forall\, g, h \in G.$$

It is evident that every ideal is a hereditary Fell sub-bundle.

A very important source of examples of hereditary Fell sub-bundles is given by the process of restriction of global actions to (not necessarily invariant) ideals, as we will now show:

Proposition 21.11. *Let β be a global C^*-algebraic action of a group G on a C^*-algebra B. Given a closed two-sided ideal A of B, let*

$$\alpha = \big(\{D_g\}_{g\in G}, \{\alpha_g\}_{g\in G}\big)$$

be the restriction of β to A, as defined in (3.2). *Denoting by $\mathscr{A}$ and $\mathscr{B}$ the semi-direct product bundles relative to α and β, respectively, one has that $\mathscr{A}$ is a hereditary*[27] *Fell sub-bundle of $\mathscr{B}$.*

[27]One should not be misled into thinking that $\mathscr{A}$ is an ideal in $\mathscr{B}$, despite the fact that A is an ideal in B.

PROOF. Recall that D_g is defined by

$$D_g = A \cap \beta_g(A),$$

so D_g is a closed subspace of B, and we may then see $D_g\delta_g$ is a closed subspace of $B\delta_g$. We leave it for the reader to verify that the multiplication and adjoint operations of $\mathscr{A}$ are precisely the restrictions of the corresponding operations on $\mathscr{B}$, so that $\mathscr{A}$ is indeed a Fell sub-bundle of $\mathscr{B}$, as desired.

In order to prove that $\mathscr{A}$ is hereditary, suppose that we are given g, h and k in G, a in D_g, b in B, and c in D_k. Then

$$(a\delta_g)(b\delta_h)(c\delta_k) = a\beta_g(b)\beta_{gh}(c)\delta_{ghk}. \tag{21.11.1}$$

Observing that each D_g is an ideal in B, notice that the coefficient of δ_{ghk} above satisfies

$$\begin{aligned} a\beta_g(b)\beta_{gh}(c) &\in D_g \cap \beta_{gh}(D_k) \\ &= A \cap \beta_g(A) \cap \beta_{gh}\big(A \cap \beta_k(A)\big) \\ &= A \cap \beta_g(A) \cap \beta_{gh}(A) \cap \beta_{ghk}(A) \\ &\subseteq A \cap \beta_{ghk}(A) \\ &= D_{ghk}, \end{aligned}$$

so the element mentioned in (21.11.1) in fact lies in $D_{ghk}\delta_{ghk}$, proving $\mathscr{A}$ to be hereditary. □

Proposition 21.12. *Let $\mathscr{B}$ be a Fell bundle and let $\mathscr{A}$ be a hereditary sub-bundle (resp. ideal) of $\mathscr{B}$. Then $C^*_{\text{red}}(\mathscr{A})$ is a hereditary subalgebra (resp. ideal) of $C^*_{\text{red}}(\mathscr{B})$.*

PROOF. We first observe that $C^*_{\text{red}}(\mathscr{A})$ may indeed be seen as a subalgebra of $C^*_{\text{red}}(\mathscr{B})$ by (21.3). It is then easy to see that

$$C_c(\mathscr{A})C_c(\mathscr{B})C_c(\mathscr{A}) \subseteq C_c(\mathscr{A}),$$

in the hereditary case, and

$$C_c(\mathscr{A})C_c(\mathscr{B}) \subseteq C_c(\mathscr{A}), \quad \text{and} \quad C_c(\mathscr{B})C_c(\mathscr{A}) \subseteq C_c(\mathscr{A}),$$

in the ideal case, from where the result follows. □

As seen in (21.8), the canonical map ι of (21.6) is not always injective. However under more restrictive conditions such pathologies disappear:

Theorem 21.13. *Let $\mathscr{B}$ be a Fell bundle and let $\mathscr{A}$ be a hereditary Fell sub-bundle. Then the canonical map*

$$\iota : C^*(\mathscr{A}) \to C^*(\mathscr{B})$$

of (21.6) is injective, so $C^*(\mathscr{A})$ is naturally isomorphic to the range on ι, which is a hereditary *-subalgebra of $C^*(\mathscr{B})$.

PROOF. Consider the diagram

$$\begin{array}{ccc} C^*(\mathscr{A}) & \xrightarrow{\ \mathscr{S}'\ } & C^*_{\text{red}}(\mathscr{A}) \otimes_{\max} C^*_{\text{red}}(G) \\ \big\downarrow{\scriptstyle \iota} & & \big\downarrow{\scriptstyle \iota_{\text{red}} \otimes id} \\ C^*(\mathscr{B}) & \xrightarrow{\ \mathscr{S}\ } & C^*_{\text{red}}(\mathscr{B}) \otimes_{\max} C^*_{\text{red}}(G) \end{array}$$

where $\mathscr{S}'$ and $\mathscr{S}$ are given by (18.9) for the respective bundles. By checking on the dense set $C_c(\mathscr{A}) \subseteq C^*(\mathscr{A})$, it is easy to see that this is a commutative diagram.

Employing [**19**, Proposition 3.6.4] we conclude that $\iota_{\rm red}\otimes id$ is injective because $\iota_{\rm red}$ is injective by the last sentence of (21.3), and $C^*_{\rm red}(\mathscr{A})$ is a hereditary subalgebra of $C^*_{\rm red}(\mathscr{B})$, by (21.12). Since $\mathscr{S}'$ and $\mathscr{S}$ are also injective, we conclude that ι is injective. That $C^*(\mathscr{A})$ is a hereditary subalgebra may be proved easily by the same argument used in the proof of (21.12). □

Given an ideal $\mathcal{J}=\{J_g\}_{g\in G}$ of a Fell bundle $\mathscr{B}=\{B_g\}_{g\in G}$, consider, for each g in G, the quotient space B_g/J_g. It is clear that the operations of $\mathscr{B}$ drop to the quotient, giving multiplication operations

$$\cdot : \frac{B_g}{J_g}\times\frac{B_h}{J_h}\to\frac{B_{gh}}{J_{gh}}$$

and involutions

$$* : \frac{B_g}{J_g}\to\frac{B_{g^{-1}}}{J_{g^{-1}}}.$$

One then easily verifies that the collection

$$\mathscr{B}/\mathcal{J}:=\{B_g/J_g\}_{g\in G}$$

is a Fell bundle with the above operations.

Definition 21.14. Given an ideal $\mathcal{J}$ of a Fell bundle $\mathscr{B}$, the Fell bundle $\mathscr{B}/\mathcal{J}$ constructed above will be called the quotient of $\mathscr{B}$ by $\mathcal{J}$.

Let us now compare the cross sectional C*-algebra of the quotient Fell bundle with the quotient of the corresponding cross sectional C*-algebras:

Proposition 21.15. *If $\mathcal{J}$ is an ideal in the Fell bundle $\mathscr{B}$, then $C^*(\mathcal{J})$ is an ideal in $C^*(\mathscr{B})$. Moreover the quotient of $C^*(\mathscr{B})$ by $C^*(\mathcal{J})$ is isomorphic to $C^*(\mathscr{B}/\mathcal{J})$, thus giving an exact sequence of C*-algebras*

$$0\longrightarrow C^*(\mathcal{J})\overset{\iota}{\to}C^*(\mathscr{B})\overset{q}{\to}C^*(\mathscr{B}/\mathcal{J})\longrightarrow 0.$$

PROOF. Since ideals are necessarily hereditary sub-bundles, (21.13) applies and hence $C^*(\mathcal{J})$ is naturally isomorphic to a subalgebra of $C^*(\mathscr{B})$, which the reader may easily prove to be an ideal using the same argument employed in the proof of (21.12).

For each g in G, let us denote by

$$q_g : B_g\to B_g/J_g$$

the quotient map. It is then obvious that $q=\{q_g\}_{g\in G}$ is a morphism from $\mathscr{B}$ to $\mathscr{B}/\mathcal{J}$, which therefore induces via (21.2) a clearly surjective *-homomorphism,

$$q : C^*(\mathscr{B})\to C^*(\mathscr{B}/\mathcal{J}),$$

still denoted by q by abuse of language, such that

$$q(\hat{\jmath}_g(b))=\hat{\jmath}_g(b+J_g),\quad\forall g\in G,\quad\forall b\in B_g,$$

where we rely on the context to determine which universal representation $\hat{\jmath}$ is meant in each case. Notice that q vanishes on $C^*(\mathcal{J})$, so

$$C^*(\mathcal{J})\subseteq \mathrm{Ker}(q).$$

The proof will then be complete once we show that these sets are in fact equal. For each g in G, define

$$\pi^0_g : b\in B_g\mapsto\hat{\jmath}_g(b)+C^*(\mathcal{J})\in C^*(\mathscr{B})/C^*(\mathcal{J}).$$

Since π_g^0 clearly vanishes on J_g, it factors through B_g/J_g, giving a linear mapping

$$\pi_g : B_g/J_g \to C^*(\mathscr{B})/C^*(\mathscr{J}).$$

One may now check that $\{\pi_g\}_{g\in G}$ is a representation of $\mathscr{B}/\mathscr{J}$, the integrated form of which, according to (16.29), is a *-homomorphism

$$\varphi : C^*(\mathscr{B}/\mathscr{J}) \to C^*(\mathscr{B})/C^*(\mathscr{J}),$$

such that

$$\varphi\big(\hat{\jmath}_g(b+J_g)\big) = \hat{\jmath}_g(b) + C^*(\mathscr{J}), \quad \forall\, g \in G, \quad \forall\, b \in B_g,$$

where again the context should be enough to determine the appropriate versions of $\hat{\jmath}$. The composition

$$C^*(\mathscr{B}) \xrightarrow{q} C^*(\mathscr{B}/\mathscr{J}) \xrightarrow{\varphi} C^*(\mathscr{B})/C^*(\mathscr{J})$$

thus sends each $\hat{\jmath}_g(b)$ to $\hat{\jmath}_g(b)+C^*(\mathscr{J})$, which means that it is precisely the quotient map modulo $C^*(\mathscr{J})$. This immediately implies that

$$\mathrm{Ker}(q) \subseteq C^*(\mathscr{J}),$$

thus concluding the proof. □

Remark 21.16. It is interesting to observe that *reduced* cross sectional algebras behave well with respect to *Fell sub-bundles* by (21.6), while *full* cross sectional algebras behave well with respect to *exact sequences* by (21.15).

Interchanging the roles of *full* and *reduced*, we have already obtained a partial result regarding *Fell sub-bundles* and *full* cross sectional algebras in (21.13), so we will now discuss the behavior of *exact sequences* under *reduced* cross sectional algebras. We begin with a technical result:

Lemma 21.17. *Given a Fell bundle $\mathscr{B} = \{B_g\}_{g\in G}$, there exists a bounded completely positive linear map*

$$\Psi : C^*(\mathscr{B}) \underset{min}{\otimes} C^*_{red}(G) \to C^*_{red}(\mathscr{B}),$$

such that, for every g and h in G, and every b in B_g one has that

$$\Psi\big(\hat{\jmath}_g(b)\otimes\lambda^G_h\big) = \begin{cases}\lambda_g(b), & \text{if } g=h,\\ 0, & \text{if } g\neq h.\end{cases}$$

Consequently $\Psi\circ\tau$ coincides with the identity of $C^*_{red}(\mathscr{B})$, where τ is given by (18.6).

Proof. Let π be any *-representation of $\mathscr{B}$ on a Hilbert space H such that π_1 is faithful. This can easily be obtained by composing a faithful representation of $C^*(\mathscr{B})$ with the universal representation $\hat{\jmath}$, for example. Also let

$$\varphi : C^*(\mathscr{B}) \to \mathscr{L}\left(H\otimes \ell^2(G)\right)$$

be built from π, as in (18.2), so that

$$\varphi\big(\hat{\jmath}_g(b)\big) = \pi_g(b)\otimes\lambda^G_g, \quad \forall\, g\in G, \quad \forall\, b \in B_g.$$

Recalling that $C^*_{\rm red}(G)$ is the closed *-subalgebra of $\mathscr{L}\left(\ell^2(G)\right)$ generated by the range of the left-regular representation of G, consider the representation

$$\varphi\otimes id : C^*(\mathscr{B}) \underset{\rm min}{\otimes} C^*_{\rm red}(G) \to \mathscr{L}\big(H\otimes\ell^2(G)\otimes\ell^2(G)\big),$$

characterized by the fact that

$$(\varphi \otimes id)(b \otimes \lambda_h^G) = \pi_g(b) \otimes \lambda_g^G \otimes \lambda_h^G,$$

for any g and h in G, and any b in B_g. Let K be the closed subspace of $H \otimes \ell^2(G) \otimes \ell^2(G)$, given by

$$K = \bigoplus_{g \in G} H \otimes e_g \otimes e_g,$$

and let P be the orthogonal projection onto K. Also consider the completely positive map

$$\Psi : C^*(\mathscr{B}) \underset{\min}{\otimes} C^*_{\text{red}}(G) \to \mathscr{L}(K)$$

given by $\Psi(x) = P\Big((\varphi \otimes id)(x)\Big)P$.

There is an obvious isometric isomorphism between K and $H \otimes \ell_2(G)$ under which a vector of the form $\xi \otimes e_g \otimes e_g$ is mapped to $\xi \otimes e_g$. If we identify $\mathscr{L}(K)$ with $\mathscr{L}\big(H \otimes \ell_2(G)\big)$ under this map, one sees that,

$$\Psi(\hat{\jmath}_g(b) \otimes \lambda_h^G) = \begin{cases} \pi_g(b) \otimes \lambda_g^G, & \text{if } g = h, \\ 0, & \text{if } g \neq h, \end{cases}$$

for every g and h in G, and every b in B_g.

The range of Ψ is therefore the closed linear span of the set of all $\pi_g(b) \otimes \lambda_g^G$, for b in B_g. This is also the range of φ, which in turn is the same as the range of the map ψ of (18.4).

Since ψ is a faithful representation by the last sentence of (18.4), given that π_1 is faithful by construction, we may then view Ψ as taking values in $C^*_{\text{red}}(\mathscr{B})$, thus providing the desired map. □

Let us now discuss an important consequence of the exactness of the base group to Fell bundles:

Theorem 21.18. *Let G be an exact group, and let $\mathscr{B}$ be a Fell bundle over G. If $\mathscr{J}$ is an ideal in $\mathscr{B}$, then the quotient of $C^*_{red}(\mathscr{B})$ by $C^*_{red}(\mathscr{J})$ is isomorphic to $C^*_{red}(\mathscr{B}/\mathscr{J})$, thus yielding an exact sequence of C^*-algebras*

$$0 \longrightarrow C^*_{red}(\mathscr{J}) \xrightarrow{\iota_{red}} C^*_{red}(\mathscr{B}) \xrightarrow{q_{red}} C^*_{red}(\mathscr{B}/\mathscr{J}) \longrightarrow 0.$$

PROOF. Recall from (21.12) that ι_{red} is a natural isomorphism from $C^*_{\text{red}}(\mathscr{J})$ onto an ideal in $C^*_{\text{red}}(\mathscr{B})$. On the other hand, the map q_{red} referred to above is the one given by (21.3) in terms of the quotient morphism $q = \{q_g\}_{g \in G}$.

As in (21.15), it is immediate to check that q_{red} is surjective and that the range of ι_{red} is contained in the kernel of q_{red}. Therefore the only point requiring our attention is the proof that the range of ι_{red} contains the kernel of q_{red}. So, let us pick z in $C^*_{\text{red}}(\mathscr{B})$, such that

$$q_{\text{red}}(z) = 0.$$

Temporarily turning to full cross sectional algebras, recall from (21.15) that

$$0 \longrightarrow C^*(\mathscr{J}) \xrightarrow{\iota} C^*(\mathscr{B}) \xrightarrow{q} C^*(\mathscr{B}/\mathscr{J}) \longrightarrow 0$$

is an exact sequence. Since $C^*_{\text{red}}(G)$ is an exact C*-algebra by hypothesis, the middle row of the diagram below is exact at $C^*(\mathscr{B}) \otimes_{\min} C^*_{\text{red}}(G)$.

$$
\begin{array}{ccccc}
C^*_{\text{red}}(\mathcal{J}) & \xrightarrow{\iota_{\text{red}}} & C^*_{\text{red}}(\mathscr{B}) & & \\
\Psi_1 \uparrow & & \Psi_2 \uparrow & & \\
C^*(\mathcal{J}) \underset{\min}{\otimes} C^*_{\text{red}}(G) & \xrightarrow{\iota\otimes 1} & C^*(\mathscr{B}) \underset{\min}{\otimes} C^*_{\text{red}}(G) & \xrightarrow{q\otimes 1} & C^*(\mathscr{B}/\mathcal{J}) \underset{\min}{\otimes} C^*_{\text{red}}(G) \\
& & \tau_2 \uparrow & & \tau_3 \uparrow \\
& & C^*_{\text{red}}(\mathscr{B}) & \xrightarrow{q_{\text{red}}} & C^*_{\text{red}}(\mathscr{B}/\mathcal{J})
\end{array}
$$

In this diagram we have also marked the maps Ψ_1 and Ψ_2 given by (21.17) for $\mathcal{J}$ and $\mathscr{B}$, respectively, as well as the maps τ_2 and τ_3 given by (18.6) for $\mathscr{B}$ and $\mathscr{B}/\mathcal{J}$. The usual method of checking on the appropriate dense subalgebras easily shows that the diagram is commutative.

Since $q_{\text{red}}(z) = 0$, we have that $(q \otimes 1)\big(\tau_2(z)\big) = 0$, so by exactness there exists x in $C^*(\mathcal{J}) \otimes_{\min} C^*_{\text{red}}(G)$ such that

$$(\iota \otimes 1)(x) = \tau_2(z).$$

Therefore

$$\iota_{\text{red}}\big(\psi_1(x)\big) = \psi_2\big((\iota \otimes 1)(x)\big) = \psi_2\big(\tau_2(z)\big) \overset{(21.17)}{=} z,$$

proving z to be in the range of ι_{red}, as desired. □

The above result should be seen from the point of view of (21.16). That is, even though *reduced* cross sectional algebras and *exact sequences* are not the best friends, under the assumption that the group is exact, we get a satisfactory result.

We will now go back to discussing another slightly disgruntled relationship, namely that of *full* cross sectional algebras and *Fell sub-bundles*, a relationship that has already appeared in (21.13), where heredity proved to be the crucial hypothesis. Instead of heredity we will now work under the existence of conditional expectations, as defined below.

Definition 21.19. Let $\mathscr{B} = \{B_g\}_{g\in G}$ be a Fell bundle and $\mathscr{A} = \{A_g\}_{g\in G}$ be a Fell sub-bundle. A *conditional expectation* from $\mathscr{B}$ to $\mathscr{A}$ is a collection of maps

$$P = \{P_g\}_{g\in G},$$

where each

$$P_g : B_g \to B_g$$

is a bounded, idempotent linear mapping, with range equal to A_g, such that P_1 is a conditional expectation from B_1 to A_1 and, for every g and h in G, every b in B_g, and every c in B_h, one has that

(i) $P_g(b)^* = P_{g^{-1}}(b^*)$,
(ii) $P_{gh}(bc) = bP_h(c)$, provided $b \in A_g$,
(iii) $P_{gh}(bc) = P_g(b)c$, provided $c \in A_h$.

Observe that (21.19.iii) easily follows from (21.19.i)–(21.19.iii) by taking adjoints.

► From now on, let us assume that we are given a Fell bundle $\mathscr{B} = \{B_g\}_{g\in G}$, and a Fell sub-bundle $\mathscr{A} = \{A_g\}_{g\in G}$, admitting a conditional expectation

$$P = \{P_g\}_{g\in G}.$$

It is our next goal to show that there exists a conditional expectation from $C^*_{\text{red}}(\mathscr{B})$ to $C^*_{\text{red}}(\mathscr{A})$ extending each P_g. This result will later be used in conjunction with (18.9) to give us the desired embedability result for full crossed sectional algebras.

The method we will adopt uses a version of the Jones-Watatani basic construction [**106**]. As the first step we will construct a right Hilbert A_1-module from $C_c(\mathscr{B})$, vaguely resembling the B_1-module $\ell^2(\mathscr{B})$ defined in (17.1). Given y and z in $C_c(\mathscr{B})$, define

$$\langle y, z\rangle_P = \sum_{g\in G} P_1\big((y_g)^* z_g\big).$$

We have already seen that $C_c(\mathscr{B})$ has the structure of a right B_1-module via the standard inclusion $j_1 : B_1 \to C_c(\mathscr{B})$. Restricting this module structure to A_1, we may view $C_c(\mathscr{B})$ as a right A_1-module. It is then easy to prove that $\langle \cdot, \cdot\rangle_P$ is an A_1-valued pre-inner product, the positivity following from (16.1.k) and the assumption that P_1 is positive.

Definition 21.20. We shall denote by $\ell^2_P(\mathscr{B})$ the right Hilbert A_1-module obtained by completing $C_c(\mathscr{B})$ under the semi-norm $\|\cdot\|_{2,P}$ arising from the inner-product defined above (after modding out the subspace of vectors of length zero). For each b in any B_g, we will denote by $j^P_g(b)$ the canonical image of $j_g(b)$ in $\ell^2_P(\mathscr{B})$.

The following result is a version of (17.3) to the present situation:

Proposition 21.21. *Given b in any B_g, the operator*

$$\lambda_g(b) : \ y \in C_c(\mathscr{B}) \longmapsto j_g(b) \star y \in C_c(\mathscr{B})$$

is bounded relative to $\|\cdot\|_{2,P}$ and hence extends to a bounded operator on $\ell^2_P(\mathscr{B})$, which we will denote by $\lambda^P_g(b)$, such that $\|\lambda^P_g(b)\| \leq \|b\|$, and which moreover satisfies

$$\lambda^P_g(b)\big(j^P_h(c)\big) = j^P_{gh}(bc), \quad \forall\, h \in G, \quad \forall\, c \in B_h.$$

PROOF. The last assertion follows from the corresponding identity proved in (17.3). Addressing the boundedness of $\lambda_g(b)$, given y in $C_c(\mathscr{B})$, notice that

$$\begin{aligned}\big\langle \lambda_g(b)y, \lambda_g(b)y\big\rangle_P &= \sum_{h\in G} P_1\big((y_{g^{-1}h})^* b^* b y_{g^{-1}h}\big)\\ &= \sum_{h\in G} P_1\big((y_h)^* b^* b y_h\big)\\ &\overset{(17.2)}{\leq} \|b\|^2 \sum_{h\in G} P_1\big((y_h)^* y_h\big)\\ &= \|b\|^2 \langle y, y\rangle_P,\end{aligned}$$

from where the result follows. □

Much like we did in (17.4), one may now prove that the collection of maps $\lambda^P = \{\lambda^P_g\}_{g\in G}$ is a representation of $\mathscr{B}$ as adjointable operators on $\ell^2_P(\mathscr{B})$, and hence gives rise to an integrated form, which we denote by

$$\Lambda^P : C^*(\mathscr{B}) \to \mathscr{L}\left(\ell^2_P(\mathscr{B})\right),$$

and which in turn is characterized by the fact that

$$\Lambda^P\big(\hat{j}_g(b)\big) = \lambda^P_g(b), \quad \forall\, g \in G, \quad \forall\, b \in B_g. \tag{21.22}$$

Proposition 21.23. *The map Λ^P defined above vanishes on the kernel of the regular representation Λ of $\mathscr{B}$.*

PROOF. Mimicking (17.12) one may prove that

$$\langle j_g^P(b), \Lambda^P(z) j_h^P(c)\rangle_P = P_1\Big(b^* E_{gh^{-1}}\big(\Lambda(z)\big)c\Big),$$

for all $z \in C^*(\mathscr{B})$, $g, h \in G$, $b \in B_g$, and $c \in B_h$. Thus, if z is in the kernel of Λ, we have that the left-hand-side above vanishes identically, and this easily implies that $\Lambda^P(z) = 0$. □

As a consequence, we see that Λ^P factors through $C^*_{\text{red}}(\mathscr{B})$, producing a representation

$$\Lambda^P_{\text{red}} : C^*_{\text{red}}(\mathscr{B}) \to \mathscr{L}\left(\ell^2_P(\mathscr{B})\right),$$

such that

$$\Lambda^P_{\text{red}}\big(\lambda_g(b)\big) = \lambda_g^P(b), \quad \forall\, g \in G, \quad \forall\, b \in B_g. \tag{21.24}$$

In order to proceed, we need to prove the following key inequality:

Lemma 21.25. *For b in any B_g, one has that*

$$P_g(b)^* P_g(b) \leq P_1(b^* b).$$

PROOF. Observing that $b - P_g(b)$ belongs to B_g, and hence by (16.1.k) one has

$$\big(b - P_g(b)\big)^* \big(b - P_g(b)\big) \geq 0$$

in B_1, we deduce from the positivity of P_1 that

$$\begin{aligned}
0 &\leq P_1\Big(\big(b - P_g(b)\big)^* \big(b - P_g(b)\big)\Big)\\
&= P_1\Big(b^* b - b^* P_g(b) - P_g(b)^* b + P_g(b)^* P_g(b)\Big)\\
&= P_1(b^* b) - P_{g^{-1}}(b^*) P_g(b) - P_g(b)^* P_g(b) + P_g(b)^* P_g(b)\\
&= P_1(b^* b) - P_g(b)^* P_g(b),
\end{aligned}$$

from where the conclusion follows. □

Observe that $C_c(\mathscr{A})$ may be seen as a right A_1-sub-module of $C_c(\mathscr{B})$ in a natural way. Moreover, the inclusion of the former into the latter is clearly isometric for the usual 2-norm on $C_c(\mathscr{A}) \subseteq \ell^2(\mathscr{A})$, and the norm $\|\cdot\|_{2,P}$ on $C_c(\mathscr{B})$. Consequently this map extends to an isometric right-A_1-linear map

$$V : \ell^2(\mathscr{A}) \to \ell^2_P(\mathscr{B}).$$

Adopting the policy of using single quotes for denoting the relevant maps for $\mathscr{A}$, let us denote by

$$j'_g : A_g \to C_c(\mathscr{A})$$

the maps given by (16.21). We may then characterize V by the fact that

$$V\big(j'_g(a)\big) = j_g^P(a), \quad \forall\, a \in A_g. \tag{21.26}$$

Since bounded linear maps on Hilbert modules are not necessarily adjointable, we need a bit of work to provide an adjoint for V.

Proposition 21.27. *The mapping*

$$p : C_c(\mathscr{B}) \to C_c(\mathscr{A}) \subseteq \ell^2(\mathscr{A}),$$

defined by

$$p(y)_g = P_g(y_g), \quad \forall\, y \in C_c(\mathscr{B}), \quad \forall\, g \in G,$$

is bounded with respect to $\|\cdot\|_{2,P}$*, and hence extends to a bounded operator from* $\ell^2_P(\mathscr{B})$ *to* $\ell^2(\mathscr{A})$*, still denoted by* p*, by abuse of language. Moreover* p *is the adjoint of the map* V *defined above.*

PROOF. Given y in $C_c(\mathscr{B})$, one has

$$\big\langle p(y), p(y)\big\rangle = \sum_{g\in G} P_g(y_g)^* P_g(y_g) \overset{(21.25)}{\leq} \sum_{g\in G} P_1\big((y_g)^* y_g\big) = \langle y, y\rangle_P.$$

This proves the boundedness of p. In order to prove the last assertion in the statement, let $y \in C_c(\mathscr{A})$ and $z \in C_c(\mathscr{B})$. Then

$$\langle y, p(z)\rangle = \sum_{h\in G} y_g^* P_g(z_g) = \sum_{h\in G} P_1(y_g^* z_g) = \langle V(y), z\rangle_P.$$

□

Since $p = V^*$, we will from now on dispense with the notation "p", using "V^*" instead.

As already mentioned, V is an isometry, so V^*V is the identity operator on $\ell^2(\mathscr{A})$, while VV^* is a projection in $\mathscr{L}(\ell^2_P(\mathscr{B}))$, whose range is clearly the canonical copy of $\ell^2(\mathscr{A})$ within $\ell^2_P(\mathscr{B})$.

The following result describes what happens when we compress elements of the range of Λ^P down to $\ell^2(\mathscr{A})$ via V. In its statement we will continue with the trend of using single quotes when denoting the relevant maps for $\mathscr{A}$.

Proposition 21.28. *For* b *in any* B_g*, one has that*

$$V^*\lambda^P_g(b)V = \lambda'_g\big(P_g(b)\big).$$

PROOF. Given h in G, and a in A_h we have that

$$\begin{aligned}
\Big(V^*\lambda^P_g(b)V\Big)\Big|_{j'_h(a)} &\overset{(21.26)}{=} V^*\lambda^P_g(b)\big(j^P_h(a)\big)\\
&\overset{(21.21)}{=} V^* j^P_{gh}(ba)\\
&\overset{(21.27)}{=} j'_{gh}\big(P_{gh}(ba)\big)\\
&= j'_{gh}\big(P_g(b)a\big)\\
&\overset{(17.3)}{=} \lambda'_g\big(P_g(b)\big) j'_h(a).
\end{aligned}$$

Since the elements $j'_h(a)$ considered above span a dense subspace of $\ell^2(\mathscr{A})$, the proof is concluded. □

With this we may now prove our next main result:

Theorem 21.29. *Given a Fell bundle* $\mathscr{B} = \{B_g\}_{g\in G}$*, and a Fell sub-bundle* $\mathscr{A} = \{A_g\}_{g\in G}$ *admitting a conditional expectation* $P = \{P_g\}_{g\in G}$*, there exists a conditional expectation*

$$E : C^*_{red}(\mathscr{B}) \to C^*_{red}(\mathscr{A}),$$

such that

$$E\big(\lambda_g(b)\big) = \lambda'_g\big(P_g(b)\big), \quad \forall\, g \in G, \quad \forall\, b \in B_g,$$

where λ' *and* λ *denote the regular representations of* $\mathscr{A}$ *and* $\mathscr{B}$*, respectively.*

PROOF. Define

$$E : C^*_{\text{red}}(\mathscr{B}) \to \mathscr{L}\left(\ell^2(\mathscr{A})\right),$$

by

$$E(z) = V^*\Lambda^P_{\text{red}}(z)V, \quad \forall\, z \in C^*_{\text{red}}(\mathscr{B}).$$

In case $z = \lambda_g(b)$, for some b in B_g, notice that

$$E\big(\lambda_g(b)\big) = V^*\Lambda^P_{\text{red}}\big(\lambda_g(b)\big)V \overset{(21.24)}{=} V^*\lambda^P_g(b)V \overset{(21.28)}{=} \lambda'_g\big(P_g(b)\big),$$

proving the last assertion in the statement. Since the $\lambda_g(b)$ span $C^*_{\text{red}}(\mathscr{B})$, this also proves that the range of E is contained in $C^*_{\text{red}}(\mathscr{A})$, so we may view E as a $C^*_{\text{red}}(\mathscr{A})$-valued map, as required.

Conditional expectations are meant to be maps from an algebra to a subalgebra. So it is important for us to view $C^*_{\text{red}}(\mathscr{A})$ as a subalgebra of $C^*_{\text{red}}(\mathscr{B})$, and we of course do it through (21.6). Technically this means that we identify $\lambda'_g(a)$ and $\lambda_g(a)$, for a in any A_g. In this case, since $P_g(a) = a$, we deduce from the equation just proved that $E\big(\lambda_g(a)\big) = \lambda_g(a)$, so E is the identity on $C^*_{\text{red}}(\mathscr{A})$, and the reader may now easily prove that E is in fact a conditional expectation. □

We may now return to the discussion of *full* cross sectional algebras versus *Fell sub-bundles*:

Theorem 21.30. *Let $\mathscr{B}$ be a Fell bundle and let $\mathscr{A}$ be a Fell sub-bundle. If $\mathscr{A}$ admits a conditional expectation, then the canonical map*

$$\iota : C^*(\mathscr{A}) \to C^*(\mathscr{B})$$

of (21.6) *is injective.*

PROOF. The proof is essentially identical to the proof of (21.13), with the exception that, in order to conclude that $\iota_{\text{red}} \otimes id$ is injective, instead of invoking [**19**, Proposition 3.6.4] we rely on [**19**, Proposition 3.6.6], observing that condition [**19**, 3.6.6.1] is fulfilled by the conditional expectation E provided by (21.29). □

Let us now study amenability of Fell sub-bundles.

Proposition 21.31. *Let $\mathscr{B}$ be an amenable Fell bundle and let $\mathscr{A}$ be a Fell sub-bundle of $\mathscr{B}$. If $\mathscr{A}$ is either hereditary or admits a conditional expectation, then $\mathscr{A}$ is also amenable.*

PROOF. Consider the diagram

$$\begin{array}{ccc} C^*(\mathscr{A}) & \xrightarrow{\ \iota\ } & C^*(\mathscr{B}) \\ \Big\downarrow{\scriptstyle \Lambda'} & & \Big\downarrow{\scriptstyle \Lambda} \\ C^*_{\text{red}}(\mathscr{A}) & \xrightarrow{\ \iota_{\text{red}}\ } & C^*_{\text{red}}(\mathscr{B}) \end{array}$$

where Λ' and Λ are the regular representations of $\mathscr{A}$ and $\mathscr{B}$, respectively, and ι and ι_{red} are as in (21.6). It is elementary to check that this is a commutative diagram.

Employing either (21.13) or (21.30), as appropriate, we have that ι is injective. Since $\mathscr{B}$ is amenable, Λ is injective as well, so we conclude that Λ' is injective, whence $\mathscr{A}$ is amenable. □

Using the method employed in the proof of (21.7) it is possible to produce an amenable Fell bundle containing a non-amenable Fell sub-bundle[28]. Therefore the hypothesis that $\mathscr{A}$ is either hereditary or admits a conditional expectation is crucial for the validity of the result above.

Let us now study the approximation property for Fell sub-bundles.

Proposition 21.32. *Let $\mathscr{B} = \{B_g\}_{g\in G}$ be a Fell bundle satisfying the approximation property, and let $\mathscr{A} = \{A_g\}_{g\in G}$ be a Fell sub-bundle of $\mathscr{B}$. If A_1 is a hereditary subalgebra of B_1, then $\mathscr{A}$ also satisfies the approximation property.*

PROOF. Let $\{a_i\}_{i\in I}$ be a Cesaro net for $\mathscr{B}$, as defined in (20.4), and let $\{v_j\}_{j\in J}$ be an approximate identity for A_1. For each (i,j) in $I\times J$, consider the function

$$c_{i,j} : g\in G \mapsto v_j a_i(g) v_j \in A_1.$$

Notice that $v_j a_i(g) v_j$ indeed lies in A_1 because $A_1 B_1 A_1 \subseteq A_1$, given that A_1 is supposed to be hereditary.

Considering $I\times J$ as a directed set with coordinate-wise order, we claim that $\{c_{i,j}\}_{(i,j)\in I\times J}$ is a Cesaro net for $\mathscr{A}$.

Since each a_i is a finitely supported function on G, it is clear that so are the $c_{i,j}$. In order to prove condition (20.4.i), suppose we are given $(i,j)\in I\times J$. Then, taking into account that $v_j^* v_j \leq 1$, one has that

$$\begin{aligned}\Big\|\sum_{g\in G} c_{i,j}(g)^* c_{i,j}(g)\Big\| &= \Big\|\sum_{g\in G} v_j^* a_i(g)^* v_j^* v_j a_i(g) v_j\Big\| \\ &\leq \Big\|\sum_{g\in G} v_j^* a_i(g)^* a_i(g) v_j\Big\| \\ &\leq \Big\|\sum_{g\in G} a_i(g)^* a_i(g)\Big\|,\end{aligned}$$

so (20.4.i) follows from the corresponding property of the Cesaro net $\{a_i\}_{i\in I}$. In order to prove (20.4.ii), given b in any A_g, we must prove that

$$\sum_{h\in\mathscr{G}} c_{i,j}(gh)^* b c_{i,j}(h) \overset{(i,j)\to\infty}{\longrightarrow} b. \tag{21.32.1}$$

For each i in I, let us consider the linear operator W_i defined on A_g by

$$W_i(b) = \sum_{h\in\mathscr{G}} a_i(gh)^* b a_i(h), \quad \forall b\in A_g.$$

Since this may be seen as the restriction of the map V of (20.3) to A_g, we have that

$$\|W_i\| \leq \Big\|\sum_{h\in\mathscr{G}} a_i(h)^* a_i(h)\Big\| \leq M,$$

where M may be chosen independently of i. Focusing on the left-hand-side of (21.32.1), we have

$$\sum_{h\in G} c_{i,j}(gh)^* b c_{i,j}(h) = \sum_{h\in G} v_j^* a_i(gh)^* v_j^* b v_j a_i(h) v_j = v_j^* W_i(v_j^* b v_j) v_j,$$

so (21.32.1) translates into

$$v_j^* W_i(v_j^* b v_j) v_j \overset{(i,j)\to\infty}{\longrightarrow} b. \tag{21.32.2}$$

[28]Take an exact, non-amenable group G, acting amenably on a compact topological space. Then, as seen in the proof of (21.7), the associated semi-direct product bundle is amenable, but it contains the trivial one-dimensional Fell bundle, which is non-amenable.

Observing that the terms of an approximate identity have norm at most 1, we have that

$$\begin{aligned}\|b - v_j^* W_i(v_j^* b v_j) v_j\| &\leq \|b - v_j^* b v_j\| + \|v_j^* b v_j - v_j^* W_i(b) v_j\| \\ &\quad + \|v_j^* W_i(b) v_j - v_j^* W_i(v_j^* b v_j) v_j\| \\ &\leq \|b - v_j^* b v_j\| + \|b - W_i(b)\| + \|W_i\| \|b - v_j^* b v_j\|.\end{aligned}$$

By (16.9) we have that $bv_j \overset{j\to\infty}{\longrightarrow} b$, so (21.32.2) follows. This verifies that $\{c_{i,j}\}_{(i,j)\in I\times J}$ is indeed a Cesaro net for $\mathscr{A}$, as desired. □

Observe that the last hypothesis of the above result, namely that A_1 is a hereditary subalgebra of B_1, is evidently satisfied if $\mathscr{A}$ is a hereditary sub-bundle of $\mathscr{B}$. It might also be worth pointing out that the fact that A_1 is a hereditary subalgebra of B_1 does not imply that $\mathscr{A}$ is a hereditary Fell sub-bundle of $\mathscr{B}$. A counter example may be easily obtained by choosing any Fell bundle $\mathscr{B}$ and selecting $A_1 = B_1$, while $A_g = \{0\}$, for all $g \neq 1$. In this case there is no reason for $A_1 B_g A_1$ to be contained in A_g.

After spending a little effort, we have not been able to determine if the approximation property passes to sub-bundles under the existence of a conditional expectation, but we believe there is a good chance this is true.

With respect to amenability of quotient Fell bundles we have the following:

Proposition 21.33. *Suppose we are given Fell bundles $\mathscr{B} = \{B_g\}_{g\in G}$ and $\mathscr{A} = \{A_g\}_{g\in G}$, as well as a morphism $\varphi = \{\varphi_g\}_{g\in G}$ from $\mathscr{B}$ to $\mathscr{A}$, such that φ_g is onto A_g for every g.*

(i) If $\mathscr{B}$ is amenable and G is an exact group, then $\mathscr{A}$ is also amenable.

(ii) If $\mathscr{B}$ satisfies the approximation property, then so does $\mathscr{A}$.

Proof. Given a Cesaro net $\{a_i\}_{i\in I}$ for $\mathscr{B}$, it is elementary to check that $\{\varphi_1(a_i)\}_{i\in I}$ is a Cesaro net for $\mathscr{A}$. Thus (ii) follows.

With respect to (i), for each g in G, let J_g be the kernel of φ_g. It is then clear that $\mathcal{J} = \{J_g\}_{g\in G}$ is an ideal of $\mathscr{B}$, in the sense of (21.10.b), and one has that $\mathscr{B}/\mathcal{J}$ is isomorphic to $\mathscr{A}$.

Consider the diagram

$$\begin{array}{ccccccccc} 0 & \longrightarrow & C^*(\mathcal{J}) & \longrightarrow & C^*(\mathscr{B}) & \longrightarrow & C^*(\mathscr{A}) & \longrightarrow & 0 \\ & & \downarrow{\scriptstyle \Lambda_{\mathcal{J}}} & & \downarrow{\scriptstyle \Lambda_{\mathscr{B}}} & & \downarrow{\scriptstyle \Lambda_{\mathscr{A}}} & & \\ 0 & \longrightarrow & C^*_{\text{red}}(\mathcal{J}) & \longrightarrow & C^*_{\text{red}}(\mathscr{B}) & \longrightarrow & C^*_{\text{red}}(\mathscr{A}) & \longrightarrow & 0 \end{array}$$

whose rows are exact by (21.15) and (21.18), and the vertical arrows indicate the various regular representations. It is immediate to check that this is commutative. Moreover, $\Lambda_{\mathscr{B}}$ is one-to-one by hypothesis, and $\Lambda_{\mathcal{J}}$ is onto by definition of reduced cross sectional algebras, so one may prove that $\Lambda_{\mathscr{A}}$ is injective by a standard diagram chase, similar to one used to prove the Five Lemma. □

Notes and remarks. Theorem (21.18) was proved in [**53**]. It would be interesting to decide if (21.33.i) holds without the hypothesis that G is exact.

CHAPTER 22

Functoriality for Partial Actions

As already mentioned, some of the most important examples of Fell bundles arise from partial dynamical systems. In this chapter we will therefore present some consequences of the results proved in the above chapter to semi-direct product bundles. We will also study the relationship between the semi-direct product bundle for a partial action β and the corresponding bundle for a restriction of β.

► So, let us now fix two C*-algebraic partial dynamical systems

$$\alpha = \big(A, G, \{A_g\}_{g\in G}, \{\alpha_g\}_{g\in G}\big)$$

and

$$\beta = \big(B, G, \{B_g\}_{g\in G}, \{\beta_g\}_{g\in G}\big).$$

We will denote the semi-direct product bundles for α and β by

$$\mathscr{A} = \{A_g\delta_g\}_{g\in G}, \quad \text{and} \quad \mathscr{B} = \{B_g\delta_g\}_{g\in G},$$

respectively.

Recall from (2.7) that a *-homomorphism $\varphi : A \to B$ is said to be G-equivariant provided $\varphi(A_g) \subseteq B_g$, and $\varphi\big(\alpha_g(a)\big) = \beta_g\big(\varphi(a)\big)$, for a in any A_g.

Proposition 22.1. *Given a G-equivariant *-homomorphism $\varphi : A \to B$, for each g in G consider the mapping*

$$\varphi_g : A_g\delta_g \to B_g\delta_g,$$

given by

$$\varphi_g(a\delta_g) = \varphi(a)\delta_g, \quad \forall\, a \in A_g.$$

Then $\{\varphi_g\}_{g\in G}$ is a morphism from $\mathscr{A}$ to $\mathscr{B}$.

PROOF. Left to the reader. □

As an immediate consequence of (21.2) and (21.3), we have:

Proposition 22.2. *Given a G-equivariant *-homomorphism $\varphi : A \to B$, there are *-homomorphisms*

$$\Phi : A \rtimes G \to B \rtimes G, \quad \text{and} \quad \Phi_{red} : A \rtimes_{red} G \to B \rtimes_{red} G$$

sending $a\delta_g$ to $\varphi(a)\delta_g$, for a in any A_g, with the appropriate interpretation of δ_g in each case.

In case φ is injective one may see $\mathscr{A}$ as a Fell sub-bundle of $\mathscr{B}$. Two of the most important examples of this are given by (21.9) and (21.11).

Proposition 22.3. *Let $\beta = \big(B, G, \{B_g\}_{g\in G}, \{\beta_g\}_{g\in G}\big)$ be a C^*-algebraic partial dynamical system, and let A be a closed *-subalgebra of B. Suppose that either:*

(i) A is invariant under β, or

(*ii*) *β is a global action and A is an ideal of B.*

In either case, let $\alpha = \big(\{A_g\}_{g\in G}, \{\alpha_g\}_{g\in G}\big)$ be the restriction of β to A (defined in (2.10) *in the first case, or in* (3.2) *in the second case). Then we have a natural inclusion*

$$A \rtimes_{red} G \subseteq B \rtimes_{red} G.$$

In situation (ii) one moreover has that $A \rtimes_{red} G$ is a hereditary subalgebra of $B \rtimes_{red} G$.

PROOF. We have that $\mathscr{A}$ is a Fell sub-bundle of $\mathscr{B}$ by (21.9) or (21.11). That $A \rtimes_{\text{red}} G$ is naturally a subalgebra of $B \rtimes_{\text{red}} G$ then follows from the last assertion in (21.3). Under hypothesis (ii), we have by (21.11) that $\mathscr{A}$ is a hereditary Fell sub-bundle of $\mathscr{B}$ so the last sentence in the statement follows from (21.12). □

For the case of full crossed products we have the following:

Proposition 22.4. *Let β be a global action of a group G on a C^*-algebra B, and let A be an ideal of B. Considering the partial action α of G on A given by restriction, as defined in* (3.2), *the canonical mapping*

$$\iota : A \rtimes G \to B \rtimes G$$

is injective. Moreover the range of ι is a hereditary *-subalgebra of $B \rtimes G$.

PROOF. We have seen in (21.11) that the semi-direct product bundle for α is hereditary in the one for β. So the conclusion follows immediately from (21.13). □

In order to take advantage of (21.30), namely the injectivity of the canonical mapping under the existence of conditional expectations, we must understand the relationship between partial actions and conditional expectations. This is the goal of our next result.

Proposition 22.5. *Suppose we are given a C^*-algebraic partial dynamical system*

$$\beta = \big(B, G, \{B_g\}_{g\in G}, \{\beta_g\}_{g\in G}\big),$$

and let A be a closed *-subalgebra of B. Suppose moreover that there is a G-equivariant conditional expectation F from B onto A. Then

(*i*) *A is invariant under β,*

(*ii*) *the semi-direct product bundle for the action α obtained by restricting β to A, henceforth denoted by $\mathscr{A}$, seen as a Fell sub-bundle of the semi-direct product bundle for β, henceforth denoted by $\mathscr{B}$, admits a conditional expectation $P = \{P_g\}_{g\in G}$, where*

$$P_g(b\delta_g) = F(b)\delta_g, \quad \forall\, b \in B_g.$$

(*iii*) *the canonical mapping*

$$\iota : A \rtimes G \to B \rtimes G$$

is injective.

PROOF. Given g in G, we have that

$$\beta_g(A \cap B_{g^{-1}}) = \beta_g\big(F(A \cap B_{g^{-1}})\big) \overset{(2.7.\text{ii})}{=} F\big(\beta_g(A \cap B_{g^{-1}})\big) \subseteq A,$$

proving that A is indeed invariant under β.

Notice that if b is in any B_g, then $F(b) \in B_g \cap A =: A_g$, by invariance, so indeed $F(b)\delta_g$ lies in $A_g\delta_g$, as needed. To prove (21.19.i), we compute

$$\begin{aligned}
P_g(b\delta_g)^* &= \big(F(b)\delta_g\big)^* \\
&= \beta_{g^{-1}}\big(F(b^*)\big)\delta_{g^{-1}} \\
&= F\big(\beta_{g^{-1}}(b^*)\big)\delta_{g^{-1}} \\
&= P_{g^{-1}}\big(\beta_{g^{-1}}(b^*)\delta_{g^{-1}}\big) \\
&= P_{g^{-1}}\big((b\delta_g)^*\big).
\end{aligned}$$

If we now take c in some A_h, we have

$$\begin{aligned}
P_{gh}\big((b\delta_g)(c\delta_h)\big) &= P_{gh}\Big(\beta_g\big(\beta_{g^{-1}}(b)c\big)\delta_{gh}\Big) \\
&= F\Big(\beta_g\big(\beta_{g^{-1}}(b)c\big)\Big)\delta_{gh} \\
&= \beta_g\Big(F\big(\beta_{g^{-1}}(b)c\big)\Big)\delta_{gh} \\
&= \beta_g\Big(F\big(\beta_{g^{-1}}(b)\big)c\Big)\delta_{gh} \\
&= \beta_g\Big(\beta_{g^{-1}}\big(F(b)\big)c\Big)\delta_{gh} \\
&= \big(F(b)\delta_g\big)\big(c\delta_h\big) \\
&= P_g\big(b\delta_g\big)\big(c\delta_h\big),
\end{aligned}$$

proving (21.19.iii) and then (21.19.ii) follows by taking adjoints. The fact that $P_g(B_g\delta_g) = A\delta_g$, or equivalently that $F(B_g) = A_g$, may be easily proved based on the G-equivariance of F. Point (iii) now follows from (21.30). □

Let us now study the behavior of crossed products relative to ideals and quotients.

▶ We therefore fix, for the time being, a C*-algebraic partial action

$$\alpha = \big(\{A_g\}_{g\in G}, \{\alpha_g\}_{g\in G}\big)$$

of a group G on a C*-algebra A, and an α-invariant closed two-sided ideal $J \trianglelefteq A$. The restriction of α to J, defined according to (2.10), will henceforth be denoted by

$$\tau = \big(\{J_g\}_{g\in G}, \{\tau_g\}_{g\in G}\big).$$

Denoting by B the quotient C*-algebra, we obtain the exact sequence

$$0 \to J \xrightarrow{i} A \xrightarrow{q} B \to 0, \tag{22.6}$$

where ι denoted the inclusion, and q the quotient mapping. Letting

$$B_g = q(A_g), \quad \forall\, g \in G,$$

it is then easy to see that B_g is a closed two-sided ideal in B and that B_g is *-isomorphic to $A_g/(A_g \cap J)$. Moreover, since each α_g maps $A_{g^{-1}}$ to A_g in such a way that

$$\alpha_g(A_{g^{-1}} \cap J) \subseteq A_g \cap J$$

by invariance of J, we see that α_g drops to the quotient providing a *-homomorphism

$$\beta_g : B_{g^{-1}} \to B_g,$$

such that

$$\beta_g\big(q(a)\big) = q\big(\alpha_g(a)\big), \quad \forall g \in G, \quad \forall a \in A_{g^{-1}}.$$

Proposition 22.7. *One has that*

$$\beta := \big(\{B_g\}_{g\in G}, \{\beta_g\}_{g\in G}\big)$$

is a partial action of G on B.

PROOF. Since (2.1.i) is trivially true, it suffices to verify (2.5.i)–(2.5.ii). Given $g, h \in G$, we claim that

$$B_g \cap B_h = q(A_g \cap A_h). \tag{22.7.1}$$

By the Cohen-Hewitt Theorem every ideal in a C*-algebra is idempotent. In particular, given any two ideals I and J, one has

$$I \cap J = (I \cap J)(I \cap J) \subseteq IJ.$$

Since IJ is obviously contained in $I \cap J$, we deduce that $I \cap J = IJ$. Thus, for every $g, h \in G$, we have

$$B_g \cap B_h = B_g B_h = q(A_g)q(A_h) = q(A_g A_h) = q(A_g \cap A_h),$$

proving (22.7.1). In order to prove (2.5.i), we then compute

$$\begin{aligned}\beta_g\big(B_{g^{-1}} \cap B_h\big) &= \beta_g\big(q(A_{g^{-1}} \cap A_h)\big)\\ &= q\big(\alpha_g(A_{g^{-1}} \cap A_h)\big)\\ &\subseteq q(A_{gh})\\ &= B_{gh}.\end{aligned}$$

Regarding (2.5.ii), pick x in $B_{h^{-1}} \cap B_{(gh)^{-1}}$, and write $x = q(a)$, with a in $A_{h^{-1}} \cap A_{(gh)^{-1}}$, by (22.7.1). Then

$$\begin{aligned}\beta_g\big(\beta_h(x)\big) &= \beta_g\big(\beta_h(q(a))\big)\\ &= \beta_g\big(q(\alpha_h(a))\big)\\ &= q\big(\alpha_g(\alpha_h(a))\big)\\ &= q\big(\alpha_{gh}(a)\big)\\ &= \beta_{gh}\big(q(a)\big)\\ &= \beta_{gh}(x).\end{aligned}$$

This completes the proof. □

Regarding the corresponding semi-direct product bundles we have:

Proposition 22.8. *Let $\alpha = \big(\{A_g\}_{g\in G}, \{\alpha_g\}_{g\in G}\big)$ be a C^*-algebraic partial action of a group G on a C^*-algebra A, and let $J \trianglelefteq A$ be an α-invariant closed two-sided ideal. Also let $\mathcal{J}$, $\mathscr{A}$ and $\mathscr{B}$ be the semi-direct product bundles relative to the partial actions τ, α and β, above. Then $\mathcal{J}$ is an ideal in $\mathscr{A}$, and $\mathscr{B}$ is naturally isomorphic to the quotient Fell bundle $\mathscr{A}/\mathcal{J}$.*

PROOF. Left for the reader. □

We thus have the following consequences of our study of ideals in Fell bundles in the previous chapter:

Theorem 22.9. *Let*

$$0 \to J \xrightarrow{i} A \xrightarrow{q} B \to 0,$$

be an exact sequence of C^*-algebras and let $\alpha = \big(\{A_g\}_{g\in G}, \{\alpha_g\}_{g\in G}\big)$ be a partial action of G on A, relative to which J is invariant. Then the corresponding sequence

$$0 \to J \rtimes G \to A \rtimes G \to B \rtimes G \to 0$$

is also exact. Moreover, if

(i) *G is an exact group, or*
(ii) *the semi-direct product bundle associated to α satisfies the approximation property, then the sequence*

$$0 \to J \rtimes_{red} G \to A \rtimes_{red} G \to B \rtimes_{red} G \to 0$$

is exact as well.

PROOF. The first assertion is a direct application of (21.15).

Under hypothesis (i), the second assertion follows immediately from (21.18), so we need only prove the second assertion under hypothesis (ii).

Let $\mathcal{J}$, $\mathscr{A}$ and $\mathscr{B}$ be the semi-direct product bundles relative to the partial actions τ, α and β, as in (22.8).

By hypothesis we have that $\mathscr{A}$ satisfies the approximation property and hence so does $\mathcal{J}$, by (21.32), as well as $\mathscr{B}$, by (21.33.ii). Therefore all three Fell bundles in sight are amenable by (20.6), so the above sequence of reduced crossed products coincides with the corresponding one for full crossed products whose exactness has already been verified. □

This result has a useful application to the study of ideals in the crossed product generated by subsets of the coefficient algebra. We state it only for full crossed products since we only envisage applications of it in this case.

Proposition 22.10. *Let*

$$\alpha = \big(A, G, \{A_g\}_{g\in G}, \{\alpha_g\}_{g\in G}\big)$$

be a C^*-algebraic partial dynamical system. Given any subset $W \subseteq A$,

(i) *let K be the ideal of $A \rtimes G$ generated by $\iota(W)$, where ι is the map defined in (11.13), and*
(ii) *let J be the smallest[29] α-invariant ideal of A containing W.*

Then K coincides with $J \rtimes G$ (or rather, its canonical image within $A \rtimes G$). In addition, there exists a *-isomorphism

$$\varphi : \frac{A \rtimes G}{K} \to \Big(\frac{A}{J}\Big) \rtimes G,$$

such that

$$\varphi(a\delta_g + K) = (a + J)\delta_g, \quad \forall\, g \in G, \quad \forall\, a \in A_g.$$

[29]It is easy to see that an arbitrary intersection of invariant ideals is again invariant, so the smallest invariant ideal always exists.

PROOF. We first claim that $\iota^{-1}(K)$ is an α-invariant ideal. While it is evident that $\iota^{-1}(K)$ is an ideal, we still need to prove it to be α-invariant. Given $a \in \iota^{-1}(K) \cap A_{g^{-1}}$, we must check that $\alpha_g(a) \in \iota^{-1}(K)$, which is to say that $\iota\big(\alpha_g(a)\big) \in K$. Using Cohen-Hewitt, write $a = bc$, where both b and c lie in $\iota^{-1}(K) \cap A_{g^{-1}}$. Then

$$\iota\big(\alpha_g(a)\big) = \alpha_g(a)\delta_1 = \alpha_g(bc)\delta_1 \overset{(8.14)}{=} \big(\alpha_g(b)\delta_g\big)(c\delta_1) = \big(\alpha_g(b)\delta_g\big)\iota(c) \in K.$$

So $\iota^{-1}(K)$ is indeed an invariant ideal, which evidently contains W. Since J is the smallest among such ideals, we have that $J \subseteq \iota^{-1}(K)$, which is to say that $\iota(J) \subseteq K$, whence

$$\iota(W) \subseteq \iota(J) \subseteq K. \tag{22.10.1}$$

This implies that K coincides with the ideal of $A \rtimes G$ generated by $\iota(J)$. On the other hand notice that, by (16.27), the ideal generated by $\iota(J)$ is $J \rtimes G$, thus proving that $K = J \rtimes G$. The second and last assertion in the statement is now an immediate consequence of (22.9). □

CHAPTER 23

Ideals in Graded Algebras

Let B be a graded C*-algebra with grading $\{B_g\}_{g\in G}$. If J is an ideal (always assumed to be closed and two-sided) in B, there might be no relationship between J and the grading of B. It is even possible that $J \cap B_g$ is trivial for every g in G.

For example, let $\mathscr{B}$ be the group bundle over $\mathbb{Z}$, so that $C^*(\mathscr{B})$ is isomorphic to $C(\mathbb{T})$, where $\mathbb{T}$ denotes the unit circle. Fixing $z_0 \in \mathbb{T}$, the ideal

$$J = \{f \in C(\mathbb{T}) : f(z_0) = 0\}$$

has trivial intersection with every homogeneous subspace $B_n = \mathbb{C}z^n$. This is because a nonzero element in B_n is invertible, and hence cannot belong to any proper ideal.

The purpose of this chapter is thus to study the relationship between ideals in graded algebras and the grading itself. For this we shall temporarily

▶ fix a graded C*-algebra B, with grading $\{B_g\}_{g\in G}$, and a closed two-sided ideal $J \trianglelefteq B$.

Proposition 23.1. *The closed two-sided ideal of B generated by $J \cap B_1$ coincides with the closure of $\bigoplus_{g\in G} J \cap B_g$.*

PROOF. Given g and h in G, notice that

$$(J \cap B_g)B_h \subseteq (JB_h) \cap (B_gB_h) \subseteq J \cap B_{gh}.$$

Therefore we see that

$$K := \overline{\bigoplus_{g\in G} J \cap B_g}$$

is invariant under right multiplication by elements of B_h, and a similar reasoning shows invariance under left multiplication as well. Since K is closed by definition, we then deduce that K is a two-sided ideal. Consequently, noticing that $J \cap B_1$ is contained in K, we have that

$$\langle J \cap B_1\rangle \subseteq K,$$

where the angle brackets above indicate the closed two-sided ideal generated by $J \cap B_1$.

In order to prove the reverse inclusion, given g in G, and $x \in J \cap B_g$, notice that $x^*x \in J \cap B_1$. Using (15.3) (which is stated for Hilbert modules, and hence also holds for C*-algebras) we then have

$$x = \lim x(x^*x)^{1/n} \in \langle J \cap B_1\rangle.$$

We therefore conclude that

$$J \cap B_g \subseteq \langle J \cap B_1\rangle,$$

from where it follows that $K \subseteq \langle J \cap B_1\rangle$, as desired. □

This justifies the introduction of the following concept:

Definition 23.2. We shall say that J is an *induced ideal*[30] (sometimes also called a *graded ideal*) provided any one of the following equivalent conditions hold:

(a) J coincides with the ideal generated by $J \cap B_1$,

(b) $\bigoplus_{g \in G} J \cap B_g$ is dense in J.

For topologically graded algebras there is a lot more to be said, so we shall assume from now on that B is topologically graded. Recall from (19.6) that in this case B admits *Fourier coefficient operators*

$$F_g : B \to B_g, \quad \forall\, g \in G,$$

such that

$$F_g(b) = \delta_{g,h} b, \quad \forall\, g, h \in G, \quad \forall\, b \in B_h.$$

Given an arbitrary ideal $J \trianglelefteq B$, let us consider the following subsets of B:

$$\begin{aligned} J' &= \langle J \cap B_1 \rangle, \\ J'' &= \{x \in B : F_g(x) \in J,\ \forall g \in G\}, \\ J''' &= \{x \in B : F_1(x^*x) \in J\}, \end{aligned} \tag{23.3}$$

where the angle brackets in the definition of J' are again supposed to mean the closed two-sided ideal generated.

Proposition 23.4. *Given any ideal J in a topologically graded C^*-algebra B, one has that the sets J', J'' and J''' defined above are closed two-sided ideals in B, and moreover*

$$J' \subseteq J'' = J'''.$$

PROOF. It is evident that these are closed subspaces of B, and moreover that J' is an ideal.

In order to prove that J'' is an ideal, let $b \in B$, and $x \in J''$, and let us prove that $bx \in J''$. Since the B_g span a dense subspace of B, we may assume that $b \in B_h$, for some $h \in G$. Then

$$F_g(bx) \overset{(19.6)}{=} bF_{h^{-1}g}(x) \in J,$$

proving that $bx \in J''$, and hence that J'' is a left ideal. One similarly proves that J'' is a right ideal. Since we will eventually prove that $J''' = J''$, we skip the proof that J''' is an ideal for now.

Observing that $J \cap B_1$ is contained in J'', and since we now know that J'' is an ideal, the ideal generated by $J \cap B_1$, namely J', is also contained in J''.

Given any $x \in B$, notice that by Parseval's identity (17.15), we have that

$$\sum_{g \in G} F_g(z)^* F_g(z) = F_1(z^*z).$$

In fact (17.15) refers to the E_g, but since $F_g = E_g \circ \psi$ (see the proof of (19.6)), our identity follows easily from (17.15). Since ideals are hereditary, we then have that

$$F_1(z^*z) \in J \ \Leftrightarrow\ F_g(z)^* F_g(z) \in J,\ \forall g \in G,$$

and we notice that the condition in the right-hand side above is also equivalent to $F_g(z) \in J$. This proves that $J'' = J'''$. □

[30] This terminology is inspired by [**85**, 1.3].

Having seen how the ideals defined in (23.3) relate to each other, let us also discuss how do they relate to J, itself. It is elementary to see that J always contains J', but the relationship between J and J'' is not straightforward. We will see below that $J' = J''$ under certain conditions, in which case it will follow that $J'' \subseteq J$. However there are examples in which J'' is not a subset of J, and in fact it may occur that, on the contrary, J is a proper subset of J''.

This is the case, for example, if B is the full group C*-algebra of a non-amenable group G, and $J = \{0\}$. One may then prove that J'' is the kernel of the regular representation, hence J'' is strictly larger than J.

One might suspect that the culprit for this anomaly is the failure of faithfulness of the standard conditional expectation on $C^*(G)$, but examples may also be found in topologically graded C*-algebras with faithful conditional expectations. Take, for example, a group G and a C*-algebra B. One may then prove that $B\otimes_{\min}C^*_{\text{red}}(G)$ is isomorphic to the reduced crossed product of B by G under the trivial action, so the former is a topologically G-graded C*-algebra with faithful conditional expectation by (17.13).

Assuming that G is a non-exact group, one may find a short exact sequence of C*-algebras

$$0 \to J \xrightarrow{i} B \xrightarrow{\pi} A \to 0,$$

for which

$$0 \to J \underset{\min}{\otimes} C^*_{\text{red}}(G) \xrightarrow{i\otimes 1} B \underset{\min}{\otimes} C^*_{\text{red}}(G) \xrightarrow{\pi\otimes 1} A \underset{\min}{\otimes} C^*_{\text{red}}(G) \to 0,$$

is not exact. It is well known that the only place where exactness may fail is at the mid point of this sequence, meaning that the range of $i \otimes 1$ is properly contained in the kernel of $\pi \otimes 1$. Letting J be the range of $i \otimes 1$, one may prove that J'' is the kernel of $\pi \otimes 1$, whence J is properly contained in J'', as claimed. Consequently we have that

$$J' \subseteq J \subsetneq J'', \tag{23.5}$$

so this also produces an example in which J'' is strictly larger than J'.

Proposition 23.6. *Let B be a topologically graded C*-algebra with grading $\{B_g\}_{g\in G}$, and assume that the associated Fell bundle has the approximation property. Then, for every ideal $J \trianglelefteq B$, one has that the ideals J' and J'' defined in* (23.3) *are equal.*

PROOF. It clearly suffices to prove that $J'' \subseteq J'$. Given $x \in J''$, we have by definition that each $F_g(x) \in J$, and we claim that $F_g(x) \in J'$. In order to see this, notice that

$$F_g(x)^*F_g(x) \in J \cap B_1 \subseteq J',$$

so we have that $F_g(x)^*F_g(x) \equiv 0 \pmod{J'}$. Since B/J' is a C*-algebra, we have that $F_g(x) \equiv 0 \pmod{J'}$, as well, meaning that $F_g(x) \in J'$, thus proving our claim.

Let $\{a_i\}_i$ be a Cesaro net for $\mathscr{B}$, and let $\{S_i\}_i$ be the net of summation processes provided by (20.10). A glimpse at the formula defining S_i is enough to convince ourselves that $S_i(x)$ is also in J', hence also

$$x = \lim_i S_i(x) \in J'.$$

□

There is another situation in which we may guarantee the coincidence of the ideals J' and J''.

Theorem 23.7. *Let G be a discrete group and let B be a topologically G-graded C^*-algebra. Suppose that G is exact and that the standard conditional expectation $F : B \to B_1$ is faithful. Then for every ideal J of B one has that the ideals J' and J'' defined in (23.3) coincide.*

PROOF. Denote by $\mathscr{B} = \{B_g\}_{g\in G}$ the underlying Fell bundle and note that B is isomorphic to $C^*_{\text{red}}(\mathscr{B})$ by (19.8). For each g in G, let $J_g = J \cap B_g$ so that $\mathcal{J} := \{J_g\}_{g\in G}$ is an ideal in $\mathscr{B}$. Employing (21.18) we have that the sequence

$$0 \to C^*_{\text{red}}(\mathcal{J}) \xrightarrow{\iota_{\text{red}}} B \xrightarrow{q_{\text{red}}} C^*_{\text{red}}(\mathscr{B}/\mathcal{J}) \to 0 \tag{23.7.1}$$

is exact. We next claim that

$$J' = \iota_{\text{red}}\big(C^*_{\text{red}}(\mathcal{J})\big), \quad \text{and} \quad J'' = \text{Ker}(q_{\text{red}}). \tag{23.7.2}$$

Given g in G, notice that

$$J_g^* J_g \subseteq B_1 \cap J \subseteq J',$$

so by (16.12) one has that

$$J_g = [J_g J_g^* J_g] \subseteq [J_g J'] \subseteq J',$$

so $\iota_{\text{red}}\big(C^*_{\text{red}}(\mathcal{J})\big) \subseteq J'$. Since the reverse inclusion is evident, we have proven the first identity in (23.7.2).

On the other hand, denoting by E the faithful standard conditional expectation of $C^*_{\text{red}}(\mathscr{B}/\mathcal{J})$, it is easy to see that $E \circ q_{\text{red}} = q_{\text{red}} \circ F$, so for any b in B we have that

$$q_{\text{red}}(b) = 0 \;\Leftrightarrow\; E\big(q_{\text{red}}(b^*b)\big) = 0 \;\Leftrightarrow\; q_{\text{red}}\big(F(b^*b)\big) = 0 \;\Leftrightarrow\; F(b^*b) \in J_1,$$

where the last step is a consequence of the fact that $F(b^*b)$ is in B_1, and that the behavior of q on B_1 is that given by (21.3). This shows that $\text{Ker}(q_{\text{red}}) = J''$, concluding the verification of (23.7.2).

Since the sequence (23.7.1) is exact, the proof follows. □

Definition 23.8. Let B be a topologically graded C*-algebra with grading $\{B_g\}_{g\in G}$ and Fourier coefficient operators F_g. We will say that a closed, two-sided ideal $J \trianglelefteq B$ is a *Fourier ideal*, if $F_g(J) \subseteq J$, for every g in G.

Thus J is a Fourier ideal if and only if $J \subseteq J''$, while J is induced if and only if $J = J'$. We may thus reinterpret (23.4), (23.6) and (23.7) as follows.

Proposition 23.9. *Let B be a topologically graded C^*-algebra. Then every induced ideal of B is a Fourier ideal. Moreover, the converse holds if either*

(i) *the associated Fell bundle has the approximation property, or*
(ii) *G is exact and the standard conditional expectation of B onto B_1 is faithful.*

For an example of a Fourier ideal which is not induced, see (23.5).

The next result is stated for Fourier ideals but, because of the reasoning above, it also holds for induced ones.

Proposition 23.10. *Let B be a topologically graded C^*-algebra with grading $\{B_g\}_{g\in G}$, and let J be a Fourier ideal of B. Then B/J is topologically graded by the spaces $q(B_g)$, where q is the quotient map.*

PROOF. Since J is invariant under each Fourier coefficient operator F_g, we have that F_g passes to the quotient giving a well defined bounded map on B/J, namely

$$\tilde{F}_g(x+J) = F_g(x) + J, \quad \forall x \in B.$$

Notice that

$$q(B_g) = \tilde{F}_g(B/J) = \text{Ker}(id - \tilde{F}_g),$$

the last step holding thanks to the fact that $\tilde{F}_g$ is idempotent. As a consequence we deduce that $q(B_g)$ is a closed subspace of B/J.

It is now immediate to verify that the collection $\{q(B_g)\}_{g\in G}$ satisfies (19.1.i)–(19.1.iii), and that $\tilde{F}_1$ fills in the rest of the hypothesis there to allow us to conclude that this is in fact a topological grading for B/J. □

All of the above results have their versions in the setting of partial crossed product algebras, since these are graded algebras. The next simple result, which we will use later, has no counterpart for graded algebras since its conclusion explicitly mentions the partial action.

Proposition 23.11. *Let $\theta = \big(A, G, \{D_g\}_{g\in G}, \{\theta_g\}_{g\in G}\big)$ be a C^*-algebraic partial dynamical system. Given an ideal J of either $A \rtimes G$ or $A \rtimes_{red} G$, let $K = J \cap A$. Then K is a θ-invariant ideal of A.*

PROOF. Of course we are identifying A with its copy $A\delta_1$ in the crossed product algebra. Regardless of whether we are working with the full or reduced crossed product, the proof is the same: given a in $K \cap D_{g^{-1}}$, choose an approximate identity $\{v_i\}_i$ for D_g. Then

$$J \ni (v_i\delta_g)(a\delta_1)(v_i\delta_g)^* \overset{(8.14)}{=} v_i\theta_g(a)v_i^*\delta_1 \overset{i\to\infty}{\longrightarrow} \theta_g(a)\delta_1,$$

so $\theta_g(a)$ is in K, proving the statement. □

Notes and remarks. The motivation for this chapter comes from Nica's work on induced ideals of algebras of Wiener-Hopf operators [**85**, Section 6], which in turn is inspired by Strătilă and Voiculescu's work on AF-algebras [**104**]. Propositions (23.4) and (23.6) have been proven in [**49**], while Theorem (23.7) is from [**52**, Theorem 5.1][31].

[31] Please note that [**52**] is the preprint version of [**53**].

CHAPTER 24

Pre-Fell-Bundles

In this short chapter we will develop some algebraic aspects of the theory of Fell bundles. We will begin by introducing the notion of an *algebraic* Fell bundle and the main question we shall analyze is whether or not they admit a norm with which one may obtain a classical Fell bundle.

The motivation for this is the study of tensor products of Fell bundles by C*-algebras which we will do in the next chapter.

Definition 24.1. An *algebraic Fell bundle* over a group G is a collection

$$\mathscr{C} = \{C_g\}_{g\in G}$$

of complex vector spaces, such that the disjoint union of all the C_g's, which we also denote by $\mathscr{C}$, by abuse of language, is moreover equipped with a multiplication operation and an involution

$$\cdot : \mathscr{C} \times \mathscr{C} \to \mathscr{C}, \qquad\qquad * : \mathscr{C} \to \mathscr{C},$$

satisfying the following properties for all g and h in G, and all b and c in $\mathscr{C}$:

(a) $C_g C_h \subseteq C_{gh}$,
(b) multiplication is bi-linear from $C_g \times C_h$ to C_{gh},
(c) multiplication on $\mathscr{C}$ is associative,

(e) $(C_g)^* \subseteq C_{g^{-1}}$,
(f) involution is conjugate-linear from C_g to $C_{g^{-1}}$,
(g) $(bc)^* = c^* b^*$,
(h) $b^{**} = b$.

Compared to (16.1) observe that axioms (d) and (i–k) are missing since these refer to norms which are absent in the present case. Our goal here is to furnish a norm on each C_g with respect to which the completions form a Fell bundle.

When $G = \{1\}$, an algebraic Fell bundle consists of a single *-algebra and the reader might know that providing a C*-norm on a *-algebra is not a problem with a straightforward solution. In the case of a general group G, the above axioms imply that C_1 is a *-algebra so we should therefore expect our construction of norms on the C_g to require at least an initial choice of a C*-norm on C_1.

If we are indeed given a C*-norm on C_1, we may consider the completion

$$B_1 := \overline{C}_1$$

with respect to this norm, which will therefore be a C*-algebra. Given c in any C_g, we may view c^*c as an element of B_1, however without further hypothesis there is no reason why c^*c is positive in B_1. Without this positivity condition there is clearly no hope of turning our algebraic Fell bundle into a Fell bundle.

Still assuming that $c \in C_g$, consider the mapping

$$a \in C_1 \mapsto c^*ac \in C_1.$$

Again there seems to be no reason why this is continuous with respect to the given norm on C_1 and again this continuity is a necessary condition for us to proceed.

Definition 24.2. A *pre-Fell-bundle* over a group G is an algebraic Fell bundle

$$\mathscr{C} = \{C_g\}_{g\in G}$$

equipped with a C*-norm $\|\cdot\|$ on C_1 such that, for every c in any C_g, one has that

(i) c^*c is a positive element in the C*-algebra B_1 obtained by completing C_1 relative to the norm given above,

(ii) the mapping

$$Ad_c : a \in C_1 \mapsto c^*ac \in C_1,$$

is continuous with respect to $\|\cdot\|$.

Should one prefer to avoid any reference to the completed algebra B_1 in (24.2.i), above, one could instead require that for every $c \in C_g$, there exists a sequence $\{a_n\}_n \subseteq C_1$, such that $\|a_n^*a_n - c^*c\| \overset{i\to\infty}{\longrightarrow} 0$.

► From now on we will fix a pre-Fell-bundle $\mathscr{C} = \{C_g\}_{g\in G}$, and we will look for suitable norms $\|\cdot\|_g$ on the other C_g's with the intent of obtaining a Fell bundle. In view of (16.1.j), we have no choice but to define

$$\|c\|_g = \|c^*c\|^{\frac{1}{2}}, \quad \forall c \in C_g,$$

and our task is then to check the remaining axioms. We begin with some technical results.

Proposition 24.3. *For every c in any C_g, one has that $\|cc^*\| = \|c^*c\|$.*

Proof. Recall that a self-adjoint element x in a C*-algebra satisfies $\|x^n\| = \|x\|^n$, for every $n \in \mathbb{N}$. This identity also holds for our C*-norm on C_1, as it follows from the corresponding fact applied to the completed algebra B_1. So

$$\begin{aligned} \|cc^*\|^{n+1} &= \|(cc^*)^{n+1}\| \\ &= \|c(c^*c)^n c^*\| \\ &= \|Ad_c\big((c^*c)^n\big)\| \\ &\le \|Ad_c\|\|(c^*c)^n\| \\ &\le \|Ad_c\|\|c^*c\|^n. \end{aligned}$$

Taking n^{th} root leads to

$$\|cc^*\|^{\frac{n+1}{n}} \le \|Ad_c\|^{\frac{1}{n}}\|c^*c\|,$$

and when $n \to \infty$, we get

$$\|cc^*\| \le \|c^*c\|.$$

The reverse inequality follows by replacing c with c^*. □

Let us now prove a technical result, reminiscent of (16.9).

Lemma 24.4. *Let $\{v_i\}_i$ be an approximate identity for C_1. Then, for every c in any C_g, one has that*

$$\lim_i c^*v_ic = c^*c.$$

PROOF. Using that $\|\cdot\|$ is a C*-norm, we have

$$\begin{aligned}\|c^*v_ic - c^*c\|^2 &= \|(c^*v_ic - c^*c)^*(c^*v_ic - c^*c)\| \\ &= \|c^*v_i^*cc^*v_ic - c^*v_i^*cc^*c - c^*cc^*v_ic + c^*cc^*c\| \\ &= \|Ad_c(v_i^*cc^*v_i - v_i^*cc^* - cc^*v_i + cc^*)\| \overset{i\to\infty}{\longrightarrow} 0.\end{aligned}$$

□

With this we may prove a version of (17.2).

Lemma 24.5. *Given c in any C_g, one has that*

$$c^*xc \leq \|x\|c^*c, \quad \forall x \in C_1 \cap B_{1+},$$

where B_{1+} denotes the set of positive elements of the C^*-algebra B_1 obtained by completing C_1, and the order relation "$\leq$" is that of B_1.

PROOF. Given c in C_g, we claim that

$$x \in C_1 \cap B_{1+} \quad \Rightarrow \quad c^*xc \in B_{1+}. \tag{24.5.1}$$

In fact, for each x in $C_1 \cap B_{1+}$, write $x = y^*y$, with $y \in B_1$, and choose a sequence $\{z_n\}_n \subseteq C_1$, converging to y. Then

$$c^*xc = Ad_c\big(\lim_{n\to\infty} z_n^*z_n\big) \overset{(24.2.ii)}{=} \lim_{n\to\infty} Ad_c(z_n^*z_n) = \lim_{n\to\infty} c^*z_n^*z_nc \overset{(24.2.i)}{\in} B_{1+},$$

proving (24.5.1). Next choose an approximate identity $\{v_i\}_i$ for B_1, contained in C_1. Then, for $x \in C_1 \cap B_{1+}$, and for any i we have that

$$v_i^*(\|x\| - x)v_i \in C_1 \cap B_{1+},$$

whence by (24.5.1) we have

$$B_{1+} \ni c^*v_i^*(\|x\| - x)v_ic = \|x\|c^*v_i^*v_ic - c^*v_i^*xv_ic. \tag{23.5.2}$$

Observing that $\{v_i^*v_i\}_i$ is also an approximate identity for B_1, we deduce from (24.4) that

$$c^*v_i^*v_ic \overset{i\to\infty}{\longrightarrow} c^*c,$$

and from (24.2.ii) we have

$$c^*v_i^*xv_ic \overset{i\to\infty}{\longrightarrow} c^*xc.$$

Taking the limit as $i \to \infty$ in (23.5.2) we finally obtain

$$\|x\|c^*c - c^*xc \in B_{1+}.$$

□

We now have all of the necessary tools in order to prove the main result of this chapter:

Theorem 24.6. *Given a pre-Fell-bundle $\mathscr{C} = \{C_g\}_{g\in G}$, there is a unique family of seminorms $\|\cdot\|_g$ on the C_g's, such that $\|\cdot\|_1$ is the given norm on C_1, and for $g, h \in G$, $b \in C_g$, and $c \in C_h$, one has that*

(d) $\|bc\|_{gh} \leq \|b\|_g\|c\|_h$,
(i) $\|b^*\|_{g^{-1}} = \|b\|_g$,
(j) $\|b^*b\|_1 = \|b\|_g^2$.

PROOF. Define $\|\cdot\|_1$ to coincide with the given norm on C_1, and for all $g \neq 1$, and all $b \in C_g$, set

$$\|b\|_g = \|b^*b\|^{\frac{1}{2}}.$$

If $b \in C_g$ and $c \in C_h$, we have

$$(bc)^*bc = c^*b^*bc \overset{(24.5)}{\leq} \|b^*b\|c^*c,$$

whence

$$\|bc\|_{gh}^2 = \|(bc)^*bc\| \leq \|b^*b\|\|c^*c\| = \|b\|_g^2\|c\|_h^2,$$

proving (d). Notice that (i) follows from (24.3), while (j) follows by definition when $g \neq 1$, and otherwise from the fact that the norm on C_1 is assumed to be a C*-norm.

To conclude we prove the triangle inequality: given $b, c \in C_g$, we have

$$\begin{aligned}
\|b + c\|_g^2 &= \|(b + c)^*(b + c)\| \\
&= \|b^*b + b^*c + c^*b + c^*c\| \\
&\leq d\|b^*\|_{g^{-1}}\|b\|_g + \|b^*\|_{g^{-1}}\|c\|_g + \|c^*\|_{g^{-1}}\|b\|_g + \|c^*\|_{g^{-1}}\|c\|_g \\
&= i\|b\|_g\|b\|_g + \|b\|_g\|c\|_g + \|c\|_g\|b\|_g + \|c\|_g\|c\|_g \\
&= (\|b\|_g + \|c\|_g)^2.
\end{aligned}$$

This concludes the proof of the existence part, while uniqueness follows easily from (j). □

Given a pre-Fell-bundle $\mathscr{C} = \{C_g\}_{g\in G}$, one may then consider the completion of each C_g under $\|\cdot\|_g$, say

$$B_g = \overline{C}_g,$$

and, after extending the operations by continuity, the verification of axioms (16.1.i)–(16.1.k) becomes routine, especially since they are already known to hold on dense sets. We thus obtain a Fell bundle $\mathscr{B} = \{B_g\}_{g\in G}$.

Definition 24.7. The Fell bundle $\mathscr{B}$, obtained as above from a given pre-Fell-bundle $\mathscr{C}$, will be called the *completion* of $\mathscr{C}$.

The question of extending representations from a pre-Fell-bundle to its completion is an important one which we will discuss next.

Proposition 24.8. *Let $\mathscr{C} = \{C_g\}_{g\in G}$ be a pre-Fell-bundle and let A be a C*-algebra. Suppose we are given a collection of linear maps $\pi = \{\pi_g\}_{g\in G}$, where*

$$\pi_g : C_g \to A,$$

such that

(i) $\pi_g(b)\pi_h(c) = \pi_{gh}(bc)$,
(ii) $\pi_g(b)^* = \pi_{g^{-1}}(b^*)$,

for all $g, h \in G$, and all $b \in C_g$, and $c \in C_h$. Suppose, moreover, that π_1 is continuous relative to the norm on C_1. Then each π_g is continuous relative to the norm $\|\cdot\|_g$ on C_g, and hence it extends to a bounded linear map

$$\tilde{\pi}_g : B_g \to A,$$

where $\mathscr{B} = \{B_g\}_{g\in G}$ is the Fell bundle completion of $\mathscr{C}$. In addition the collection of maps $\tilde{\pi} = \{\tilde{\pi}_g\}_{g\in G}$ is a representation of $\mathscr{B}$ in A.

PROOF. For c in any C_g we have

$$\|\pi_g(c)\|^2 = \|\pi_g(c)^*\pi_g(c)\| \overset{\text{(i\&ii)}}{=} \|\pi_1(c^*c)\| \leq \|c^*c\|_1 = \|c\|_g^2,$$

proving π_g to be continuous. The remaining statements are verified in a routine way. □

CHAPTER 25

Tensor Products of Fell Bundles

Tensor products form a very important part of the theory of C*-algebras and, given the very close relationship between C*-algebras and Fell bundles, no treatment of Fell bundles is complete without a careful study of their tensor products.

The most general form of this theory would start by considering two Fell bundles $\mathscr{B}$ and $\mathscr{B}'$, over two groups G and G', respectively, and one would then attempt to construct a Fell bundle $\mathscr{B}\otimes\mathscr{B}'$ over the group $G\times G'$. However, given the applications we have in mind, we will restrict ourselves to the special case in which G' is a trivial group. In other words, we will restrict our study to tensor products of Fell bundles by single C*-algebras.

Among the main aspects of tensor products we plan to analyze is the relationship between the corresponding versions of spatial and maximal norms.

► Let us fix a C*-algebra A and a Fell bundle $\mathscr{B}=\{B_g\}_{g\in G}$. For each g in G, let us consider the vector space tensor product

$$C_g := A\odot B_g.$$

Given g and h in G, it is easy to see that there exists a bi-linear operation $C_g\times C_h\to C_{gh}$, such that

$$(a\otimes b)(a'\otimes b')=(aa')\otimes(bb'),\quad \forall\, a,a'\in A,\quad \forall\, b\in B_g,\quad \forall\, b'\in B_h.$$

Likewise one may show the existence of a conjugate-linear map * from C_g to $C_{g^{-1}}$, such that

$$(a\otimes b)^*=a^*\otimes b^*,\quad \forall\, a\in A,\quad \forall\, b\in B_g.$$

One then easily checks that the collection

$$A\odot\mathscr{B}=\{C_g\}_{g\in G} \tag{25.1}$$

is an algebraic Fell bundle with the above operations.

The next result is intended to aid the verification of (24.2.i), once we have a C*-norm on C_1.

Proposition 25.2. *Given c in any C_g, there exist $x_1,\ldots,x_n\in C_1$, such that*

$$c^*c=\sum_{i=1}^n x_i^*x_i.$$

PROOF. Writing $c=\sum_{i=1}^n a_i\otimes b_i$, with $a_i\in A$, and $b_i\in B_g$, we claim that the $n\times n$ matrix

$$m=\{b_i^*b_j\}_{i,j}\in M_n(B_1),$$

is positive. To prove it notice that, viewed as a matrix over the cross sectional C*-algebra of $\mathscr{B}$, we have that $m=y^*y$, where y is the row matrix

$$y=[b_1\ b_2\ \ldots\ b_n].$$

Thus m is positive as a matrix over $C^*(\mathscr{B})$, but since B_1 is a subalgebra of $C^*(\mathscr{B})$ by (17.9.iv), and since the coefficients of m lie in B_1, we have that m is positive as a matrix over B_1. So there exists h in $M_n(B_1)$, such that $m = h^*h$, which translates into

$$b_i^*b_j = \sum_{k=1}^{n}(h^*)_{i,k}h_{k,j} = \sum_{k=1}^{n}(h_{k,i})^*h_{k,j}, \quad \forall\, i,j = 1,\ldots,n.$$

We then have

$$\begin{aligned} c^*c &= \Big(\sum_{i=1}^{n} a_i \otimes b_i\Big)^*\Big(\sum_{j=1}^{n} a_j \otimes b_j\Big) \\ &= \sum_{i,j=1}^{n} a_i^*a_j \otimes b_i^*b_j \\ &= \sum_{i,j,k=1}^{n} a_i^*a_j \otimes (h_{ki})^*h_{kj} \\ &= \sum_{k=1}^{n}\sum_{i,j=1}^{n} (a_i\otimes h_{ki})^*(a_j\otimes h_{kj}) \\ &= \sum_{k=1}^{n}\Big(\sum_{i=1}^{n} a_i\otimes h_{ki}\Big)^*\Big(\sum_{j=1}^{n} a_j\otimes h_{kj}\Big) \\ &= \sum_{k=1}^{n} x_k^*x_k, \end{aligned}$$

where $x_k = \sum_{i=1}^{n} a_i \otimes h_{ki}$. □

We then see that axiom (24.2.i) will hold for any choice of C*-norm on C_1.

Proposition 25.3. *Let $\|\cdot\|_{max}$ and $\|\cdot\|_{min}$ be the maximal and minimal C*-norms on the algebraic tensor product $A\odot B_1$, respectively. Then $A\odot\mathscr{B}$ is a pre-Fell-bundle with either one of these norms.*

PROOF. After the remark in the paragraph just before the statement, it is now enough to prove that, for every a and a' in A, and every b and b' in any B_g, the mapping

$$\rho\ :\ t\in C_1 \mapsto (a\otimes b)^*t(a'\otimes b') \in C_1,$$

is continuous with respect to both the maximal and minimal C*-norms. If $x \in A$, and $y \in B_1$, notice that

$$\rho(x\otimes y) = a^*xa' \otimes b^*yb',$$

thus ρ is seen to be the tensor product of the maps φ and ψ given by

$$\varphi : x\in A \mapsto a^*xa' \in A, \quad \text{and} \quad \psi : y \in B_1 \mapsto b^*yb' \in B_1.$$

By the polarization identity we may write φ as a linear combination of four maps of the form

$$\varphi_i : x\in A\mapsto a_i^*xa_i \in A,$$

with $a_i \in A$, and likewise ψ may be written as a linear combination of four maps of the form

$$\psi_j : y\in B_1 \mapsto b_j^*yb_j \in B_1,$$

with $b_j \in B_g$. Consequently ρ may be written as a linear combination of the sixteen maps $\varphi_i\odot\psi_j$, and it is therefore enough to show that these are continuous. Since both the φ_i and the ψ_j are completely positive maps, the result follows from [**19**, Theorem 3.5.3]. □

Definition 25.4. Given a C*-algebra A and a Fell bundle $\mathscr{B} = \{B_g\}_{g\in G}$, let τ be any C*-norm on $A \odot B_1$, with respect to which $A \odot \mathscr{B}$ is a pre-Fell-bundle. Then its completion, according to (24.7), will be denoted by $A \otimes_\tau \mathscr{B}$. For each g in G, we will denote the corresponding fiber of $A \otimes_\tau \mathscr{B}$ by $A \otimes_\tau B_g$. In the case of the maximal and minimal C*-norms on $A \odot B_1$, we will respectively denote the corresponding completions by

$$A \underset{\max}{\otimes} \mathscr{B}, \quad \text{and} \quad A \underset{\min}{\otimes} \mathscr{B},$$

with fibers

$$A \underset{\max}{\otimes} B_g, \quad \text{and} \quad A \underset{\min}{\otimes} B_g,$$

for each g in G.

We are in fact only interested in the maximal and minimal norms, but nevertheless there is a lot of current interest in studying *exotic norms* [**21**].

Employing similar methods one could also study the tensor product of a Fell bundle $\mathscr{B} = \{B_g\}_{g\in G}$ over a group G, by a Fell bundle $\mathscr{C} = \{C_h\}_{h\in H}$ over a group H, obtaining a Fell bundle

$$\mathscr{B} \otimes \mathscr{C} = \{\overline{B_g \otimes C_h}\}_{(g,h)\in G\times H}$$

over the group $G \times H$. This will of course require an appropriate choice of norm on $B_1 \odot C_1$. We will however not pursue these ideas here.

Proposition 25.5. *Given a Fell bundle $\mathscr{B} = \{B_g\}_{g\in G}$, let C and D be C*-algebras and suppose that τ and ν are C*-norms on $C\odot B_1$ and $D\odot B_1$, respectively, making $C \odot \mathscr{B}$ and $D \odot \mathscr{B}$ into pre-Fell-bundles. Suppose moreover that $\varphi : C \to D$ is a *-homomorphism such that*

$$\varphi \otimes id : C \odot B_1 \to D \odot B_1$$

is continuous relative to τ and ν. Then there exists a morphism $\tilde{\varphi} = \{\varphi_g\}_{g\in G}$ from $C \otimes_\tau \mathscr{B}$ to $D \otimes_\nu \mathscr{B}$, such that

$$\varphi_g(c \otimes b) = \varphi(c) \otimes b,$$

for every b in any B_g.

Proof. Given g in G, define

$$\varphi_g^0 : C \odot B_g \to D \odot B_g$$

by $\varphi_g^0 = \varphi \otimes id$. It is then easy to see that the collection of maps

$$\widetilde{\varphi^0} = \{\varphi_g^0\}_{g\in G}$$

satisfy (21.1.i) & (21.1.ii), relative to the algebraic Fell bundle structures of $C \odot B_g$ and $D \odot B_g$ given in (25.1). We moreover claim that each φ_g^0 is continuous relative to the norms $\|\cdot\|_g^\tau$ and $\|\cdot\|_g^\nu$, given by (24.6) on $C \odot B_g$ and $D \odot B_g$, respectively.

In order to verify the claim, notice that the case $g = 1$ is granted by hypothesis. For an arbitrary g in G, pick any x in $C \odot B_g$, and observe that

$$\left(\|\varphi_g^0(x)\|_g^\nu\right)^2 = \|\varphi_g^0(x)^*\varphi_g^0(x)\|_1^\nu = \|\varphi_1^0(x^*x)\|_1^\nu \leq \|x^*x\|_1^\tau = \left(\|x\|_g^\tau\right)^2,$$

where the inequality above is a consequence of the already verified case $g = 1$. We may therefore extend each φ_g^0 to a continuous mapping

$$\varphi_g : C \otimes_\tau B_g \to D \otimes_\nu B_g,$$

and the reader may then easily verify that the resulting collection of maps $\tilde{\varphi} = \{\varphi_g\}_{g\in G}$ satisfies the required conditions. □

Recall that any Fell bundle admits a *universal* representation in its full cross sectional C*-algebra by (16.26), as well a *regular* representation in its reduced cross sectional C*-algebra by (17.4). In the case of $A \otimes_{\max} \mathscr{B}$ and $A \otimes_{\min} \mathscr{B}$, we believe it is best not to introduce any special notation for these representations, instead relying on (17.9.iv)–(17.9.v), which allows us to identify the fibers of our bundles as subspaces of both the full and reduced cross sectional C*-algebras. In principle this has a high risk of confusion, since when a is in A and b is in some B_g, the expression $a \otimes b$ may be interpreted as elements of eleven different spaces, namely $A \odot B_g$, $A \otimes_{\max} B_g$, $A \otimes_{\min} B_g$, $C^*(A \otimes_{\max} \mathscr{B})$, $C^*(A \otimes_{\min} \mathscr{B})$, $C^*_{\text{red}}(A \otimes_{\max} \mathscr{B})$, $C^*_{\text{red}}(A \otimes_{\min} \mathscr{B})$, besides the maximal and minimal tensor products of A by either $C^*(\mathscr{B})$ or $C^*_{\text{red}}(\mathscr{B})$. Fortunately, as we will see, the context will always suffice to determine the correct interpretation of $a \otimes b$.

Proposition 25.6. *For every a in A, and b in any B_g, one has that*

$$\|a \otimes b\| = \|a\|\|b\|,$$

for any one of the various interpretations of $a \otimes b$, except, of course, for $A \odot B_g$ which is a space devoid of a norm.

PROOF. We begin by treating the interpretation of $a \otimes b$ in $A \otimes_{\max} B_g$. By (24.6.j) we have

$$\begin{aligned}\|a \otimes b\|_g^2 &= \|(a \otimes b)^*(a \otimes b)\|_1 \\ &= \|a^*a \otimes b^*b\|_1 \\ &= \|a^*a \otimes b^*b\|_{\max} \\ &= \|a^*a\|\|b^*b\| \\ &= \|a\|^2\|b\|^2,\end{aligned}$$

where, in the penultimate step above, we have used that the norm on $A \otimes_{\max} B_1$ is a cross-norm (i.e., satisfies the identity in the statement).

A very similar argument proves the result in the case of $A \otimes_{\min} B_g$, and then (17.9.iv) takes care of $C^*(A \otimes_{\max} \mathscr{B})$ and $C^*(A \otimes_{\min} \mathscr{B})$, while (17.9.v) does it for $C^*_{\text{red}}(A \otimes_{\max} \mathscr{B})$ and $C^*_{\text{red}}(A \otimes_{\min} \mathscr{B})$.

Again by (17.9.iv)–(17.9.v), we know that the norm of b, as interpreted within $C^*(\mathscr{B})$ or $C^*_{\text{red}}(\mathscr{B})$, coincide with its norm as an element of B_g. Thus, the result relative to the maximal or minimal tensor products of A by either $C^*(\mathscr{B})$ or $C^*_{\text{red}}(\mathscr{B})$ follow, since both the maximal or minimal norms are cross-norms. □

Theorem 25.7. *Given a C^*-algebra A and a Fell bundle $\mathscr{B}$, one has that*

$$C^*(A \underset{max}{\otimes} \mathscr{B}) \simeq A \underset{max}{\otimes} C^*(\mathscr{B}),$$

via an isomorphism which sends $a \otimes b \to a \otimes \hat{\jmath}_g(b)$, for b in any B_g, and all a in A.

PROOF. For each g in G, define

$$\pi_g : A \odot B_g \to A \underset{\max}{\otimes} C^*(\mathscr{B})$$

by

$$\pi_g(a \otimes b) = a \otimes \hat{\jmath}_g(b), \quad \forall\, a \in A, \quad \forall\, b \in B_g,$$

where $\hat{\jmath}$ is the universal representation of $\mathscr{B}$ in $C^*(\mathscr{B})$.

It is clear that the π_g satisfy (24.8.i)–(24.8.ii). In addition, by the universal property of the maximal norm [**19**, Theorem 3.3.7], we have that π_1 is continuous for $\|\cdot\|_{\max}$. It then follows from (24.8) that the π_g extend to a representation $\tilde{\pi} = \{\tilde{\pi}_g\}_{g\in G}$ of $A \otimes_{\max} \mathscr{B}$ in $A \otimes_{\max} C^*(\mathscr{B})$. We then conclude from (16.29) that there exists a unique *-homomorphism

$$\varphi : C^*(A \underset{\max}{\otimes} \mathscr{B}) \to A \underset{\max}{\otimes} C^*(\mathscr{B}) \tag{25.7.1}$$

such that φ composed with the universal representation of $A \otimes_{\max} \mathscr{B}$ in $C^*(A \otimes_{\max} \mathscr{B})$ yields $\tilde{\pi}$. Employing the identifications discussed in the paragraph before (25.6), this means that

$$\varphi(a \otimes b) = a \otimes \hat{\jmath}_g(b), \quad \forall\, a \in A, \quad \forall\, g \in G, \quad \forall\, b \in B_g.$$

In order to prove that φ is an isomorphism we will construct an inverse for it. This will involve understanding a representation of $\mathscr{B}$ in the multiplier algebra of $C^*(A \otimes_{\max} \mathscr{B})$, which we will now describe. Given b in any B_g, we claim that there are bounded linear operators

$$L_b, R_b : C^*(A \underset{\max}{\otimes} \mathscr{B}) \to C^*(A \underset{\max}{\otimes} \mathscr{B}),$$

such that

$$L_b(a \otimes c) = a \otimes bc, \quad \text{and} \quad R_b(a \otimes c) = a \otimes cb,$$

for any a in A, and any c in the total space of $\mathscr{B}$.

To see this let $\{v_i\}_i$ be an approximate identity for A, and consider the operators $L_{v_i \otimes b}$ on $C^*(A \otimes_{\max} \mathscr{B})$ given by left-multiplication by $v_i \otimes b$. We then have that

$$L_{v_i \otimes b}(a \otimes c) = v_i a \otimes bc \overset{i\to\infty}{\longrightarrow} a \otimes bc.$$

Note that this convergence is guaranteed by (25.6).

Therefore the net $\{L_{v_i \otimes b}\}_i$ converges pointwise on $C_c(A \otimes_{\max} \mathscr{B})$ and, being a uniformly bounded net, it actually converges pointwise everywhere to a bounded operator which we denote by L_b, and which clearly satisfies the required conditions. The existence of R_b is proved similarly, and it is then easy to see that the pair (L_b, R_b) is a multiplier of $C^*(A \otimes_{\max} \mathscr{B})$. Moreover the mappings

$$\mu_g : b \in B_g \mapsto (L_b, R_b) \in \mathcal{M}\big(C^*(A \underset{\max}{\otimes} \mathscr{B})\big),$$

once put together, form a representation

$$\mu = \{\mu_g\}_{g\in G}$$

of $\mathscr{B}$ in $\mathcal{M}\big(C^*(A \otimes_{\max} \mathscr{B})\big)$. Feeding this representation into (16.29) provides a *-homomorphism

$$\psi : C^*(\mathscr{B}) \to \mathcal{M}\big(C^*(A \underset{\max}{\otimes} \mathscr{B})\big),$$

such that $\psi \circ \hat{\jmath}_g = \mu_g$, for all g in G. In particular

$$\psi\big(\hat{\jmath}_g(b)\big)(a \otimes c) = a \otimes bc, \quad \forall\, b \in B_g, \quad \forall\, c \in \mathscr{B}, \quad \forall\, a \in A,$$

where $\mathscr{B}$ is being used here to denote total space.

We now prove the existence of *-homomorphism

$$\chi : A \to \mathcal{M}\big(C^*(A \underset{\max}{\otimes} \mathscr{B})\big),$$

such that

$$\chi(a)(a' \otimes c) = aa' \otimes c, \quad \forall\, a, a' \in A, \quad \forall\, c \in \mathscr{B}.$$

Since the reasoning is similar to the above, we limit ourselves to sketching it. Fixing a in A and an approximate identity $\{v_i\}_i$ for B_1, one has that the left-multiplication operators $L_{a\otimes v_i}$ satisfy

$$L_{a\otimes v_i}(a'\otimes c) = aa'\otimes v_ic \overset{i\to\infty}{\longrightarrow} aa'\otimes c, \quad \forall\, a'\in A, \quad \forall\, c\in \mathscr{B},$$

by (16.9). The strong limit of the $L_{a\otimes v_i}$ therefore exists and gives the first coordinate of a multiplier pair (L_a, R_a) which we set to be $\chi(a)$.

Once the existence of χ is established, observing that the ranges of ψ and χ commute, we may employ once again the universal property of the maximal norm [**19**, Theorem 3.3.7], obtaining a *-homomorphism $\chi\times\psi$ from $A\otimes_{\max} C^*(\mathscr{B})$ to the above multiplier algebra, such that

$$(\chi\times\psi)(a\otimes y) = \chi(a)\psi(y), \quad \forall\, a\in A, \quad \forall\, y\in C^*(\mathscr{B}).$$

In particular, for $a, a'\in A$, $b\in B_g$, and $c\in\mathscr{B}$ (total space), we have

$$\Big((\chi\times\psi)\big(a\otimes \hat{\jmath}_g(b)\big)\Big)(a'\otimes c) = aa'\otimes bc.$$

This implies that

$$(\chi\times\psi)\big(a\otimes\hat{\jmath}_g(b)\big) = a\otimes b,$$

so the range of $\chi\times\psi$ is actually contained in $C^*(A\otimes_{\max}\mathscr{B})$ (or rather, in its canonical copy inside the multiplier algebra). It is now easy to see that $\chi\times\psi$ is the inverse of the map φ in (25.7.1), so the proof is concluded. □

The following is the reduced/minimal version of the result above.

Theorem 25.8. *Given a C^*-algebra A and a Fell bundle $\mathscr{B}$, one has that*

$$C^*_{red}(A \underset{min}{\otimes} \mathscr{B}) \simeq A \underset{min}{\otimes} C^*_{red}(\mathscr{B}),$$

via an isomorphism sending $a\otimes b\to a\otimes\lambda_g(b)$, for b in any B_g, and all a in A.

Proof. The first few steps of this proof are very similar to the proof of (25.7), but significant differences will appear along the way. For each g in G, define

$$\pi_g : A\odot B_g \to A \underset{\min}{\otimes} C^*_{\rm red}(\mathscr{B})$$

by

$$\pi_g(a\otimes b) = a\otimes\lambda_g(b), \quad \forall\, a\in A, \quad \forall\, b\in B_g,$$

where λ is the regular representation of $\mathscr{B}$ in $C^*_{\rm red}(\mathscr{B})$ given in (17.4).

It is clear that the π_g satisfy (24.8.i)–(24.8.ii). In addition, by [**19**, Proposition 3.6.1], we have that π_1 is isometric for the minimal norm on $A\odot B_1$, and hence also continuous. It then follows from (24.8) that the π_g extend to a representation $\tilde\pi = \{\tilde\pi_g\}_{g\in G}$ of $A\otimes_{\min}\mathscr{B}$ in $A\otimes_{\min} C^*_{\rm red}(\mathscr{B})$. Note that $\tilde\pi_1$ is then isometric on $A\otimes_{\min} B_1$.

Let us denote the integrated form of $\tilde\pi$ by

$$\rho : C^*(A \underset{\min}{\otimes} \mathscr{B}) \to A \underset{\min}{\otimes} C^*_{\rm red}(\mathscr{B})$$

so that by (16.29) we have

$$\rho(a\otimes b) = a\otimes\lambda_g(b), \quad \forall\, a\in A, \quad \forall\, b\in B_g.$$

We next apply (18.5) for $\tilde\pi$, obtaining a *-homomorphism

$$\psi : C^*_{\rm red}\Big(A \underset{\min}{\otimes} \mathscr{B}\Big) \to A \underset{\min}{\otimes} C^*_{\rm red}(\mathscr{B}) \underset{\min}{\otimes} C^*_{\rm red}(G),$$

such that for all $a \otimes b$ in any $A \otimes_{\min} B_g$, one has

$$\psi(a \otimes b) = a \otimes \lambda_g(b) \otimes \lambda_g^G.$$

Since $\tilde{\pi}_1$ is faithful, we have by (18.5) that ψ is faithful. Using the above maps, in addition to the map σ provided by (18.7), we build the diagram

$$\begin{array}{ccc} C^*(A \otimes_{\min} \mathscr{B}) & \xrightarrow{\quad\rho\quad} & A \otimes_{\min} C^*_{\text{red}}(\mathscr{B}) \\ \big\downarrow{\scriptstyle \Lambda} & & \big\downarrow{\scriptstyle id\otimes\sigma} \\ C^*_{\text{red}}(A \otimes_{\min} \mathscr{B}) & \xrightarrow{\psi} & A \otimes_{\min} C^*_{\text{red}}(\mathscr{B}) \otimes_{\min} C^*_{\text{red}}(G) \end{array}$$

which the reader may easily prove to be commutative.

Since σ is faithful, we have by [**19**, Proposition 3.6.1], that $id \otimes \sigma$ is faithful. We have also seen above that ψ is faithful, so one deduces that ρ and Λ must have the same null spaces. Consequently ρ factors through the kernel of Λ, producing the required isomorphism. □

Let us now study the approximation property for tensor products of C*-algebras by Fell bundles.

Proposition 25.9. *Let A be a C*-algebra and $\mathscr{B} = \{B_g\}_{g \in G}$ be a Fell bundle. Also let τ be any C*-norm on $A \odot B_1$, with respect to which $A \odot \mathscr{B}$ is a pre-Fell-bundle. If $\mathscr{B}$ satisfies the approximation property described in* (20.4), *then so does $A \otimes_\tau \mathscr{B}$.*

PROOF. Let $\{v_i\}_{i \in I}$ be an approximate identity for A, and choose a Cesaro net $\{a_j\}_{j \in J}$ for $\mathscr{B}$. Considering $I \times J$ as an ordered set with coordinate-wise order relation, it is clear that $I \times J$ is a directed set. For each $(i,j) \in I \times J$, let

$$\alpha_{i,j} : G \to A \otimes_\tau B_1,$$

be defined by

$$\alpha_{i,j}(g) = v_i \otimes a_j(g), \quad \forall g \in G.$$

We then claim that α is a Cesaro net for $A \otimes_\tau \mathscr{B}$. Indeed, given $(i,j) \in I \times J$, we have

$$\begin{aligned} \Big\| \sum_{g \in G} \alpha_i(g)^* \alpha_i(g) \Big\| &= \Big\| \sum_{g \in G} v_i^* v_i \otimes a_i(g)^* a_i(g) \Big\| \\ &= \|v_i^* v_i\| \, \Big\| \sum_{g \in G} a_i(g)^* a_i(g) \Big\|, \end{aligned}$$

where we have denoted τ by the usual norm symbol, noting that, as well as any C*-norm on $A \odot B_1$, τ is a cross norm [**19**, Lemma 3.4.10]. This proves that the α_i satisfy (20.4.i).

Given a in A, and b in any B_g, we have for all $(i,j) \in I \times J$, that

$$\begin{aligned} \sum_{h \in \mathscr{G}} \alpha_i(gh)^* (a \otimes b) \alpha_i(h) &= \sum_{h \in G} \big(v_i \otimes a_j(gh)\big)^* (a \otimes b) \big(v_i \otimes a_j(h)\big) \\ &= \sum_{h \in G} v_i^* a v_i \otimes a_j(gh)^* b a_j(h) \\ &= v_i^* a v_i \otimes \sum_{h \in G} a_j(gh)^* b a_j(h) \overset{i,j \to \infty}{\longrightarrow} a \otimes b. \end{aligned}$$

The conclusion now follows from (20.5). □

Recall from [**19**, Theorem 3.8.7] that a C*-algebra A is *nuclear* if, for every C*-algebra B, there is a unique C*-norm on $A \odot B$. This is clearly the same as saying that the maximal and minimal tensor norms on $A \odot B$ coincide.

The next result gives sufficient conditions for the nuclearity of cross sectional C*-algebras.

Proposition 25.10. *Suppose that the Fell bundle $\mathscr{B}$ satisfies the approximation property and that B_1 is nuclear. Then $C^*(\mathscr{B})$, which is necessarily isomorphic to $C^*_{red}(\mathscr{B})$, is also nuclear.*

PROOF. We must check that, for any C*-algebra A, there is only one C*-norm on $A \odot C^*(\mathscr{B})$. This is equivalent to showing that the canonical map from the maximal to the minimal tensor product, shown in the first row of the diagram below, is an isomorphism.

$$\begin{array}{ccccc} A \otimes_{\max} C^*(\mathscr{B}) & & \longrightarrow & & A \otimes_{\min} C^*(\mathscr{B}) \\ \uparrow & & & & \uparrow \\ C^*\big(A \otimes_{\max} \mathscr{B}\big) & = & C^*(A \otimes_{\min} \mathscr{B}) & \xrightarrow{\Lambda} & C^*_{\text{red}}(A \otimes_{\min} \mathscr{B}) \end{array}$$

Consider the isomorphisms of (25.7) and (25.8) in place of the left and right-hand vertical arrows above, respectively. Observe, in addition, that since B_1 is nuclear by hypothesis, the maximal and minimal norms on $A \odot B_1$ coincide, so that $A \otimes_{\max} \mathscr{B}$ and $A \otimes_{\min} \mathscr{B}$ are in fact equal. Moreover $A \otimes_{\min} \mathscr{B}$ satisfies the approximation property by (25.9), and hence is amenable by (20.6), so its regular representation, marked as Λ in the above diagram, is an isomorphism. Finally, since all maps above are essentially the identity on the various dense copies of $A \odot C_c(\mathscr{B})$, we deduce that the diagram commutes, which implies that the arrow in the first row of the diagram is an isomorphism, as desired. □

The following is a a partial converse of the above result:

Theorem 25.11. *If the reduced cross-sectional C^*-algebra of a Fell bundle $\mathscr{B}$ is nuclear, then $\mathscr{B}$ is amenable.*

PROOF. Consider the diagram

$$\begin{array}{ccc} C^*(\mathscr{B}) & \xrightarrow{\mathscr{S}} & C^*_{\text{red}}(\mathscr{B}) \otimes_{\max} C^*_{\text{red}}(G) \\ {\scriptstyle\Lambda}\big\downarrow & & \big\downarrow{\scriptstyle q} \\ C^*_{\text{red}}(\mathscr{B}) & \xrightarrow{\sigma} & C^*_{\text{red}}(\mathscr{B}) \otimes_{\min} C^*_{\text{red}}(G) \end{array}$$

where σ is provided by (18.7), $\mathscr{S}$ by (18.9), and q is the natural map from the maximal to the minimal tensor product. By checking on elements of the form $\hat{\jmath}_g(b)$, it is easy to see that the diagram commutes. Assuming that $C^*_{\text{red}}(\mathscr{B})$ is nuclear, we have that q is injective by [**19**, 3.6.12], and hence Λ is injective as well. □

Our next result gives sufficient conditions for the reduced cross sectional C*-algebra of a Fell bundle to be exact.

Proposition 25.12. *Let $\mathscr{B} = \{B_g\}_{g \in G}$ be a Fell bundle over an exact group G, such that B_1 is an exact C^*-algebra. Then $C^*_{red}(\mathscr{B})$ is exact.*

PROOF. Let

$$0 \to J \xrightarrow{\iota} A \xrightarrow{\pi} Q \to 0$$

be an exact sequence of C*-algebras. We need to prove that

$$0 \to J \underset{\min}{\otimes} C^*_{\text{red}}(\mathscr{B}) \xrightarrow{\iota \otimes id} A \underset{\min}{\otimes} C^*_{\text{red}}(\mathscr{B}) \xrightarrow{\pi \otimes id} Q \underset{\min}{\otimes} C^*_{\text{red}}(\mathscr{B}) \to 0 \tag{25.12.1}$$

is also exact.

Considering the Fell bundles $J \otimes_{\min} \mathscr{B}$, $A \otimes_{\min} \mathscr{B}$, and $Q \otimes_{\min} \mathscr{B}$, introduced in (25.4), we will next show that $J \otimes_{\min} \mathscr{B}$ is naturally isomorphic to an ideal in $A \otimes_{\min} \mathscr{B}$, and the corresponding quotient is isomorphic to $Q \otimes_{\min} \mathscr{B}$. By [**19**, Theorem 3.5.3] we have that both

$$\iota \otimes id : J \odot B_1 \to A \odot B_1, \quad \text{and} \quad \pi \otimes id : A \odot B_1 \to Q \odot B_1$$

are continuous for the minimal tensor product norms, so we may employ (25.5) to conclude that there are morphisms

$$\tilde{\iota} = \{\iota_g\}_{g \in G}, \quad \text{and} \quad \tilde{\pi} = \{\pi_g\}_{g \in G}$$

from $J \otimes_{\min} C^*_{\text{red}}(\mathscr{B})$ to $A \otimes_{\min} C^*_{\text{red}}(\mathscr{B})$, and from there to $Q \otimes_{\min} C^*_{\text{red}}(\mathscr{B})$, respectively.

By [**19**, Theorem 3.6.1] we have that $\iota \otimes id$, also known as ι_1, is injective on $J \otimes_{\min} B_1$, so all the ι_g are isometric by (21.4.a). We may therefore identify $J \otimes_{\min} C^*_{\text{red}}(\mathscr{B})$ with a Fell sub-bundle of $A \otimes_{\min} C^*_{\text{red}}(\mathscr{B})$.

By first checking the conditions in (21.10.b). for the corresponding dense algebraic tensor products it is easy to see that $J \otimes_{\min} C^*_{\text{red}}(\mathscr{B})$ is in fact an ideal in $A \otimes_{\min} C^*_{\text{red}}(\mathscr{B})$.

It is evident that each π_g vanishes on $J \otimes_{\min} B_g$, and we claim that the null space of π_g is precisely $J \otimes_{\min} B_g$. For the special case $g = 1$ this follows from the exactness of the sequence

$$0 \to J \underset{\min}{\otimes} B_1 \xrightarrow{\iota \otimes id} A \underset{\min}{\otimes} B_1 \xrightarrow{\pi \otimes id} Q \underset{\min}{\otimes} B_1 \to 0,$$

since B_1 is assumed to be an exact C*-algebra. Given an arbitrary g in G, and given any x in $A \otimes_{\min} B_g$, with $\pi_g(x) = 0$, we have that

$$0 = \pi_g(x)^* \pi_g(x) = \pi_1(x^* x),$$

from where we conclude that x^*x lies in $J \otimes_{\min} B_1$. Then

$$x \overset{(15.3)}{=} \lim_{n \to \infty} x(x^*x)^{1/n} \in \left[(A \underset{\min}{\otimes} B_g)(J \underset{\min}{\otimes} B_1)\right] \subseteq J \underset{\min}{\otimes} B_g.$$

This shows that indeed $J \otimes_{\min} B_g = \text{Ker}(\pi_g)$.

Observe that the range of each π_g contains the dense subspace $Q \odot B_g$, so π_g is surjective by (21.4.b). Together with the above conclusion about the kernel of π_g we then conclude that $Q \otimes_{\min} B_g$ is isomorphic to the quotient of $A \otimes_{\min} B_g$ by $J \otimes_{\min} B_g$. In other words, $Q \otimes_{\min} \mathscr{B}$ is isomorphic to the quotient Fell bundle

$$\left(A \underset{\min}{\otimes} \mathscr{B}\right) \Big/ \left(J \underset{\min}{\otimes} \mathscr{B}\right).$$

We may then invoke (21.18) to obtain the exact sequence of C*-algebras

$$0 \to C^*_{\text{red}}\big(J \underset{\min}{\otimes} \mathscr{B}\big) \to C^*_{\text{red}}\big(A \underset{\min}{\otimes} \mathscr{B}\big) \to C^*_{\text{red}}\big(Q \underset{\min}{\otimes} \mathscr{B}\big) \to 0.$$

The isomorphisms provided by (25.8) may now be used to transform this sequence into (25.12.1), which is therefore also an exact sequence. This concludes the proof. □

Notes and remarks. This chapter is based on [**7**, Sections 5 and 6]. In particular, Theorem (25.11) first appeared in [**7**, Theorem 6.4] and Proposition (25.12) is from [**7**, Proposition 5.2].

CHAPTER 26

Smash Product

The title of this chapter, as well as the terminology for the main concept to be introduced here, is taken from the theory of Hopf algebras [**82**], where one defines the *smash product* of an algebra by the *co-action* of a Hopf algebra.

In the special case of the Hopf C*-algebra $C^*(G)$, where G is a discrete group, Quigg has shown that co-actions correspond to Fell bundles [**93**]. Incidentally, our study of Fell bundles may therefore be seen as a special case of the theory of co-actions of Hopf C*-algebras. Having decided to concentrate on Fell bundles, rather than more general co-actions, we will likewise introduce the notion of smash product, below, in an ad-hoc way, avoiding the whole apparatus of Hopf algebras. Our limited view of smash products will nevertheless have many important applications in the sequel.

Given a group G, we will denote the algebra of all compact operators on $\ell^2(G)$ by $\mathcal{K}\big(\ell^2(G)\big)$. Given g in G, we will let e_g be the canonical basis vector of $\ell^2(G)$, and for each $g, h \in G$, we will denote by $e_{g,h}$ the rank-one operator on $\ell^2(G)$ defined by

$$e_{g,h}(\xi) = \langle \xi, e_h \rangle e_g, \quad \forall\, \xi \in \ell^2(G),$$

so that

$$e_{g,h}(e_k) = \delta_{h,k} e_g, \quad \forall\, k \in G.$$

It is well known that $\mathcal{K}\big(\ell^2(G)\big)$ is the closed linear span of the set formed by all the $e_{g,h}$.

Since $\mathcal{K}\big(\ell^2(G)\big)$ is a nuclear algebra, the minimal tensor product of a C*-algebra A by $\mathcal{K}\big(\ell^2(G)\big)$ is isomorphic to its maximal version, so we will use the symbol "$\otimes$" in

$$A \otimes \mathcal{K}\big(\ell^2(G)\big),$$

indistinctly meaning "$\otimes_{\min}$" or "$\otimes_{\max}$".

Proposition 26.1. *If $\mathscr{B} = \{B_g\}_{g\in G}$ is a Fell bundle, then the subset of $C^*(\mathscr{B}) \otimes \mathcal{K}\big(\ell^2(G)\big)$ given by*

$$\mathscr{B} \sharp_0 G = \sum_{g,h\in G} B_{g^{-1}h} \otimes e_{g,h}$$

is a *-subalgebra.

Proof. It is enough to notice that, for every $g, h, k, l \in G$, one has that

$$(B_{g^{-1}h} \otimes e_{g,h})(B_{k^{-1}l} \otimes e_{k,l}) \subseteq \delta_{h,k}(B_{g^{-1}h}B_{h^{-1}l} \otimes e_{g,l}) \subseteq B_{g^{-1}l} \otimes e_{g,l},$$

and also that

$$(B_{g^{-1}h} \otimes e_{g,h})^* = B_{h^{-1}g} \otimes e_{h,g}.$$

□

Definition 26.2. The *smash product*[32] of the Fell bundle $\mathscr{B}$ by G, denoted $\mathscr{B}\sharp G$, is the closed *-subalgebra of $C^*(\mathscr{B})\otimes\mathcal{K}\big(\ell^2(G)\big)$ given by the closure of $\mathscr{B}\overset{\sharp}{_0}G$.

Remark 26.3. If B is any graded C*-algebra whose grading coincides with our Fell bundle $\mathscr{B}$, we could define a variant of $\mathscr{B}\overset{\sharp}{_0}G$ by looking at

$$\sum_{g,h\in G} B_{g^{-1}h}\otimes e_{g,h}$$

as a subalgebra of $B\otimes\mathcal{K}\big(\ell^2(G)\big)$, as opposed to $C^*(\mathscr{B})\otimes\mathcal{K}\big(\ell^2(G)\big)$. Its closure within $B\otimes\mathcal{K}\big(\ell^2(G)\big)$ could then be taken as an alternative definition of $\mathscr{B}\sharp G$. However it is not difficult to see that this alternative smash product is isomorphic to the one defined in (26.2). The reason is that for all finite subsets $F\subseteq G$, the set

$$S_F := \sum_{g,h\in F} B_{g^{-1}h}\otimes e_{g,h},$$

(no closure) seen within $B\otimes\mathcal{K}\big(\ell^2(G)\big)$, is a closed *-subalgebra whose isomorphism class clearly does not depend on the way $\mathscr{B}$ is represented in B. Since the smash product (any variant of it) is the inductive limit of the S_F, as F ranges over the finite subsets of G, we then see that the smash product itself does not depend on the graded algebra B. Our choice of $C^*(\mathscr{B})$ in the definition of $\mathscr{B}\sharp G$, above, is therefore arbitrary.

Faithfully representing $C^*(\mathscr{B})$ on a Hilbert space H, we have that

$$C^*(\mathscr{B})\otimes\mathcal{K}\big(\ell^2(G)\big)\subseteq\mathscr{L}\left(H\otimes\ell^2(G)\right), \tag{26.4}$$

(where we think of the tensor product sign in the left-hand-side above as the spatial tensor product). Hence we may also view $\mathscr{B}\sharp G$ as an algebra of operators on $H\otimes\ell^2(G)$.

Observe that we have for all $g,h,k\in G$, that

$$(1\otimes e_{g,g})(B_{h^{-1}k}\otimes e_{h,k})=\delta_{gh}(B_{h^{-1}k}\otimes e_{h,k})\subseteq\mathscr{B}\sharp G, \tag{26.5}$$

from where we see that $(1\otimes e_{g,g})(\mathscr{B}\sharp G)\subseteq\mathscr{B}\sharp G$, and similarly one has that $(\mathscr{B}\sharp G)(1\otimes e_{g,g})\subseteq\mathscr{B}\sharp G$. In other words, $1\otimes e_{g,g}$ is a multiplier of $\mathscr{B}\sharp G$.

It should be noticed, however, that the same is not true for $1\otimes e_{g,h}$, when $g\neq h$.

Proposition 26.6. *Given any w in $\mathscr{B}\sharp G$, and given g and h in G, there is a unique $w_{g,h}\in B_{g^{-1}h}$ such that*

$$(1\otimes e_{g,g})w(1\otimes e_{h,h})=w_{g,h}\otimes e_{g,h}.$$

PROOF. This is obvious for w in $\mathscr{B}\overset{\sharp}{_0}G$. So the result follows from the density of the latter in $\mathscr{B}\sharp G$. □

We will soon be dealing with numerous ideals in $\mathscr{B}\sharp G$. In preparation for this we present the following simple criterion for deciding when an element of $\mathscr{B}\sharp G$ belongs to a given ideal.

[32]One may prove that $C^*(\mathscr{B})$ admits a co-action of the Hopf algebra $C^*(G)$, and a construction based on the usual Hopf algebra concept of smash product leads to the notion presently being introduced.

Proposition 26.7. *Let J be a closed subspace of $\mathscr{B}\sharp G$. Given w in $\mathscr{B}\sharp G$, consider the statements:*

(*i*) $w_{g,h} \otimes e_{g,h} \in J$, *for all g and h in G,*
(*ii*) $w \in J$.

Then (i) implies (ii). In case J satisfies

$$(1 \otimes \mathcal{K})J(1 \otimes \mathcal{K}) \subseteq J,$$

in particular if J is a two-sided ideal, the converse also holds.

Proof. For every finite subset $F \subseteq G$, let

$$P_F = \sum_{g \in F} 1 \otimes e_{g,g}.$$

We then claim that,

$$w = \lim_{F \uparrow G} P_F w P_F, \quad \forall w \in \mathscr{B}\sharp G,$$

where we think of the collection of all finite subsets $F \subseteq G$ as a directed set relative to set inclusion. The reason for the claim is that it clearly holds for $w \in \mathscr{B}\sharp_0 G$, which is dense in $\mathscr{B}\sharp G$, while the P_F are uniformly bounded.

Observe that for F as above, we have

$$P_F w P_F = \sum_{g,h \in F} (1 \otimes e_{g,g}) w (1 \otimes e_{h,h}) \overset{(26.6)}{=} \sum_{g,h \in F} w_{g,h} \otimes e_{g,h}.$$

Thus, assuming that w satisfies (i), we deduce that $P_F w P_F$ is in J, and hence also that w is in J, because J is closed under taking limits.

The last assertion in the statement is obvious, observing that closed two-sided ideals are invariant under left and right multiplication by multipliers of the ambient algebra. □

Definition 26.8. The *restricted smash product* of the Fell bundle $\mathscr{B}$ by G, denoted $\mathscr{B}\flat G$, is defined to be

$$\mathscr{B}\flat G = \overline{\sum_{g,h \in G} [B_{g^{-1}} B_h] \otimes e_{g,h}},$$

where the closure is taken within $C^*(\mathscr{B}) \otimes \mathcal{K}(\ell^2(G))$.

Observe that $\mathscr{B}\flat G \subseteq \mathscr{B}\sharp G$, because $[B_{g^{-1}} B_h] \subseteq B_{g^{-1}h}$. Moreover, unless $\mathscr{B}$ is saturated, this is a proper inclusion.

Proposition 26.9. *One has that $\mathscr{B}\flat G$ is a closed two-sided ideal of $\mathscr{B}\sharp G$.*

Proof. Given $g, h, k, l \in G$, notice that

$$\begin{aligned}(B_{g^{-1}h} \otimes e_{g,h})(B_{k^{-1}} B_l \otimes e_{k,l}) &\subseteq \delta_{h,k}(B_{g^{-1}h} B_{h^{-1}} B_l \otimes e_{g,l}) \\ &\subseteq B_{g^{-1}} B_l \otimes e_{g,l} \\ &\subseteq \mathscr{B}\flat G,\end{aligned}$$

from where we deduce that $\mathscr{B}\flat G$ is a left ideal of $\mathscr{B}\sharp G$, and a similar reasoning proves it to be a right ideal as well. □

We may now use our membership criterion (26.7) to characterize elements of $\mathscr{B}\flat G$ as follows:

Proposition 26.10. *Let w be in $\mathscr{B}\sharp G$. Then the following are equivalent:*

(*i*) $w_{g,h} \in [B_{g^{-1}} B_h]$ *(closed linear span), for every g and h in G,* (*ii*) $w \in \mathscr{B}\flat G$.

PROOF. (i) ⇒ (ii) Follows from (26.7).

(ii) ⇒ (i) Evident. □

Recalling that λ^G denotes the left-regular representation of G on $\ell^2(G)$, notice that

$$\lambda^G_g e_{h,k} = e_{gh,k}, \quad \text{and} \quad e_{h,k}\lambda^G_g = e_{h,g^{-1}k}, \quad \forall\, g,k,h \in G.$$

Consequently,

$$\lambda^G_g e_{h,k}\lambda^G_{g^{-1}} = e_{gh,gk},$$

from where we see that

$$\begin{aligned}(1\otimes\lambda^G_g)(B_{h^{-1}k}\otimes e_{h,k})(1\otimes\lambda^G_{g^{-1}}) &= B_{h^{-1}k}\otimes e_{gh,gk}\\ &= B_{(gh)^{-1}(gk)}\otimes e_{gh,gk}\\ &\subseteq \mathscr{B}\sharp G.\end{aligned}$$

This implies that $\mathscr{B}\sharp G$ is invariant under conjugation by $1\otimes\lambda^G_g$, so we may define an action Γ of G on $\mathscr{B}\sharp G$ by

$$\Gamma_g : w \in \mathscr{B}\sharp G \mapsto (1\otimes\lambda^G_g)w(1\otimes\lambda^G_{g^{-1}}) \in \mathscr{B}\sharp G. \tag{26.11}$$

Notice that the ideal $\mathscr{B}\flat G$ is not necessarily invariant under Γ, since

$$(1\otimes\lambda^G_g)(B_{h^{-1}}B_k\otimes e_{h,k})(1\otimes\lambda^G_{g^{-1}}) = B_{h^{-1}}B_k\otimes e_{gh,gk},$$

and there is no reason why

$$B_{h^{-1}}B_k \overset{?}{\subseteq} [B_{h^{-1}g^{-1}}B_{gk}],$$

even though both of these are subsets of $B_{h^{-1}k}$. In case $\mathscr{B}$ is saturated then $[B_{h^{-1}}B_k]$ and $[B_{h^{-1}g^{-1}}B_{gk}]$ both coincide with $B_{h^{-1}k}$, and hence the above inclusion would hold, but in general it does not.

Regardless of this lack of invariance, we may still restrict Γ to a *partial action* Δ of G on $\mathscr{B}\flat G$, according to (3.2).

Definition 26.12. Let $\mathscr{B}$ be a Fell bundle.

(a) The global action Γ of G on $\mathscr{B}\sharp G$, defined above, will be called the *dual global action* for $\mathscr{B}$.
(b) The partial action Δ of G, obtained by restricting Γ to $\mathscr{B}\flat G$, will be called the *dual partial action* for $\mathscr{B}$.

The justification for this terminology is as follows: if $\mathscr{B}$ is the Fell bundle formed by the spectral subspaces for an action θ of a compact abelian group K on a C*-algebra B, one may show that the crossed product $B\rtimes_\theta K$ is isomorphic to $\mathscr{B}\sharp G$, and that dual action of $\widehat{K}$ on $B\rtimes_\theta K$ is equivalent to the *dual global action* defined above.

The dual partial action will play a central role from now on, so it pays to describe the ideals involved:

Proposition 26.13. *For each g in G, let $E_g = \Gamma_g(\mathscr{B}\flat G)\cap(\mathscr{B}\flat G)$. Then*

$$E_g = \overline{\sum_{h,k\in G}[B_{h^{-1}}D_gB_k]\otimes e_{h,k}},$$

where $D_g = [B_gB_{g^{-1}}]$.

PROOF. Given w in E_g, in particular w is in $\mathscr{B}\flat G$, so we have by (26.10) that $w_{h,k} \in [B_{h^{-1}}B_k]$, for all h and k in G. Since w is also in $\Gamma_g(\mathscr{B}\flat G)$, then $y := \Gamma_{g^{-1}}(w) \in \mathscr{B}\flat G$, so

$$[B_{h^{-1}g}B_{g^{-1}k}] \ni y_{g^{-1}h,g^{-1}k} = w_{h,k}.$$

Choosing approximate identities $\{u_i\}_i$ and $\{v_j\}_j$, for the ideals $[B_{h^{-1}}B_h]$ and $[B_{k^{-1}}B_k]$, respectively, we have by (16.11) that

$$\begin{aligned} w_{g,h} &= \lim_{i,j} u_i w_{g,h} v_j \\ &\in [B_{h^{-1}}B_h B_{h^{-1}g}B_{g^{-1}k}B_{k^{-1}}B_k] \\ &\subseteq [B_{h^{-1}}B_g B_{g^{-1}}B_k] \\ &= [B_{h^{-1}}D_g B_k]. \end{aligned}$$

That w is in the set on the right-hand side in the statement then follows from (26.7), therefore proving "$\subseteq$".

Conversely, given $g, h, k \in G$, notice that $D_g \subseteq B_1$, so $D_g B_k \subseteq B_k$, whence

$$B_{h^{-1}}D_g B_k \otimes e_{h,k} \subseteq B_{h^{-1}}B_k \otimes e_{h,k} \subseteq \mathscr{B}\flat G.$$

On the other hand we also have that

$$[B_{h^{-1}}D_g B_k] = [B_{h^{-1}}B_g B_{g^{-1}}B_k] \subseteq [B_{h^{-1}g}B_{g^{-1}k}],$$

so $B_{h^{-1}}D_g B_k \otimes e_{g^{-1}h,g^{-1}k}$ is also contained in $\mathscr{B}\flat G$. We thus have that

$$\begin{aligned} B_{h^{-1}}D_g B_k \otimes e_{h,k} &= (1\otimes \lambda^G_g)(B_{h^{-1}}D_g B_k \otimes e_{g^{-1}h,g^{-1}k})(1\otimes \lambda^G_{g^{-1}}) \\ &\subseteq \Gamma_g(\mathscr{B}\flat G), \end{aligned}$$

proving that each $B_{h^{-1}}D_g B_k \otimes e_{h,k}$ is contained in E_g. This shows the remaining inclusion "$\supseteq$", and hence the proof is concluded. □

Since Δ is a restriction of Γ, it is interesting to ask what exactly is the globalization of Δ. The answer could not be other than the corresponding dual global action!

Proposition 26.14. *The dual global action for a Fell bundle is the globalization of the corresponding dual partial action.*

PROOF. Calling our bundle $\mathscr{B} = \{B_g\}_{g\in G}$, all we must do is prove that

$$\sum_{g\in G} \Gamma_g(\mathscr{B}\flat G)$$

is dense in $\mathscr{B}\sharp G$. For each g and h in G, observe that $B_{g^{-1}h} \otimes e_{1,g^{-1}h}$ is contained in $\mathscr{B}\flat G$. Moreover

$$\Gamma_g(B_{g^{-1}h} \otimes e_{1,g^{-1}h}) = B_{g^{-1}h} \otimes e_{g,h},$$

so $B_{g^{-1}h} \otimes e_{g,h}$ is contained in the orbit of $\mathscr{B}\flat G$ under Γ, from where the proof follows. □

Notes and remarks. Algebras resembling the restricted smash product in the context of partial actions first appeared in [**45**]. In the case of partial actions of continuous groups, the smash product was used by Abadie in [**1**] and [**2**].

CHAPTER 27

Stable Fell Bundles as Partial Crossed Products

As we have seen in (17.11.vi) & (17.11.vii), a partial crossed product is always a graded C*-algebra. In this chapter we will present o Mne of the most important results of the theory of partial actions, proving a converse of the above statement under quite broad hypotheses.

The most general form of such a converse is unfortunately not true, meaning that not all graded C*-algebras are partial crossed products. This may be seen from the following example: consider the $\mathbb{Z}$-grading of $M_3(\mathbb{C})$ given by

$$B_0 = \begin{bmatrix} \bigstar & 0 & 0 \\ 0 & \bigstar & \bigstar \\ 0 & \bigstar & \bigstar \end{bmatrix}, \quad B_1 = \begin{bmatrix} 0 & 0 & 0 \\ \bigstar & 0 & 0 \\ \bigstar & 0 & 0 \end{bmatrix}, \quad B_{-1} = \begin{bmatrix} 0 & \bigstar & \bigstar \\ 0 & 0 & 0 \\ 0 & 0 & 0 \end{bmatrix},$$

while $B_n = \{0\}$, for all $n \in \mathbb{Z} \setminus \{0, \pm 1\}$. In order to see that this grading does not arise from a partial action, let us argue by contradiction and suppose that there is a partial action

$$\theta = \big(\{D_n\}_{n\in\mathbb{Z}}, \{\theta_n\}_{n\in\mathbb{Z}}\big)$$

of $\mathbb{Z}$ on B_0 whose associated semi-direct product bundle coincides with the Fell bundle given by the above grading of $M_3(\mathbb{C})$. In this case, one would have

$$[B_{-1}B_1] \overset{(8.14.f)}{=} D_{-1} \simeq D_1 = [B_1 B_{-1}], \tag{27.1}$$

(brackets meaning closed linear span) but notice that

$$[B_{-1}B_1] = \begin{bmatrix} \bigstar & 0 & 0 \\ 0 & 0 & 0 \\ 0 & 0 & 0 \end{bmatrix}, \quad \text{and} \quad [B_1 B_{-1}] = \begin{bmatrix} 0 & 0 & 0 \\ 0 & \bigstar & \bigstar \\ 0 & \bigstar & \bigstar \end{bmatrix},$$

which are not isomorphic algebras, hence bringing about a contradiction.

Further analyzing this contradiction, notice that even though $[B_{-1}B_1]$ and $[B_1 B_{-1}]$ fail to be isomorphic, they are Morita-Rieffel-equivalent, given that B_1 is an imprimitivity bimodule. More generally, if B is any graded C*-algebra with grading $\{B_g\}_{g\in G}$, then for every g in G, one has that

$$D_g := [B_g B_{g^{-1}}]$$

is an ideal in B_1.

Assuming that the grading arises from a partial action, as in (17.11), then $D_{g^{-1}}$ is isomorphic to D_g by the reasoning used in (27.1). But, regardless of this assumption, $D_{g^{-1}}$ is always Morita-Rieffel-equivalent to D_g, with B_g playing the role of the imprimitivity bimodule. Thus, considering Morita-Rieffel-equivalence as a weak form of isomorphism, we see that the rudiments of a partial action are present in any graded C*-algebra. Furthermore, in case B_1 is a separable stable

C*-algebra, then by [**17**] the above Morita-Rieffel-equivalence actually implies that $D_{g^{-1}}$ is isomorphic to D_g, so we may choose, for each g, an isomorphism

$$\theta_g : D_{g^{-1}} \to D_g,$$

getting us even closer to obtaining a partial action.

In this chapter we will carefully explore these ideas in order to prove that every separable Fell bundle with stable unit fiber algebra is isomorphic to the semi-direct product bundle for a suitable partial action of the base group on B_1.

Recall that a C*-algebra A is said to be *stable* if A is isomorphic to the tensor product of some other C*-algebra B by the algebra $\mathcal{K}$ of all compact operators on a separable, infinite dimensional Hilbert space. In symbols

$$A \simeq B \otimes \mathcal{K}.$$

We again refrain from specifying either "$\otimes_{\min}$" or "$\otimes_{\max}$", since $\mathcal{K}$ is a nuclear C*-algebra, whence the minimal and maximal norms are equal.

Since $\mathcal{K} \otimes \mathcal{K} \simeq \mathcal{K}$, in case A is stable we have

$$A \simeq B \otimes \mathcal{K} \simeq B \otimes \mathcal{K} \otimes \mathcal{K} \simeq A \otimes \mathcal{K},$$

which is to say that the algebra B, referred to above, might as well be taken to be A itself.

We will now present a useful criterion for the stability of C*-algebras.

Lemma 27.2. *A C*-algebra A is stable if and only if there exists a non-degenerate *-homomorphism from $\mathcal{K}$ to the multiplier algebra $\mathcal{M}(A)$.*

PROOF. Assuming that A is stable, write $A = B \otimes \mathcal{K}$, where B is a C*-algebra. Supposing without loss of generality that B is faithfully represented on a Hilbert space H, we may view $B \otimes \mathcal{K}$ as an algebra of operators on $H \otimes \ell^2$. Defining

$$\gamma : k \in \mathcal{K} \mapsto 1 \otimes k \in \mathscr{L}(H \otimes \ell^2),$$

it is easy to see that the range of γ is contained in the algebra of multipliers of $B \otimes \mathcal{K}$ and that, seen as a map from $\mathcal{K}$ to $\mathcal{M}(B \otimes \mathcal{K}) = \mathcal{M}(A)$, one has that γ is non-degenerate.

Conversely, given a non-degenerate *-homomorphism from $\mathcal{K}$ to $\mathcal{M}(A)$, let $\{e_{i,j}\}_{i,j \in \mathbb{N}}$ be the standard matrix units of $\mathcal{K} = \mathcal{K}(\ell^2)$, and let

$$B = \gamma(e_{1,1}) A \gamma(e_{1,1}).$$

For each $n \in \mathbb{N}$ one may easily prove that the map

$$\varphi_n \ : \ \{b_{i,j}\}_{i,j=1}^n \ \in \ M_n(B) \ \longmapsto \ \sum_{i,j=1}^n \gamma(e_{i,1}) b_{i,j} \gamma(e_{1,j}) \ \in \ A,$$

is an injective *-homomorphism. After also checking that these maps are compatible with the usual inductive limit structure of

$$B \otimes \mathcal{K} \simeq \varinjlim M_n(B),$$

we obtain a *-homomorphism

$$\varphi : B \otimes \mathcal{K} \to A,$$

which is easily seen to be injective. In order to prove that φ is also surjective, first notice that

$$a = \lim_{n\to\infty} \sum_{i,j=1}^{n} \gamma(e_{i,i}) a \gamma(e_{j,j}), \quad \forall a \in A, \tag{27.2.1}$$

since γ is assumed to be non-degenerate. Given $a \in A$, and setting

$$b_{i,j} = \gamma(e_{1,i}) a \gamma(e_{j,1}), \quad \forall i, j \leq n,$$

one has that $b := \{b_{i,j}\}_{i,j=1}^{n}$ is an element of $M_n(B)$, and that

$$\varphi_n(b) = \sum_{i,j=1}^{n} \gamma(e_{i,1})\gamma(e_{1,i}) a \gamma(e_{j,1})\gamma(e_{1,j}) = \sum_{i,j=1}^{n} \gamma(e_{i,i}) a \gamma(e_{j,j}) \xrightarrow{(27.2.1)} a,$$

proving that φ is surjective. Therefore $A \simeq B \otimes \mathcal{K}$, and hence A is stable. ☐

Our first use of this tool will be in proving the stability of $\mathscr{B}\flat G$ under appropriate hypotheses.

Lemma 27.3. *Let $\mathscr{B} = \{B_g\}_{g\in G}$ be a Fell bundle over a countable group G. If B_1 is stable, then so is $\mathscr{B}\flat G$.*

PROOF. Recall that if $C^*(\mathscr{B})$ is faithfully represented on a Hilbert space H, then

$$\mathscr{B}\flat G \subseteq C^*(\mathscr{B}) \otimes \mathcal{K} \subseteq \mathscr{L}\big(H \otimes \ell^2(G)\big),$$

where $\mathcal{K} = \mathcal{K}\big(\ell^2(G)\big)$. Considering the representation of B_1 on $H \otimes \ell^2(G)$ given by

$$\pi(b) = b \otimes 1, \quad \forall b \in B_1,$$

it is easy to see that the range of π is contained in the multiplier algebra of $\mathscr{B}\flat G$, so we may view π as a *-homomorphism

$$\pi : B_1 \to \mathcal{M}(\mathscr{B}\flat G),$$

(which should not be confused with another rather canonical mapping from B_1 to $\mathscr{B}\flat G$, namely $b \to b \otimes e_{1,1}$).

We claim that π is non-degenerate. In order to see this, let $\{v_i\}_{i\in I}$ be an approximate identity for B_1. Then, for every $g, h \in G$, and every $b \in B_{g^{-1}}B_h$, one has that

$$\pi(v_i)(b \otimes e_{g,h}) = (v_i \otimes 1)(b \otimes e_{g,h}) = v_i b \otimes e_{g,h} \overset{i\to\infty}{\longrightarrow} b \otimes e_{g,h},$$

by (16.9). This shows that π is non-degenerate, as desired.

It is well known that if C and D are C*-algebras, and $\varphi : C \to \mathcal{M}(D)$ is a given non-degenerate *-homomorphism, then φ admits a unique extension to a unital *-homomorphism $\tilde{\varphi} : \mathcal{M}(C) \to \mathcal{M}(D)$. Applying this to the present situation, let

$$\tilde{\pi} : \mathcal{M}(B_1) \to \mathcal{M}(\mathscr{B}\flat G)$$

be the extension of π thus obtained. Since B_1 is stable, we may use (27.2) to get a non-degenerate *-homomorphism $\gamma : \mathcal{K} \to \mathcal{M}(B_1)$, and we may then consider the composition

$$\mathcal{K} \xrightarrow{\gamma} \mathcal{M}(B_1) \xrightarrow{\tilde{\pi}} \mathcal{M}(\mathscr{B}\flat G),$$

which we denote by σ. Notice that

$$\begin{aligned}[\sigma(\mathcal{K})(\mathscr{B}\flat G)] &= [\sigma(\mathcal{K})\pi(B_1)(\mathscr{B}\flat G)]\\ &= [\tilde{\pi}\big(\gamma(\mathcal{K})\big)\pi(B_1)(\mathscr{B}\flat G)]\\ &= [\tilde{\pi}\big(\gamma(\mathcal{K})B_1\big)(\mathscr{B}\flat G)]\\ &= [\pi(B_1)(\mathscr{B}\flat G)]\\ &= \mathscr{B}\flat G,\end{aligned}$$

so σ is non-degenerate, and hence $\mathscr{B}\flat G$ is stable. □

The reader could use the above method to show that, if B_1 is stable, then $C^*(\mathscr{B})$ and $C^*_{\text{red}}(\mathscr{B})$ are also stable.

The following is a slight improvement on [**16**, Lemma 2.5], using ideas from the proof of [**17**, Theorem 3.4]:

Lemma 27.4. *Let A be a separable stable C^*-algebra. If p is a full projection*[33] *in $\mathcal{M}(A)$ such that pAp is also stable, then there is $v \in \mathcal{M}(A)$ such that $v^*v = 1$, and $vv^* = p$.*

PROOF. Let e be any minimal projection in $\mathcal{K}$, for example $e = e_{1,1}$. We first claim that there exists $v_{A} \in \mathcal{M}(A\otimes\mathcal{K})$ such that

$$v_{A}^*v_{A} = 1\otimes 1, \quad \text{and} \quad v_{A}v_{A}^* = 1\otimes e. \tag{27.4.1}$$

To see this, recall that any two separable infinite dimensional Hilbert spaces are isometrically isomorphic to each other, so there is u in $\mathcal{M}(\mathcal{K}\otimes\mathcal{K})$ such that $u^*u = 1\otimes 1$, and $uu^* = 1\otimes e$. Letting $\varphi : A\otimes\mathcal{K} \to A$ be any *-isomorphism, notice that

$$\varphi\otimes id : A\otimes\mathcal{K}\otimes\mathcal{K} \to A\otimes\mathcal{K}$$

is also a *-isomorphism, which therefore extends to the respective multiplier algebras. Defining $v_{A} = (\varphi\otimes id)(1\otimes u)$, observe that

$$\begin{aligned}v_{A}^*v_{A} &= (\varphi\otimes id)\big((1\otimes u)^*(1\otimes u)\big)\\ &= (\varphi\otimes id)(1\otimes u^*u)\\ &= (\varphi\otimes id)(1\otimes 1\otimes 1)\\ &= 1\otimes 1,\end{aligned}$$

while

$$\begin{aligned}v_{A}v_{A}^* &= (\varphi\otimes id)\big((1\otimes u)(1\otimes u)^*\big)\\ &= (\varphi\otimes id)(1\otimes uu^*)\\ &= (\varphi\otimes id)(1\otimes 1\otimes e)\\ &= 1\otimes e,\end{aligned}$$

so the claim is proved.

Since $B := pAp$ is also supposed to be stable, the same argument applies to produce $v_{B} \in \mathcal{M}(B\otimes\mathcal{K})$, such that

$$v_{B}^*v_{B} = p\otimes 1, \quad \text{and} \quad v_{B}v_{B}^* = p\otimes e,$$

where the slight difference between the above equations and (27.4.1) is due to the fact that the unit of $\mathcal{M}(B)$ is called p.

[33] A projection p in $\mathcal{M}(A)$ is said to be *full*, when $A = [ApA]$ (closed linear span).

We now observe that separable C*-algebras possess strictly positive elements, so we may apply [**16**, Lemma 2.5] to show the existence of an element $w \in \mathcal{M}(A \otimes \mathcal{K})$ such that $w^*w = 1 \otimes 1$, and $ww^* = p \otimes 1$. We then define $u = v_B w v_A^*$,

$$\begin{array}{ccc} 1 \otimes 1 & \xrightarrow{v_A} & 1 \otimes e \\ \downarrow{\scriptstyle w} & & \\ p \otimes 1 & \xrightarrow{v_B} & p \otimes e. \end{array}$$

Strictly speaking the product $v_B w$ makes no sense, since v_B belongs to $\mathcal{M}(B \otimes \mathcal{K})$, while w is in $\mathcal{M}(A \otimes \mathcal{K})$. However, since

$$B \otimes \mathcal{K} = (p \otimes 1)(A \otimes \mathcal{K})(p \otimes 1),$$

we may naturally embed $\mathcal{M}(B \otimes \mathcal{K})$ in $\mathcal{M}(A \otimes \mathcal{K})$. A simple calculation then shows that

$$u^*u = 1 \otimes e, \quad \text{and} \quad uu^* = p \otimes e.$$

Therefore u is a partial isometry and we claim that $(1 \otimes e)u(1 \otimes e) = u$. In fact, we have

$$\begin{aligned} (1 \otimes e)u(1 \otimes e) = (1 \otimes e)uu^*u &= (1 \otimes e)(p \otimes e)u \\ &= (p \otimes e)u \\ &= uu^*u \\ &= u. \end{aligned}$$

The claim proved, and since e is a minimal projection, we see that u must be of the form

$$u = v \otimes e,$$

where $v \in \mathcal{M}(A)$ satisfies $v^*v = 1$, and $vv^* = p$. □

From now on our study of Fell bundles will rely on (27.4), so we will have to restrict our attention to Fell bundles satisfying suitable separability conditions.

Definition 27.5. A Fell bundle $\mathscr{B} = \{B_g\}_{g \in G}$ is said to be *separable* if

(i) G is a countable group,
(ii) every B_g is a separable Banach space.

▶ From now on we will fix a separable Fell bundle $\mathscr{B} = \{B_g\}_{g \in G}$. It is then easy to see that most constructions originating from $\mathscr{B}$, such as $C^*(\mathscr{B})$, $C^*_{\text{red}}(\mathscr{B})$, $\mathscr{B} \sharp G$ and $\mathscr{B} \flat G$, lead to separable C*-algebras.

Recall from (26.5) that, for every g in G, one has that $1 \otimes e_{g,g}$ is a multiplier of $\mathscr{B} \sharp G$, and hence also of $\mathscr{B} \flat G$. By (26.10) the corresponding *corner* of $\mathscr{B} \flat G$ is then

$$(1 \otimes e_{g,g})(\mathscr{B} \flat G)(1 \otimes e_{g,g}) = [B_{g^{-1}} B_g] \otimes e_{g,g},$$

and in particular

$$(1 \otimes e_{1,1})(\mathscr{B} \flat G)(1 \otimes e_{1,1}) = B_1 \otimes e_{1,1}. \tag{27.6}$$

Proposition 27.7. $1 \otimes e_{1,1}$ *is a a full projection in* $\mathcal{M}(\mathscr{B} \flat G)$.

PROOF. Given g in G, notice that $B_g \otimes e_{1,g}$ is contained in $\mathscr{B}\flat G$. The result then follows immediately from the identity

$$(B_g \otimes e_{1,g})^*(1 \otimes e_{1,1})(B_h \otimes e_{1,h}) = B_{g^{-1}} B_h \otimes e_{g,h}.$$

□

► From now on we will assume, in addition, that B_1 is stable. We then have from (27.3) that $\mathscr{B}\flat G$ is a separable stable C*-algebra, and in its multiplier algebra one finds the full projection $1 \otimes e_{1,1}$, whose associated corner is the stable algebra $B_1 \otimes e_{1,1}$. We are then precisely in the situation of (27.4), so we conclude that there is a $v \in \mathcal{M}(\mathscr{B}\flat G)$, such that

$$v^*v = 1 \otimes 1, \quad \text{and} \quad vv^* = 1 \otimes e_{1,1}. \tag{27.8}$$

In particular the mapping

$$\Phi_1 : b \in B_1 \mapsto v^*(b \otimes e_{1,1})v \in \mathscr{B}\flat G \tag{27.9}$$

is an isomorphism. It is our goal to prove that Φ_1 is but one ingredient of an isomorphism

$$\Phi = \{\Phi_g\}_{g \in G}$$

between $\mathscr{B}$ and the semi-direct product bundle arising from the dual partial action of G on $\mathscr{B}\flat G$ introduced in (26.12). We will eventually define each Φ_g by the formula

$$\Phi_g(b) = v^*(b \otimes e_{1,g})\Gamma_g(v)\delta_g, \quad \forall\, b \in B_g,$$

and in preparation for this we first prove the following technical facts:

Lemma 27.10. *For every g in G, one has that*
 (i) $B_g \otimes e_{1,g} \subseteq E_g$,
 (ii) *if $y \in E_g$, then* $\Delta_{g^{-1}}(vy)v^* \in B_g \otimes e_{g^{-1},1}$.

PROOF. By (26.13) we have that

$$B_{h^{-1}} D_g B_k \otimes e_{h,k} \subseteq E_g,$$

for every h and k in G. So, plugging in $h = 1$, and $k = g$, and using (16.12), we see that (i) follows.

In order to prove (ii), recall that Δ is the restriction of Γ to $\mathscr{B}\flat G$, and that, in turn, Γ is the adjoint action relative to the unitary representation $1 \otimes \lambda^G$, as defined in (26.11). Therefore Γ may be seen as an action of G on the whole algebra of bounded operators on $H \otimes \ell^2(G)$ (see (26.4)).

Given y in E_g, let $x = \Delta_{g^{-1}}(vy)v^*$, so

$$\begin{aligned} x &= \Delta_{g^{-1}}(vv^*vy)v^*vv^* \\ &\overset{(27.8)}{=} \Gamma_{g^{-1}}(1 \otimes e_{1,1})\Delta_{g^{-1}}(vy)v^*(1 \otimes e_{1,1}) \\ &= (1 \otimes e_{g^{-1},g^{-1}})x(1 \otimes e_{1,1}) \\ &\overset{(26.6)}{=} x_{g^{-1},1} \otimes e_{g^{-1},1}. \end{aligned}$$

Using (26.10), we have that $x_{g^{-1},1} \in B_g$, proving (ii). □

We may now prove the main result of this chapter:

Theorem 27.11. *Let $\mathscr{B}$ be a separable Fell bundle whose unit fiber algebra B_1 is stable. Then there exists a C^*-algebraic partial action of G of B_1 whose associated semi-direct product bundle is isomorphic to $\mathscr{B}$.*

PROOF. As already mentioned we will prove that $\mathscr{B}$ is isomorphic to the semi-direct product bundle for the dual partial action of G on $\mathscr{B}\flat G$. One may then transfer this action over to B_1 via the isomorphism of (27.9), arriving at the conclusion in the precise form stated above.

Working with the multiplier v of $\mathscr{B}\flat G$, as in (27.8), recall that v also acts as a multiplier of any ideal of $\mathscr{B}\flat G$, such as E_g and $E_{g^{-1}}$. Consequently $\Gamma_g(v)$ may be seen as a multiplier of $\Gamma_g(E_{g^{-1}}) = E_g$, so we see that[34]

$$v^* E_g \Gamma_g(v) \subseteq E_g.$$

Using (27.10.i), we then have that

$$\varphi_g : b \in B_g \mapsto v^*(b \otimes e_{1,g})\Gamma_g(v) \in E_g,$$

is a well defined map for each g in G. On the other hand, by (27.10.ii) we see that there is a map

$$\psi_g : E_g \to B_g,$$

such that

$$\psi_g(y) \otimes e_{g^{-1},1} = \Delta_{g^{-1}}(vy)v^*, \quad \forall y \in E_g. \tag{27.11.1}$$

We will now prove that φ_g and ψ_g are each other's inverse. For this, notice that if $y \in E_g$, then

$$\begin{aligned}
\varphi_g\big(\psi_g(y)\big) &= v^*(\psi_g(y) \otimes e_{1,g})\Gamma_g(v) \\
&= v^*\Gamma_g\big((\psi_g(y) \otimes e_{g^{-1},1})v\big) \\
&\overset{(27.11.1)}{=} v^*\Gamma_g\big(\Delta_{g^{-1}}(vy)v^*v\big) \\
&\overset{(27.8)}{=} y.
\end{aligned}$$

On the other hand, given b in B_g, we have

$$\begin{aligned}
\psi_g\big(\varphi_g(b)\big) \otimes e_{g^{-1},1} &= \Delta_{g^{-1}}(v\varphi_g(b))v^* \\
&= \Delta_{g^{-1}}\big(vv^*(b \otimes e_{1,g})\Gamma_g(v)\big)v^* \\
&\overset{(27.8)}{=} \Delta_{g^{-1}}\big((1 \otimes e_{1,1})(b \otimes e_{1,g})\Gamma_g(v)\big)v^* \\
&= \Delta_{g^{-1}}\big((b \otimes e_{1,g})\Gamma_g(v)\big)v^* \\
&= (b \otimes e_{g^{-1},1})vv^* \\
&= (b \otimes e_{g^{-1},1})(1 \otimes e_{1,1}) \\
&= b \otimes e_{g^{-1},1},
\end{aligned}$$

[34]Observe that if $u = (L_u, R_u)$ and $v = (L_v, R_v)$ are multipliers of E_g, then for every a in E_g, the expression uav stands for either $L_u(R_v(a))$ or $R_v(L_u(a))$, which coincide with each other thanks to (7.9), and to the fact that E_g, being a C*-algebra, is both non-degenerate and idempotent.

from where we see that $\psi_g\big(\varphi_g(b)\big) = b$, whence ψ_g is indeed the inverse of φ_g. Since both of these maps are contractive, then both are in fact isometries. We then define

$$\Phi_g : B_g \to E_g\delta_g,$$

by

$$\Phi_g(b) = \varphi_g(b)\delta_g, \quad \forall\, b \in B_g,$$

and we will prove that the collection of maps $\Phi = \{\Phi_g\}_{g\in G}$ gives an isomorphism from $\mathscr{B}$ to the semi-direct product bundle relative to Δ.

Since we already know that the Φ_g are bijective, it now suffices to prove that Φ is a morphism of Fell bundles. We begin by proving (21.1.i). Given $b \in B_g$, and $c \in B_h$, let $x = \varphi_g(b)$, and $y = \varphi_h(c)$. We then have

$$\begin{aligned}
\Phi_g(b)\Phi_h(c) &= (x\delta_g)(y\delta_h)\\
&= \Delta_g\big(\Delta_{g^{-1}}(x)y\big)\delta_{gh}\\
&= x\Gamma_g(y)\delta_{gh}\\
&= v^*(b\otimes e_{1,g})\Gamma_g(v)\Gamma_g\big(v^*(c\otimes e_{1,h})\Gamma_h(v)\big)\delta_{gh}\\
&= v^*(b\otimes e_{1,g})\Gamma_g(1\otimes e_{1,1})(c\otimes e_{g,gh})\Gamma_{gh}(v)\delta_{gh}\\
&= v^*(b\otimes e_{1,g})(1\otimes e_{g,g})(c\otimes e_{g,gh})\Gamma_{gh}(v)\delta_{gh}\\
&= v^*(bc\otimes e_{1,gh})\Gamma_{gh}(v)\delta_{gh}\\
&= \Phi_{gh}(bc).
\end{aligned}$$

Referring to (21.1.ii), pick b in any B_g, and notice that

$$\begin{aligned}
\Phi_g(b)^* &= \big(\varphi_g(b)\delta_g\big)^*\\
&= \Delta_{g^{-1}}\big(\varphi_g(b)^*\big)\delta_{g^{-1}}\\
&= \Gamma_{g^{-1}}\big(\Gamma_g(v^*)(b^*\otimes e_{g,1})v\big)\delta_{g^{-1}}\\
&= v^*(b^*\otimes e_{1,g^{-1}})\Gamma_{g^{-1}}(v)\delta_{g^{-1}}\\
&= \Phi_{g^{-1}}(b^*).
\end{aligned}$$

This concludes the proof. □

As an application of this to graded algebras we present the following:

Corollary 27.12. *Suppose we are given a countable group G, and a separable, topologically G-graded C*-algebra B. Suppose moreover that B_1 is stable and the canonical conditional expectation onto B_1 is faithful. Then there exists a partial action of G on B_1 such that*

$$B \simeq B_1 \rtimes_{red} G.$$

PROOF. All of the conditions of (27.11) are clearly fulfilled for the associated Fell bundle $\mathscr{B}$, so we may assume that $\mathscr{B}$ is the semi-direct product bundle for a partial action of G on B_1. By (19.8) we then have that B is isomorphic to the reduced cross-sectional C*-algebra of $\mathscr{B}$, also known as $B_1 \rtimes_{\text{red}} G$. □

Notes and remarks. The fact that every separable, stable Fell bundle arises from a partial crossed product is known for almost 20 years [**48**, Theorem 7.3], except that the partial action might be twisted by a cocycle. On the other hand, the so called Packer-Raeburn trick [**86**, Theorem 3.4] asserts that, up to stabilization, every twisted global action is *exterior equivalent* to a genuine (untwisted) action.

Therefore it has been widely suspected that the cocycle in [**48**, Theorem 7.3] could be eliminated. Theorem (27.11) does precisely that. It has been proven by Sehnem in her Masters Thesis [**100**]. A purely algebraic version of [**48**, Theorem 7.3] may be found in [**35**, Theorem 8.5].

CHAPTER 28

Globalization in the C*-Context

The question of globalization is one of the richest parts of the theory of partial actions. Recall that partial actions on sets always have a unique globalization (3.5), and so do partial actions on topological spaces (5.5), although not always on a Hausdorff space (5.6). In the case of algebraic partial actions, existence (6.14) and uniqueness (6.7) may fail, except when the corresponding ideals are unital (6.13).

With such a track record, when it comes to C*-algebras we clearly shouldn't expect a smooth ride. In order to be able to properly discuss globalization for C*-algebraic partial actions we will need a large part of the material developed so far and it is for this reason that the present chapter has been postponed until now.

The concept of restriction, as defined in (3.2), is perfectly suitable for the category of C*-algebras, requiring no further adaptation. In other words, if η is a global C*-algebraic action of a group G on an algebra B, and if we are given a closed two-sided ideal $A \trianglelefteq B$, the restriction of η to A gives a bona fide C*-algebraic partial action of G on A. However the concept of globalization given in (6.6) requires some fine tuning if it is to be of any use when working with C*-algebras. The difference between (6.6) and the following definition is essentially the occurrence of the word *closure* below.

Definition 28.1. Let η be a C*-algebraic global action of a group G on an algebra B, and let A be a closed two-sided ideal of B. Also let θ be the partial action obtained by restricting η to A. If B is the *closure* of

$$\sum_{g\in G} \eta_g(A),$$

we will say that η is a *C*-algebraic globalization* of θ.

It is not difficult to adapt the example given in (6.14) to produce a C*-algebraic partial action admitting no globalization. However, example (6.7), exploiting a somewhat grotesque algebraic structure (identically zero product) to show the lack of uniqueness for globalizations of algebraic partial actions, has no counterpart for C*-algebras: our next result shows that if a C*-algebraic partial action admits a globalization, then that globalization is necessarily unique.

Proposition 28.2. *Let $\theta = (\{D_g\}_{g\in G}, \{\theta_g\}_{g\in G})$ be a C*-algebraic partial action of the group G on the algebra A, and suppose that for each $k = 1, 2$, we are given a globalization η^k of θ, acting on a C*-algebra B^k. Then there exists an equivariant *-isomorphism*

$$\varphi : B^1 \to B^2$$

which is the identity on the respective copies of A within B^1 and B^2.

PROOF. As a first step we claim that, if a and b are in A, then

$$\eta^1_g(a)b = \eta^2_g(a)b, \quad \forall g \in G. \tag{28.2.1}$$

Choosing an approximate identity $\{v_i\}_i$ for $D_{g^{-1}}$, notice that $\{\theta_g(v_i)\}_i$ is an approximate identity for D_g. Also, since

$$\eta_g^k(a)b \in \eta_g^k(A) \cap A = D_g, \quad \forall k = 1, 2,$$

we have

$$\eta_g^k(a)b = \lim_{i\to\infty} \theta_g(v_i)\eta_g^k(a)b = \lim_{i\to\infty} \eta_g^k(v_i a)b = \lim_{i\to\infty} \theta_g(v_i a)b,$$

where the last step is justified by the fact that $v_i a$ is in $D_{g^{-1}}$. Since the right-hand-side above does not depend on k, the claim is proved.

Recall that each B^k is the closure of the sum of the ideals $\eta_g^k(A)$, for g in G. This suggests the following slightly more general situation: suppose we are given a C*-algebra B, which is the closure of the sum of a family $\{J_i\}_{i\in I}$ of closed two-sided ideals. Then every b in B acts as a multiplier of each J_i by left and right multiplication, thus providing a canonical *-homomorphism

$$\mu_i : B \to \mathcal{M}(J_i).$$

Setting

$$\mu = \prod_{i\in I} \mu_i : B \to \prod_{i\in I} \mathcal{M}(J_i),$$

(here the product on the right-hand-side is defined to be the C*-algebra formed by all *bounded* families $x = (x_i)_{i\in I}$, with each x_i in $\mathcal{M}(J_i)$, equipped with coordinate-wise operations and the supremum norm), we claim that μ is injective. In fact, if $b \in B$ is such that $\mu_i(b) = 0$, for all $i \in I$, then

$$bx = 0, \quad \forall x \in \textstyle\bigcup_{i\in I} J_i.$$

Since the set of all such x's span a dense subspace of B, we conclude that $b = 0$. Being injective, μ is also necessarily isometric, so for each b in B, one has that

$$\|b\| = \|\mu(b)\| = \sup_{i\in I} \|\mu_i(b)\| = \sup_{i\in I} \sup_{\substack{x\in J_i \\ \|x\|\leq 1}} \|bx\|.$$

Returning to the above setting, given $a_1, \ldots, a_n \in A$, and $g_1, \ldots, g_n \in G$, we may then compute the norm of the element

$$b = \sum_{i=1}^n \eta_{g_i}^1(a_i) \in B^1,$$

as follows:

$$\begin{aligned}
\|b\| &= \sup_{h\in G} \sup_{\substack{a\in A\\ \|a\|\leq 1}} \|b\eta^1_h(a)\| \\
&= \sup_{h,a} \Big\| \sum_{i=1}^n \eta^1_{g_i}(a_i)\eta^1_h(a) \Big\| \\
&= \sup_{h,a} \Big\| \eta^1_h\Big(\sum_{i=1}^n \eta^1_{h^{-1}g_i}(a_i)a\Big) \Big\| \\
&= \sup_{h,a} \Big\| \sum_{i=1}^n \eta^1_{h^{-1}g_i}(a_i)a \Big\| \\
&\overset{(28.2.1)}{=} \sup_{h,a} \Big\| \sum_{i=1}^n \eta^2_{h^{-1}g_i}(a_i)a \Big\| \\
&= \cdots = \Big\| \sum_{i=1}^n \eta^2_{g_i}(a_i) \Big\|,
\end{aligned}$$

where the ellipsis indicates the reversal of our computations with the superscript "1" replaced by "2". This implies that the correspondence

$$\sum_{i=1}^n \eta^1_{g_i}(a_i) \mapsto \sum_{i=1}^n \eta^2_{g_i}(a_i), \tag{28.2.2}$$

is well defined and extends to give an isometric linear mapping $\varphi : B^1 \to B^2$, which we claim to satisfy all of the required conditions.

We begin with the verification that φ is multiplicative, for which it clearly suffices to check that

$$\varphi\big(\eta^1_g(a)\eta^1_h(b)\big) = \eta^2_g(a)\eta^2_h(b), \quad \forall\, g,h \in G, \quad \forall\, a,b \in A.$$

We have

$$\begin{aligned}
\varphi\big(\eta^1_g(a)\eta^1_h(b)\big) &= \varphi\Big(\eta^1_h\big(\eta^1_{h^{-1}g}(a)b\big)\Big) \\
&\overset{(28.2.1)}{=} \varphi\Big(\eta^1_h\big(\eta^2_{h^{-1}g}(a)b\big)\Big) \\
&\overset{(28.2.2)}{=} \eta^2_h\big(\eta^2_{h^{-1}g}(a)b\big) \\
&= \eta^2_g(a)\eta^2_h(b).
\end{aligned}$$

The easy verification that φ is equivariant and preserves the star operation is left to the reader. □

Another important aspect of globalization is that it does not mix the commutative and the non-commutative worlds:

Proposition 28.3. *Let η be a globalization of a C^*-algebraic partial action θ. If θ acts on a commutative algebra, then so does η.*

PROOF. Let A and B be the algebras where θ and η act, respectively, so that A is a commutative ideal in B. We first claim that A is contained in the center of B.

To see this pick a in A, and b in B. Using Cohen-Hewitt, we may write $a = a_1a_2$, with a_1 and a_2 in A. We then have

$$ab = (a_1a_2)b = a_1(a_2b) = (a_2b)a_1 = a_2(ba_1) = (ba_1)a_2 = ba,$$

proving that a is in the center of B. This argument in fact shows that any commutative idempotent ideal must be central.

Since B is generated by the translates of A, it is clearly enough to prove that

$$\eta_g(a)\eta_h(b) = \eta_h(b)\eta_g(a),$$

for any a and b in A, and for any g and h in G, but this is easily proven with the following computation:

$$\eta_g(a)\eta_h(b) = \eta_g\big(a\eta_{g^{-1}h}(b)\big) = \eta_g\big(\eta_{g^{-1}h}(b)a\big) = \eta_h(b)\eta_g(a).$$

□

Since partial actions on commutative C*-algebras correspond bijectively to partial actions on LCH (locally compact Hausdorff) spaces by (11.6), we may correlate the globalization questions for commutative C*-algebras on the one hand, and for topological spaces, on the other:

Proposition 28.4. *Let θ be a partial action of a group G on a LCH space X, and denote by θ' the partial action of G on $C_0(X)$ corresponding to θ via* (11.6). *Then a necessary and sufficient condition for θ' to admit a globalization is that the globalization of θ provided by* (5.5) *take place on a Hausdorff space.*

PROOF. If (η, Y) is a globalization of θ and Y is Hausdorff, it is easy to see that the corresponding action η' of G on $C_0(Y)$ is a globalization for θ'.

Conversely, if we are given a globalization (η', B) for θ', we have by (28.3) that B is commutative, hence $B \simeq C_0(Y)$, where Y is the spectrum of B. Denoting by η the global action of G on Y corresponding to η' under (11.6), it is easy to see that η is the globalization for θ. Being the spectrum of a commutative C*-algebra, Y is Hausdorff. □

This result may easily be used to produce examples of C*-algebraic partial actions not admitting a globalization: just take a partial action on a LCH space X whose graph is not closed, so that its globalization will be non-Hausdorff by (5.6). The corresponding partial action on $C_0(X)$ will therefore admit no globalization by (28.4).

Having seen that the existence question for globalization of C*-algebraic partial actions often has a negative answer, one might try to relax the question itself, and look for Morita-Rieffel-equivalent actions admitting a globalization. The following main result shows that this can always be attained.

Theorem 28.5. *Every C*-algebraic partial action is Morita-Rieffel-equivalent to one admitting a globalization. More precisely, every C*-algebraic partial action is Morita-Rieffel-equivalent to the dual partial action Δ on the restricted smash product for the corresponding semi-direct product bundle (which admits a globalization by* (26.14)*).*

PROOF. Given a C*-algebraic partial action

$$\theta = (\{D_g\}_{g\in G},\ \{\theta_g\}_{g\in G})$$

of the group G on the C*-algebra A, denote by $\mathscr{B}$ its semi-direct product bundle. Our task is therefore to construct a Hilbert A-$\mathscr{B}\flat G$-bimodule M, and a partial action γ of G on M, satisfying the conditions required by (15.7). We begin by letting M be the subspace of $\mathscr{B}\flat G$ given by

$$M = \overline{\sum_{h\in G} B_h \otimes e_{1,h}}.$$

If the reader is used to thinking of elements of $C^*(\mathscr{B}) \otimes \mathcal{K}$ as infinite matrices with entries in $C^*(\mathscr{B})$, then M should be thought of as the space of all row matrices in $\mathscr{B}\flat G$. Observe that M is invariant under left-multiplication by $B_1 \otimes e_{1,1}$, so, upon identifying A with $B_1 \otimes e_{1,1}$ via

$$a \in A \mapsto a\delta_1 \otimes e_{1,1} \in B_1 \otimes e_{1,1}, \tag{28.5.1}$$

we may use the multiplicative structure of $\mathscr{B}\flat G$ to define the left A-module structure. Given h, k and l in G, notice that

$$(B_h \otimes e_{1,h})(B_{k^{-1}} B_l \otimes e_{k,l}) = \delta_{h,k}(B_h B_{h^{-1}} B_l \otimes e_{1,l}) \subseteq B_l \otimes e_{1,l} \subseteq M,$$

from where we see that M is a right ideal in $\mathscr{B}\flat G$, hence also a right $\mathscr{B}\flat G$-module.

Given ξ and η in M, it is easy to see that $\xi\eta^*$ (operated as elements of the C*-algebra $\mathscr{B}\flat G$) lies in $B_1 \otimes e_{1,1}$. So, under the identification (28.5.1), we may view $\xi\eta^*$ as an element of A. In other words, we define the A-valued inner-product of ξ and η to be the unique element $\langle \xi, \eta\rangle_A \in A$, such that

$$\xi\eta^* = \langle \xi, \eta\rangle_A \delta_1 \otimes e_{1,1}. \tag{28.5.2}$$

For the $\mathscr{B}\flat G$-valued inner-product we simply set

$$\langle \xi, \eta\rangle_{\mathscr{B}\flat G} = \xi^*\eta, \quad \forall\, \xi, \eta \in M.$$

Proving M to be a Hilbert A-$\mathscr{B}\flat G$-bimodule is now entirely routine. In order to construct the partial action, we let for each g in G,

$$M_g = \overline{\sum_{h\in G} [B_g B_{g^{-1}} B_h] \otimes e_{1,h}}.$$

It is easy to see that $(B_1 \otimes e_{1,1})M_g \subseteq M_g$, so that M_g is a left A-sub-module. In order to prove it to be a right $\mathscr{B}\flat G$-sub-module, notice that for each h and k in G, we have

$$(B_g B_{g^{-1}} B_h \otimes e_{1,h})(B_{h^{-1}} B_k \otimes e_{h,k}) \subseteq B_g B_{g^{-1}} B_k \otimes e_{1,k} \subseteq M_g,$$

thus concluding the verification of (15.7.i).

Recall that in (26.13) we used the notation D_g to mean $[B_g B_{g^{-1}}]$, while here D_g is supposed to mean the range of θ_g. Insisting on the present use of D_g, observe that $B_g = D_g\delta_g$, hence

$$[B_g B_{g^{-1}}] = [(D_g\delta_g)(D_{g^{-1}}\delta_{g^{-1}})] \overset{(8.14.b)}{=} [D_g\theta_g(D_{g^{-1}})\delta_1] = D_g\delta_1,$$

so the current meaning of D_g is compatible with the one used in (26.13), up to the usual identification of D_g with $D_g\delta_1$.

Given g and h in G observe that

$$[B_g B_{g^{-1}} B_h] = [D_g D_h \delta_h] = (D_g \cap D_h)\delta_h,$$

so M_g may be alternatively described as

$$M_g = \overline{\sum_{h\in G} (D_g \cap D_h)\delta_h \otimes e_{1,h}}. \tag{28.5.3}$$

Speaking of (15.7.ii), notice that for each h in G, we have that

$$\Big((D_g \cap D_h)\delta_h\Big)\Big((D_g \cap D_h)\delta_h\Big)^* \overset{(8.14.f)}{=} (D_g \cap D_h)\delta_1,$$

from where we see that $[\langle M_g, M_g\rangle_A] = D_g$. On the other hand, given h and k in G, we have

$$\begin{aligned}[\big(B_gB_{g^{-1}}B_h \otimes e_{1,h}\big)^*\big(B_gB_{g^{-1}}B_k \otimes e_{1,k}\big)] &= [B_{h^{-1}}B_gB_{g^{-1}}B_gB_{g^{-1}}B_k \otimes e_{h,k}] \\ &= [B_{h^{-1}}D_gB_k \otimes e_{h,k}],\end{aligned}$$

so we deduce from (26.13) that $[\langle M_g, M_g\rangle_{\mathscr{B}\flat G}] = E_g$, as desired.

We next claim that for each g in G, there exists a bounded linear mapping

$$\gamma_g : M_{g^{-1}} \to M_g,$$

such that

$$\gamma_g(a\delta_h \otimes e_{1,h}) = \theta_g(a)\delta_{gh} \otimes e_{1,gh}, \tag{28.5.4}$$

for all g and h in G, and for all a in $D_{g^{-1}} \cap D_h$. In order to see this, first notice that if a is as above, then

$$\theta_g(a) \in \theta_g(D_{g^{-1}} \cap D_h) \overset{(2.5.i)}{\subseteq} D_g \cap D_{gh},$$

from where we see that the right-hand-side of (28.5.4) indeed lies in M_g, as needed. Denoting by M_g^0 the dense subspace of M_g defined as in (28.5.3), but without taking closures, it is therefore immediate to prove the existence of a map $\gamma_g^0 : M_{g^{-1}}^0 \to M_g^0$, satisfying (28.5.4). Given ξ and η in $M_{g^{-1}}^0$, write them as finite sums

$$\xi = \sum_{h\in G} a_h\delta_h \otimes e_{1,h}, \quad \text{and} \quad \eta = \sum_{h\in G} b_h\delta_h \otimes e_{1,h},$$

with each a_h and b_h in $D_{g^{-1}} \cap D_h$, and notice that

$$\begin{aligned}\gamma_g^0(\xi)\gamma_g^0(\eta)^* &= \Big(\sum_{h\in G} \theta_g(a_h)\delta_{gh} \otimes e_{1,gh}\Big)\Big(\sum_{k\in G} \theta_g(b_k)\delta_{gk} \otimes e_{1,gk}\Big)^* \\ &\overset{(8.14.f)}{=} \sum_{h\in G} \theta_g(a_hb_h^*)\delta_1 \otimes e_{1,1}.\end{aligned}$$

In view of (28.5.2), this shows that

$$\langle\gamma_g^0(\xi), \gamma_g^0(\eta)\rangle_A = \sum_{h\in G} \theta_g(a_hb_h^*) = \theta_g\big(\langle\xi, \eta\rangle_A\big),$$

from where we deduce two important facts: firstly, that γ_g^0 is an isometry, so it extends to a bounded linear map γ_g from $M_{g^{-1}}$ to M_g, proving our claim and, secondly, that γ_g satisfies (15.7.iv) relative to the A-valued inner-product. Let us now also prove (15.7.iv) for the $\mathscr{B}\flat G$-valued inner-product, so we begin by taking ξ and η in $M_{g^{-1}}$ of the form

$$\xi = a\delta_h \otimes e_{1,h}, \quad \text{and} \quad \eta = b\delta_k \otimes e_{1,k},$$

with a in $D_{g^{-1}} \cap D_h$, and b in $D_{g^{-1}} \cap D_k$. So

$$\begin{aligned}\langle \gamma_g(\xi), \gamma_g(\eta) \rangle_{\mathscr{B} \flat G} &= \big(\theta_g(a)\delta_{gh} \otimes e_{1,gh}\big)^* \big(\theta_g(b)\delta_{gk} \otimes e_{1,gk}\big) \\ &\overset{(8.14.e)}{=} \theta_{h^{-1}g^{-1}}\big(\theta_g(a^*b)\big)\delta_{h^{-1}k} \otimes e_{gh,gk} \\ &\overset{(2.5.ii)}{=} \theta_{h^{-1}}(a^*b)\delta_{h^{-1}k} \otimes e_{gh,gk} \\ &= \Delta_g\Big(\theta_{h^{-1}}(a^*b)\delta_{h^{-1}k} \otimes e_{h,k}\Big) \\ &= \Delta_g\Big(\langle \xi, \eta \rangle_{\mathscr{B} \flat G}\Big).\end{aligned}$$

Since the set of elements ξ and η considered above span a dense subspace of $M_{g^{-1}}$, we have proven (15.7.iv).

To conclude we must now prove that γ is indeed a partial action on M. Since (2.1.i) is elementary, we prove only (2.1.ii). In order to do this, we first observe that, as a consequence of (26.7), given w in $\mathscr{B} \flat G$, a necessary and sufficient condition for w to be in M_g is that

$$w_{1,k} \in (D_g \cap D_k)\delta_k, \quad \text{and} \quad w_{h,k} = 0, \tag{28.5.5}$$

for all h and k in G, with $h \neq 1$. This said, we claim that, for all g and h in G, one has that

$$\gamma_h^{-1}(M_h \cap M_{g^{-1}}) \subseteq M_{(gh)^{-1}}. \tag{28.5.6}$$

Notice that the set in the left-hand-side above is precisely the domain of $\gamma_g \circ \gamma_h$ by (2.2). We thus pick an element ξ in this set, so that

$$\xi \in M_{h^{-1}}, \quad \text{and} \quad \gamma_h(\xi) \in M_{g^{-1}}.$$

Letting $a_k \in D_{h^{-1}} \cap D_k$ be defined by $\xi_{1,k} = a_k\delta_k$, observe that

$$\gamma_h(\xi)_{1,hk} = \theta_h(a_k)\delta_{hk},$$

as a quick look at (28.5.4) will reveal. Since $\gamma_h(\xi) \in M_{g^{-1}}$, we deduce from (28.5.5) that

$$\gamma_h(\xi)_{1,hk} \in (D_{g^{-1}} \cap D_{hk})\delta_{hk},$$

which implies that

$$\theta_h(a_k) \in D_{g^{-1}} \cap D_{hk} \quad \cap D_h,$$

so

$$a_k \in \theta_{h^{-1}}(D_{g^{-1}} \cap D_{hk} \cap D_h) \overset{(2.5.i)}{\subseteq} D_{h^{-1}g^{-1}} \cap D_k \cap D_{h^{-1}}.$$

Therefore

$$\xi_{1,k} = a_k\delta_k \in (D_{h^{-1}g^{-1}} \cap D_k)\delta_k,$$

whence ξ is in $M_{(gh)^{-1}}$, by (28.5.5), proving claim (28.5.6). In particular, this shows that the domain of $\gamma_g \circ \gamma_h$ is contained in the domain of γ_{gh}, and it is now easy to see that $\gamma_{gh}(\xi) = \gamma_g\big(\gamma_h(\xi)\big)$, for all ξ in the domain of $\gamma_g \circ \gamma_h$, thus verifying (2.1.ii), and hence finishing the proof. □

Starting with an arbitrary C*-algebraic partial dynamical system

$$\theta = (A, G, \{D_g\}_{g\in G}, \{\theta_g\}_{g\in G}),$$

we are then led to considering two other actions: the dual partial action Δ (26.12) for the semi-direct product bundle, and its globalization Γ (26.14).

Each of these actions comes with its full and reduced crossed products:

θ Given action	Δ Dual action	Γ Globalization
$A \rtimes G$	$(\mathscr{B} \flat G) \rtimes G$	$(\mathscr{B} \sharp G) \rtimes G$
$A \rtimes_{\text{red}} G$	$(\mathscr{B} \flat G) \rtimes_{\text{red}} G$	$(\mathscr{B} \sharp G) \rtimes_{\text{red}} G$

TABLE 28.6. Derived actions and crossed products.

We already know that θ and Δ are Morita-Rieffel-equivalent as partial actions. Even though Δ and Γ are not necessarily Morita-Rieffel-equivalent[35], we will now show that in each row of the above table all of the three algebras are Morita-Rieffel-equivalent to each other. Our next two results are designed to justify this claim.

Theorem 28.7.[36] *Let G be a group and suppose we are given Morita-Rieffel-equivalent C^*-algebraic partial dynamical systems*

$$\alpha = \big(A, G, \{A_g\}_{g\in G}, \{\alpha_g\}_{g\in G}\big), \quad \text{and} \quad \beta = \big(B, G, \{B_g\}_{g\in G}, \{\beta_g\}_{g\in G}\big).$$

Then

(i) *$A \rtimes_{red} G$ and $B \rtimes_{red} G$ are Morita-Rieffel-equivalent, and*
(ii) *$A \rtimes G$ and $B \rtimes G$ are Morita-Rieffel-equivalent.*

PROOF. Let

$$\gamma = \big(M, G, \{M_g\}_{g\in G}, \{\gamma_g\}_{g\in G}\big)$$

be an imprimitivity system for α and β, and let L be the linking algebra of M. Consider the partial action

$$\lambda = \big(\{L_g\}_{g\in G},\ \{\lambda_g\}_{g\in G}\big),$$

of G on L given by (15.10). Identified with the upper left-hand corner of L, it is clear that A is a λ-invariant subalgebra, so we may use case (i) of (22.3) to view $A \rtimes_{\text{red}} G$ as a subalgebra of $L \rtimes_{\text{red}} G$.

Unless A is a unital algebra, there is no canonical way to view the 2×2 matrix

$$e_{1,1} = \begin{pmatrix} 1 & 0 \\ 0 & 0 \end{pmatrix}$$

as an element of L, but, even when A is non-unital, the formal left- or right-multiplication of $e_{1,1}$ by elements of L is easily seen to define a multiplier of L.

By (16.27), which is stated for the full cross-sectional algebra, but which evidently also holds for the reduced one, we have that the inclusion of L in $L \rtimes_{\text{red}} G$ is a non-degenerate *-homomorphism. We may then extend it to a *-homomorphism

$$\mathcal{M}(L) \to \mathcal{M}\big(L \rtimes_{\text{red}} G\big),$$

[35]An example of a partial action not Morita-Rieffel-equivalent to its globalization is given right after (10.3). It is easy to see that the globalization of this partial action is the action of $\mathbb{Z}$ on itself by translation. These actions are not Morita-Rieffel-equivalent because Morita-Rieffel-equivalence among commutative algebras is tantamount to isomorphism.

[36]The statement in part (ii) of this result has been anticipated in (15.11).

and we will denote the image of $e_{1,1}$ under this map by $e_{1,1}\delta_1$. Given g in G, and

$$x = \begin{pmatrix} a & m \\ n^* & b \end{pmatrix} \in L_g,$$

notice that

$$(e_{1,1}\delta_1)(x\delta_g)(e_{1,1}\delta_1) \overset{(16.17)}{=} \lambda_g\big(\lambda_{g^{-1}}(e_{1,1}x)e_{1,1}\big)\delta_g$$

$$= \lambda_g\left(\begin{pmatrix} \alpha_{g^{-1}}(a) & \gamma_{g^{-1}}(m) \\ 0 & 0 \end{pmatrix} e_{1,1}\right)\delta_g$$

$$= \begin{pmatrix} a & 0 \\ 0 & 0 \end{pmatrix}\delta_g. \tag{28.7.1}$$

With this it is easy to see that

$$(e_{1,1}\delta_1)(L \rtimes_{\text{red}} G)(e_{1,1}\delta_1) = A \rtimes_{\text{red}} G, \tag{28.7.2}$$

from where it follows that $A \rtimes_{\text{red}} G$ is a hereditary subalgebra of $L \rtimes_{\text{red}} G$.

We next claim that A is a full subalgebra. To see this, we first check that the closed two-sided ideal generated by $A \rtimes_{\text{red}} G$ coincides with

$$\big[(L \rtimes_{\text{red}} G)(e_{1,1}\delta_1)(L \rtimes_{\text{red}} G)\big].$$

Temporarily denoting

$$L^{\rtimes} := L \rtimes_{\text{red}} G, \quad \text{and} \quad A^{\rtimes} := A \rtimes_{\text{red}} G,$$

and denoting $e_{1,1}\delta_1$ by p, notice that $[L^{\rtimes}pL^{\rtimes}]$ is an ideal in $L^{\rtimes}$, hence idempotent, so

$$[L^{\rtimes}pL^{\rtimes}] = [L^{\rtimes}pL^{\rtimes}L^{\rtimes}pL^{\rtimes}] = [L^{\rtimes}pL^{\rtimes}pL^{\rtimes}] \overset{(28.7.2)}{=} [L^{\rtimes}A^{\rtimes}L^{\rtimes}],$$

proving our claim. Given $g \in G$,

$$x = \begin{pmatrix} a & m \\ n^* & b \end{pmatrix} \in L_1 = L, \quad \text{and} \quad x' = \begin{pmatrix} a' & m' \\ n'^* & b' \end{pmatrix} \in L_g,$$

we have

$$(x\delta_1)(e_{1,1}\delta_1)(x'\delta_g) \overset{(16.17)}{=} xe_{1,1}x'\delta_g = \begin{pmatrix} aa' & am' \\ n^*a' & \langle n, m'\rangle_B \end{pmatrix}\delta_g.$$

This implies that the ideal generated by $A \rtimes_{\text{red}} G$ contains

$$\begin{pmatrix} [AA_g] & [AM_g] \\ [(A_gM)^*] & [\langle M, M_g\rangle_B] \end{pmatrix}\delta_g.$$

Observing that

$$[AA_g] = A_g, \qquad [AM_g] = M_g,$$

$$M_g \overset{(15.2)}{=} [\langle M_g, M_g\rangle_A M_g] \subseteq [A_gM],$$

and

$$B_g \overset{(15.7.\text{ii})}{=} [\langle M_g, M_g\rangle_B] \subseteq [\langle M, M_g\rangle_B],$$

we therefore conclude that the ideal generated by $A \rtimes_{\text{red}} G$ contains $L_g\delta_g$, for every g, and consequently also $L \rtimes_{\text{red}} G$.

The grand conclusion is that $A \rtimes_{\text{red}} G$ is a full hereditary subalgebra of $L \rtimes_{\text{red}} G$, so these are Morita-Rieffel-equivalent C*-algebras by (15.6).

We may now rerun the whole argument above in order to prove $B \rtimes_{\text{red}} G$ to be Morita-Rieffel-equivalent to $L \rtimes_{\text{red}} G$. Since Morita-Rieffel-equivalence is well known to be an equivalence relation, the proof of (i) is concluded.

Focusing now on (ii), let us consider the semi-direct product bundles $\mathscr{A}$ and $\mathscr{L}$ associated to the partial actions α and λ, respectively. We claim that $\mathscr{A}$ and $\mathscr{L}$ satisfy the hypothesis of (21.30), namely that there exists a conditional expectation from $\mathscr{L}$ onto $\mathscr{A}$.

Since each fiber of $\mathscr{L}$ is faithfully represented in $C^*_{\text{red}}(\mathscr{L})$ by (17.9.ii), we will produce the conditional expectation needed by working within $C^*_{\text{red}}(\mathscr{L})$, or equivalently, within $L \rtimes_{\text{red}} G$. We then claim that the mapping

$$E : x \in L \rtimes_{\text{red}} G \mapsto (e_{1,1}\delta_1)x(e_{1,1}\delta_1) \in A \rtimes_{\text{red}} G,$$

sends each fiber $L_g\delta_g$ to the corresponding $A_g\delta_g$, and that the restrictions P_g thus defined form a conditional expectation $P = \{P_g\}_{g\in G}$ from $\mathscr{L}$ to $\mathscr{A}$.

An expression for P_g may be obtained from our previous calculation (28.7.1), namely

$$P_g\left(\begin{pmatrix} a & m \\ n^* & b \end{pmatrix}\delta_g\right) = \begin{pmatrix} a & 0 \\ 0 & 0 \end{pmatrix}\delta_g.$$

The verification that the P_g satisfy (21.19.i)–(21.19.iii) is now straightforward and is left as an exercise.

As claimed, the hypothesis of (21.30) is now verified, so we conclude that the canonical embedding of full cross-sectional algebras is a monomorphism, which we interpret as

$$A \rtimes G \subseteq L \rtimes G.$$

The proof of (i) now generalizes ipsis literis, after replacing reduced crossed products by their full versions. □

Returning to the context of table (28.6), recall that θ is Morita-Rieffel-equivalent to Δ by (28.5). Thus, by the above result, in each one of the last two rows of table (28.6), the algebras in the first and second columns are Morita-Rieffel-equivalent.

In order to include the algebras in the third column, we present the following:

Theorem 28.8. *Let*

$$\theta = \big(A,\ G,\ \{D_g\}_{g\in G},\ \{\theta_g\}_{g\in G}\big)$$

be a C-algebraic partial dynamical system admitting a globalization η, acting on a C*-algebra B. Then:*

- (i) *$A\rtimes_{red} G$ is isomorphic to a full hereditary subalgebra of $B\rtimes_{red} G$, in a natural way, and hence $A\rtimes_{red} G$ and $B\rtimes_{red} G$ are Morita-Rieffel-equivalent.*
- (ii) *$A \rtimes G$ is isomorphic to a full hereditary subalgebra of $B \rtimes G$, in a natural way, and hence $A \rtimes G$ and $B \rtimes G$ are Morita-Rieffel-equivalent.*

PROOF. By (22.3) we have that $A\rtimes_{\text{red}} G$ is a hereditary subalgebra of $B\rtimes_{\text{red}} G$, so let us now prove that $A\rtimes_{\text{red}} G$ is a full subalgebra. This is to say that the closed two-sided ideal generated by $A \rtimes_{\text{red}} G$, say

$$J := \big[(B \rtimes_{\text{red}} G)(A \rtimes_{\text{red}} G)(B \rtimes_{\text{red}} G)\big],$$

coincides with $B \rtimes_{\text{red}} G$. For this pick g and h in G, and notice that

$$J \supseteq \big[(B\delta_h)(A\delta_1)(B\delta_{h^{-1}g})\big] = \big[B\eta_h(AB)\delta_g\big] = \eta_h(A)\delta_g,$$

so J also contains

$$\Big(\overline{\sum_{h\in G}\eta_h(A)}\Big)\delta_g = B\delta_g,$$

for every g in G, from where it follows that $J = B \rtimes_{\text{red}} G$, proving that $A \rtimes_{\text{red}} G$ is indeed full.

As for the second point, we have by (22.4) that $A \rtimes G$ is naturally isomorphic to a hereditary subalgebra of $B \rtimes G$, and the above proof that $A \rtimes_{\text{red}} G$ is a full subalgebra of $B \rtimes_{\text{red}} G$ generalizes ipsis literis to show that $A \rtimes G$ is a full subalgebra of $B \rtimes G$.

The statements about Morita-Rieffel-equivalence now follow immediately from (15.6). □

Given a partial dynamical system

$$\theta = (A, G, \{D_g\}_{g\in G}, \ \{\theta_g\}_{g\in G}),$$

we may summarize the main results obtained in this chapter as follows:

- When a globalization for θ exists, it is unique. See (28.2).
- Regardless of whether or not θ admit a globalization, there always exists a partial dynamical system Δ, Morita-Rieffel-equivalent to θ, which admits a globalization Γ. See (28.5).
- The reduced crossed products relative to the partial dynamical systems θ, Δ, and Γ are all Morita-Rieffel-equivalent to each other. See (28.7.i) & (28.8.i).
- The full crossed products relative to the partial dynamical systems θ, Δ, and Γ are all Morita-Rieffel-equivalent to each other. See (28.7.ii) & (28.8.ii).

There are a few other important results related to the globalization of C*-algebraic partial actions which we would now like to mention without proofs.

When a partial action θ is Morita-Rieffel-equivalent to another partial action Δ, which, in turn, admits a globalization Γ, one says that Γ is a *Morita-Rieffel-globalization* of θ.

The following is [**2**, Proposition 6.3]:

Theorem 28.9. *Any two Morita-Rieffel-globalizations of a given C*-algebraic partial action are Morita-Rieffel-equivalent to each other.*

Another very interesting result proved by Abadie regards partial actions on abelian C*-algebras. In order to describe it, let θ be a partial action of a group G on a LCH space X and denote by θ' the partial action of G on $C_0(X)$ corresponding to θ via (11.6).

By (5.5), we have that θ always admits a globalization, say η, acting on a space Y. However Y may be non-Hausdorff, and in this case θ' admits no globalization by (28.4). So let β be a Morita-Rieffel-globalization of θ', acting on a C*-algebra B, which always exists by (28.5).

The action β evidently induces an action $\hat{\beta}$ on the primitive ideal space $\text{Prim}(B)$, and because any two Morita-Rieffel-globalizations are equivalent by (28.9), $\hat{\beta}$ does not depend on the specific choice of β, being thus intrinsically associated to θ.

The next result is essentially [**2**, Proposition 7.4]:

Theorem 28.10. *Let θ be a topological partial action of a group G on a LCH space X, and let β, acting on B, be a Morita-Rieffel-globalization of the corresponding partial action on $C_0(X)$. Then the action $\hat{\beta}$ of G on* $\text{Prim}(B)$ *is the (possibly non-Hausdorff) globalization of θ.*

Notes and remarks. Most of the results in this chapter are extracted from Abadie's PhD Thesis [**1**], [**2**], where in fact the more general case of partial actions of locally compact groups is considered. See also [**5**] for a study of the relationship between the amenability of a partial action and that of its Morita enveloping action.

CHAPTER 29

Topologically Free Partial Actions

In chapter (23) we have studied induced and Fourier ideals in topologically graded algebras. These may be considered as the ideals which respect the given grading. In the present chapter we will study conditions which imply that all ideals somehow take the grading into account, such as having a nonzero intersection with the unit fiber algebra or, in the best-case scenario, are induced ideals.

The graded algebra at the center of the stage here will be the crossed product of an abelian C*-algebra A by a given group G, and the main conditions we will impose relate to the corresponding partial action of G on the spectrum of A, namely that there are not too many fixed points. As an application we will give conditions for a partial crossed product to be simple.

Definition 29.1. Let $\theta = \big(\{D_g\}_{g\in G}, \{\theta_g\}_{g\in G}\big)$ be a topological partial action of a group G on a locally compact Hausdorff space X, and let $g \in G$.

(a) A *fixed point* for g is any element x in $D_{g^{-1}}$, such that $\theta_g(x) = x$.
(b) The set of all fixed points for g will be denoted by F_g.
(c) We say that θ is *free* if F_g is empty for every $g \neq 1$.
(d) We say that θ is *topologically free* if the *interior* of F_g is empty for every $g \neq 1$.

Observe that $F_g = F_{g^{-1}}$, for all g in G. Also notice that F_g is a closed subset of $D_{g^{-1}}$ (and hence also of D_g by the previous remark) in the relative topology, but it is not necessarily closed in X. Incidentally, here is a useful result of general topology:

Lemma 29.2. *Let X be a topological space, and suppose we are given subsets $F \subseteq D \subseteq X$, such that F is closed relative to D, and D is open in X. If the interior of F is empty, then the interior of $\bar{F}$ is also empty, where $\bar{F}$ denotes the closure of F relative to X.*

PROOF. We should observe that, since D is open, the interior of F relative to D is the same as the interior of F relative to X.

It suffices to prove that, if V is an open subset of X, with $V \subseteq \bar{F}$, then $V = \emptyset$. Notice that for each such V, we have

$$V \cap D \subseteq \bar{F} \cap D = F,$$

the last equality being the expression that F is closed in D. Since F has no interior, we deduce that $V \cap D = \emptyset$.

It is well known that, when two open subsets are disjoint, each one is necessarily disjoint from the other's closure, whence

$$\emptyset = V \cap \bar{D} \supseteq V \cap \bar{F} = V.$$

□

Another topological fact we will need is the following baby version of Baire's category Theorem, which incidentally holds in any topological space:

Lemma 29.3. *Let X be a topological space and let $F_1, F_2, \ldots, F_n$ be nowhere dense*[37] *subsets of X. Then $F_1 \cup F_2 \cup \ldots \cup F_n$ is nowhere dense.*

PROOF. Replacing each F_i by its closure, we may clearly assume them to be closed, and hence so is their union.

In order to prove the statement we must show that every open set

$$V \subseteq F_1 \cup F_2 \cup \ldots \cup F_n$$

is necessarily empty. Given such a V, notice that,

$$W := V \setminus (F_1 \cup F_2 \cup \ldots \cup F_{n-1})$$

is an open set contained in F_n, whence W is empty by hypothesis. Thus $V \subseteq F_1 \cup F_2 \cup \ldots \cup F_{n-1}$, and the conclusion follows by induction. □

▶ From now on we will fix a group G, a locally compact Hausdorff space X and a topological partial action

$$\theta = \big(\{D_g\}_{g\in G}, \{\theta_g\}_{g\in G}\big)$$

of G on X. Our attention will be focused on the partial action of G on $C_0(X)$ induced by θ via (11.6), which we will henceforth denote by α. More precisely

$$\alpha = \Big(\big\{C_0(D_g)\big\}_{g\in G}, \{\alpha_g\}_{g\in G}\Big),$$

where each α_g is given by

$$\alpha_g : f \in C_0(D_{g^{-1}}) \mapsto f \circ \theta_{g^{-1}} \in C_0(D_g).$$

Above all, we are interested in the reduced crossed product

$$C_0(X) \rtimes_{\text{red}} G,$$

so, given f in $C_0(D_g)$, we will always interpret the expression "$f\delta_g$" as an element of the above crossed product algebra.

The following technical Lemma is the key tool in the proof of our main result below. It is intended to *shut out* certain elements in the grading subspaces $C_0(D_g)\delta_g$, with $g \neq 1$, by *compressing* them away with positive elements lying in the unit fiber algebra.

Lemma 29.4. *Given*

(*a*) $g \in G$, *with* $g \neq 1$,
(*b*) $f \in C_0(D_g)$,
(*c*) $x_0 \in X \setminus F_g$, *and*
(*d*) $\varepsilon > 0$,

there exists h in $C_0(X)$ such that

(*i*) $h(x_0) = 1$,
(*ii*) $0 \leq h \leq 1$, *and*
(*iii*) $\|(h\delta_1)(f\delta_g)(h\delta_1)\| \leq \varepsilon$.

[37] A subset of a topological space is said to be *nowhere dense* when its closure has empty interior.

PROOF. We separate the proof in two cases, according to whether or not x_0 lies in D_g. If x_0 is not in D_g, let

$$K = \{x \in D_g : |f(x)| \geq \varepsilon\}.$$

Then K is a compact subset of D_g and, since x_0 is not in K, we may use Urysohn's Lemma obtaining a continuous, real valued function on X such that

$$0 \leq h \leq 1, \quad h(x_0) = 1, \quad \text{and} \quad h|_K = 0.$$

By temporarily working in the one-point compactification of X, and replacing K by $K \cup \{\infty\}$, we may also assume that h vanishes at infinity, meaning that h is in $C_0(X)$.

Since f is bounded by ε outside of K, it follows that $\|hf\| \leq \varepsilon$, from where (iii) easily follows, hence concluding the proof in the present case.

If x_0 is in D_g, then $\theta_{g^{-1}}(x_0)$ is defined and not equal to x_0. We may then take disjoint open sets V_1 and V_2 such that

$$x_0 \in V_1, \quad \text{and} \quad \theta_{g^{-1}}(x_0) \in V_2,$$

and we may clearly assume that $V_1 \subseteq D_g$, and $V_2 \subseteq D_{g^{-1}}$. We may further shrink V_1 by replacing it with

$$V_1' = V_1 \cap \theta_g(V_2),$$

(notice that x_0 is still in V_1'), so that $\theta_{g^{-1}}(V_1') \subseteq V_2$, and then

$$\theta_{g^{-1}}(V_1') \cap V_1' = \emptyset. \tag{29.4.1}$$

Using Urysohn's Lemma again, pick h in $C_0(X)$ such that

$$0 \leq h \leq 1, \ h(x_0) = 1, \quad \text{and} \quad h|_{X \setminus V_1'} = 0.$$

We then have

$$(h\delta_1)(f\delta_g)(h\delta_1) = \alpha_g\big(\alpha_{g^{-1}}(hf)h\big)\delta_g = 0,$$

because $\alpha_{g^{-1}}(hf)$ is supported $\theta_{g^{-1}}(V_1')$, while h is supported in V_1', and these are disjoint sets by (29.4.1). This verifies (iii) in the final case, and the proof is thus concluded. □

The following result, together with its consequences to be presented below, is gist of this chapter:

Theorem 29.5. *Let θ be a topologically free partial action of a group G on a locally compact Hausdorff space X. Then any nonzero closed two-sided ideal*

$$J \trianglelefteq C_0(X) \rtimes_{red} G$$

has a nonzero intersection with $C_0(X)$.

PROOF. We should notice that the last occurrence of $C_0(X)$, above, stands for its copy $C_0(X)\delta_1$ within $C_0(X) \rtimes_{\text{red}} G$.

Denoting the conditional expectation provided by (17.8) by

$$E_1 : C_0(X) \rtimes_{\text{red}} G \to C_0(X),$$

we claim that, for every z in $C_0(X) \rtimes_{\text{red}} G$, and for every $\varepsilon > 0$, there exists $h \in C_0(X)$, such that

(i) $0 \leq h \leq 1$,
(ii) $\|hE_1(z)h\| \geq \|E_1(z)\| - \varepsilon$, and
(iii) $\|(h\delta_1)z(h\delta_1) - hE_1(z)h\delta_1\| \leq \varepsilon$.

Assume first that z is a linear combination of the form

$$z = z_1\delta_1 + \sum_{g\in T} z_g\delta_g, \tag{29.5.1}$$

where T is a finite subset of G, with $1 \notin T$, in which case $E_1(z) = z_1$. Let

$$V = \{x \in X : |z_1(x)| > \|z_1\| - \varepsilon\},$$

which is clearly open and nonempty. By the topological freeness hypothesis and by (29.2), each F_g is nowhere dense in X. Furthermore, by (29.3) one has that $\bigcup_{g\in T} F_g$ is likewise nowhere dense, hence there exists some

$$x_0 \in V \setminus \Big(\bigcup_{g\in T} F_g\Big).$$

For each g in T we may then apply (29.4), obtaining an h_g in $C_0(X)$ satisfying

$$h_g(x_0) = 1, \quad 0 \leq h_g \leq 1, \quad \text{and} \quad \|(h_g\delta_1)(z_g\delta_g)(h_g\delta_1)\| \leq \frac{\varepsilon}{|T|}.$$

Here we are tacitly assuming that $|T| > 0$, and we leave it for the reader to treat the trivial case in which $|T| = 0$.

We will now show that $h := \prod_{g\in T} h_g$, satisfies conditions (i–iii), above. Noticing that (i) is immediate, we prove (ii) by observing that x_0 is in V, so

$$\|hE_1(z)h\| = \|hz_1h\| \geq |z_1(x_0)| > \|z_1\| - \varepsilon = \|E_1(z)\| - \varepsilon.$$

As for (iii), we have

$$\begin{aligned}\|(h\delta_1)z(h\delta_1) - hE_1(z)h\delta_1\| &= \|(h\delta_1)\big(z - E_1(z)\delta_1\big)(h\delta_1)\| \\ &= \Big\|\sum_{g\in T}(h\delta_1)(z_g\delta_g)(h\delta_1)\Big\| \\ &\leq \sum_{g\in T}\|(h\delta_1)(z_g\delta_g)(h\delta_1)\| \\ &\leq \sum_{g\in T}\|(h_g\delta_1)(z_g\delta_g)(h_g\delta_1)\| \\ &\leq \varepsilon.\end{aligned}$$

This proves (i–iii) under special case (29.5.1), but since the elements of that form are dense in $C_0(X) \rtimes_{\text{red}} G$, a standard approximation argument gives the general case.

The claim verified, let us now address the statement. Arguing by contradiction we suppose that J is a nonzero ideal such that

$$J \cap C_0(X) = \{0\}.$$

Pick a nonzero element y in J, and let $z = y^*y$. Using the claim, for each positive ε we choose h in $C_0(X)$, satisfying (i–iii) above. Let

$$q \,:\, C_0(X) \rtimes_{\text{red}} G \to \frac{C_0(X) \rtimes_{\text{red}} G}{J}$$

be the quotient map. Since z lies in J, we have that

$$q\big((h\delta_1)z(h\delta_1)\big) = 0,$$

whence

$$\begin{aligned}\left\|q\big(hE_1(z)h\delta_1\big)\right\| &= \left\|q\big((h\delta_1)z(h\delta_1) - hE_1(z)h\delta_1\big)\right\| \\ &\leq \left\|(h\delta_1)z(h\delta_1) - hE_1(z)h\delta_1\right\| \\ &\overset{\text{(iii)}}{\leq} \varepsilon.\end{aligned}$$

Since $J \cap C_0(X) = \{0\}$, we deduce that q is injective, hence isometric, on $C_0(X)$. So

$$\varepsilon \geq \|hE_1(z)h\| \overset{\text{(ii)}}{\geq} \|E_1(z)\| - \varepsilon,$$

from where we see that $\|E_1(z)\| \leq 2\varepsilon$, and since ε is arbitrary, we have

$$0 = E_1(z) = E_1(y^*y).$$

It then results from (17.13) that $y = 0$, a contradiction. This concludes the proof. □

A useful consequence to the representation theory of reduced crossed products is as follows:

Corollary 29.6. *Let θ be a topologically free partial action of a group G on a locally compact Hausdorff space X. Then a *-representation of $C_0(X) \rtimes_{red} G$ is faithful if and only if it is faithful on $C_0(X)$.*

PROOF. Apply (29.5) to the kernel of the given representation. □

Definition 29.7. We will say that θ is a *minimal partial action* if there are no θ-invariant closed subsets of X, other than X and the empty set.

The complement of an invariant set is invariant too, so minimality is equivalent to the absence of nontrivial *open* invariant subsets.

Corollary 29.8. *If, in addition to the conditions of (29.5), θ is minimal, then $C_0(X) \rtimes_{red} G$ is a simple*[38] *C^*-algebra.*

PROOF. Let J be a nonzero, closed two-sided ideal of $C_0(X) \rtimes_{\text{red}} G$. Employing (29.5), we have that

$$K := J \cap C_0(X) \neq \{0\}. \tag{29.8.1}$$

Since K is an ideal in $C_0(X)$, there is an open subset $U \subseteq X$, such that $K = C_0(U)$. Using (23.11) we have that K is α-invariant, and it is easy to see that this implies U to be θ-invariant. By minimality of θ, one has that either $U = \emptyset$, or $U = X$, but under the first case we would have $K = \{0\}$, contradicting (29.8.1). So $U = X$, and hence $K = C_0(X)$, meaning that $C_0(X) \subseteq J$.

By (16.27), which is stated for the full cross-sectional algebra, but which evidently also holds for the reduced one, we have that $C_0(X)$ (the unit fiber algebra in the semi-direct product bundle) generates $C_0(X) \rtimes_{\text{red}} G$, as an ideal, whence $J = C_0(X) \rtimes_{\text{red}} G$, and the proof is concluded. □

The upshot of this result is that when only two θ-invariant open subsets exist in X, namely $\emptyset$ and the whole space, then only two closed two-sided ideals exist in $C_0(X) \rtimes_{\text{red}} G$, namely $\{0\}$ and the whole algebra.

This raises the question as to whether a correspondence may still be found between invariant open subsets and closed two-sided ideals, in case these exist in

[38] A C*-algebra is said to be simple if there are no nontrivial closed two-sided ideals.

greater numbers. In order to find such a correspondence we need a bit more than topological freeness.

Theorem 29.9. *Let θ be a topological partial action of a group G on a locally compact Hausdorff space X, such that either:*

(i) G is an exact group, or

(ii) the semi-direct product bundle associated to θ satisfies the approximation property.

In addition we suppose that θ, as well as the restriction of θ to any closed θ-invariant subset of X, is topologically free. Then there is a one-to-one correspondence between θ-invariant open subsets $U \subseteq X$, and closed two-sided ideals in $C_0(X) \rtimes_{red} G$, given by

$$U \mapsto C_0(U) \rtimes_{red} G.$$

PROOF. Given a θ-invariant open subset $U \subseteq X$, it is easy to see that $C_0(U)$ is an α-invariant ideal in $C_0(X)$. By (22.9), one then has that $C_0(U) \rtimes_{\mathrm{red}} G$ is an ideal in $C_0(X) \rtimes_{\mathrm{red}} G$.

By first checking on elements of the algebraic crossed product, it is easy to see that the conditional expectation E_1 of (17.8) satisfies

$$E_1\big(C_0(U) \rtimes_{\mathrm{red}} G\big) = C_0(U)\delta_1.$$

From this it follows that our correspondence is injective, and we are then left with the task of proving that any ideal

$$J \trianglelefteq C_0(X) \rtimes_{\mathrm{red}} G$$

is of the above form. Given J, let $U \subseteq X$ be the unique open subset of X such that

$$J \cap C_0(X) = C_0(U).$$

By (23.11) we have that $C_0(U)$ is α-invariant, whence U is θ-invariant and hence so is

$$F := X \setminus U.$$

It is a well known fact that the map

$$q : f \in C_0(X) \mapsto f|_F \in C_0(F)$$

passes to the quotient modulo $C_0(U)$, leading up to a *-isomorphism

$$C_0(X)/C_0(U) \simeq C_0(F).$$

Observe that $C_0(X)/C_0(U)$ carries a quotient partial action of G as proved in (22.7), while the restriction of θ to F induces a partial action of G on $C_0(F)$, via (11.6). The isomorphism mentioned above may then be easily shown to be G equivariant. So, employing (22.9) we get the following exact sequence of C*-algebras and *-homomorphisms

$$0 \to C_0(U) \rtimes_{\mathrm{red}} G \xrightarrow{\iota_{\mathrm{red}}} C_0(X) \rtimes_{\mathrm{red}} G \xrightarrow{q_{\mathrm{red}}} C_0(F) \rtimes_{\mathrm{red}} G \to 0.$$

By (16.27) notice that $C_0(U) \rtimes_{\mathrm{red}} G$ coincides with the ideal generated by $C_0(U)$ within $C_0(X) \rtimes_{\mathrm{red}} G$. Since $C_0(U) \subseteq J$, we then deduce that

$$C_0(U) \rtimes_{\mathrm{red}} G \subseteq J. \tag{29.9.1}$$

We next claim that

$$q_{\mathrm{red}}(J) \cap C_0(F) = \{0\}. \tag{29.9.2}$$

In order to see this, let z be in $q_{\text{red}}(J) \cap C_0(F)$, and choose y in J such that $q_{\text{red}}(y) = z$. Since $C_0(F) = q_{\text{red}}\big(C_0(X)\big)$, we may also choose f in $C_0(X)$ such that $q_{\text{red}}(f) = z$. Therefore $q_{\text{red}}(f - y) = 0$, hence we see that

$$f - y \in \text{Ker}(q_{\text{red}}) = C_0(U) \rtimes_{\text{red}} G \subseteq J,$$

from where it follows that

$$f \in J \cap C_0(X) = C_0(U),$$

so

$$0 = q_{\text{red}}(f) = z.$$

This proves (29.9.2), and since the restriction of θ to F is topologically free by hypothesis, we may use (29.5) to deduce that $q_{\text{red}}(J) = \{0\}$, which is to say that

$$J \subseteq \text{Ker}(q_{\text{red}}) = C_0(U) \rtimes_{\text{red}} G$$

so the inclusion in (29.9.1) is in fact an equality of sets. This concludes the proof. □

As a consequence we have:

Corollary 29.10. *Under the hypotheses of Theorem* (29.9)*, every ideal of* $C_0(X) \rtimes_{red} G$ *is induced (in the sense of Definition* (23.2)*), and hence also a Fourier ideal.*

PROOF. In view of (29.9) it is enough to prove that the ideal

$$J = C_0(U) \rtimes_{\text{red}} G$$

is induced for every θ-invariant open subset $U \subseteq X$. Given any such U, we need to prove that J is generated by $J \cap C_0(X)\delta_1$. However, since it is clear that

$$C_0(U)\delta_1 \subseteq J \cap C_0(X)\delta_1,$$

we deduce that

$$J = C_0(U) \rtimes_{\text{red}} G \overset{(16.27)}{=} \langle C_0(U)\delta_1 \rangle \subseteq \langle J \cap C_0(X)\delta_1 \rangle \subseteq J,$$

so J is indeed an induced ideal. By (23.9) we then also have that J is a Fourier ideal. □

Notes and remarks. Theorem (29.5) first appeared in [**59**, Theorem 2.6]. It is a direct generalization of [**74**, Theorem 4.1] for partial actions. See also [**56**, Theorem 4.4].

A groupoid version of Theorem (29.9) is stated in [**96**, Proposition 4.5], although there is a missing hypothesis in it, without which one might have the bad behavior discussed in [**97**, Remark 4.10]. The correct statement applies for amenable groupoids and is to be found in [**97**, Corollary 4.9].

Since a notion of *exact groupoid* does not seem to exist, Theorem (29.9) under hypothesis (i) does not appear to have a groupoid counterpart.

Many authors have studied results similar to the ones we proved above for global actions of discrete groups on non-abelian C*-algebras. See, for example, [**43**], [**74**] and [**8**].

Part 3

Applications

CHAPTER 30

Dilating Partial Representations

With this chapter we start a series of applications of the theory developed so far.

Recall from (9.5) that if v is a unitary group representation of G, and if p is a self-adjoint idempotent such that $v_g p v_{g^{-1}}$ commutes with p, for every g in G, then

$$u_g := p v_g p, \quad \forall g \in G$$

defines a partial representation of G.

One may think of u as a restriction of v, a process akin to restricting a partial action to a not necessarily invariant subset. In this chapter we will study a situation in which one may revert this process, proving that partial representations of a group on a Hilbert space may always be obtained from unitary representations via the above process.

Definition 30.1. In the context of (9.5), we will say that u is the *restriction* of v and, conversely, that v is a *dilation* of u.

In order to prepare for the proof of our main dilation result, we will now prove a related dilation Lemma, applicable to *-representations of C*-algebras.

Lemma 30.2. *Let B be a C*-algebra and A be a closed *-subalgebra of B. Given a *-representation π of A on a Hilbert space H, there exists a representation $\tilde{\pi}$ of B on another Hilbert space $\tilde{H}$, and a (not necessarily surjective) isometric linear operator $j : H \to \tilde{H}$, such that*

$$j\pi(a) = \tilde{\pi}(a)j, \quad \forall a \in A.$$

Proof. By splitting π in the direct sum of cyclic representations (perhaps including an identically zero sub-representation), we may assume, without loss of generality, that π is cyclic. So, let ξ be a cyclic vector for π. Defining

$$\varphi(a) = \langle \pi(a)\xi, \xi \rangle, \quad \forall a \in A,$$

we get a positive linear functional on A. Let $\tilde{\varphi}$ be a positive linear functional on B extending φ [**89**, Proposition 3.1.6], and let $\tilde{\pi}$ be the GNS representation of B associated to $\tilde{\varphi}$, acting on a Hilbert space $\tilde{H}$, with cyclic vector $\tilde{\xi}$.

We then claim that there exists an isometric linear operator $j : H \to \tilde{H}$, such that

$$j\big(\pi(a)\xi\big) = \tilde{\pi}(a)\tilde{\xi}, \quad \forall a \in A.$$

To see this, first define j on the dense subspace $\pi(A)\xi \subseteq H$ by the above formula. This is well defined and isometric because for every a in A, we have

$$\begin{aligned} \|\tilde{\pi}(a)\tilde{\xi}\|^2 &= \langle \tilde{\pi}(a)\tilde{\xi}, \tilde{\pi}(a)\tilde{\xi} \rangle \\ &= \langle \tilde{\pi}(a^*a)\tilde{\xi}, \tilde{\xi} \rangle \\ &= \tilde{\varphi}(a^*a) \\ &= \varphi(a^*a) \\ &= \langle \pi(a^*a)\xi, \xi \rangle \\ &= \|\pi(a)\xi\|^2. \end{aligned}$$

Extending j by continuity to the whole of H, we clearly obtain the claimed operator. So, in order to conclude the proof, it is enough to verify the identity in the statement.

It η is a vector in H of the form $\eta = \pi(c)\xi$, with c in A, then, for every a in A we have

$$\begin{aligned} j\big(\pi(a)\eta\big) &= j\big(\pi(a)\pi(c)\xi\big) \\ &= j\big(\pi(ac)\xi\big) \\ &= \tilde{\pi}(ac)\tilde{\xi} \\ &= \tilde{\pi}(a)\tilde{\pi}(c)\tilde{\xi} \\ &= \tilde{\pi}(a)j\big(\pi(c)\xi\big) \\ &= \tilde{\pi}(a)j(\eta). \end{aligned}$$

Since the set of η's considered above is dense in H, the proof is concluded. □

We should notice that, in the context of the above result, the range of j is easily seen to be invariant under $\tilde{\pi}(A)$. So the projection onto the range of j, namely jj^*, commutes with $\tilde{\pi}(A)$. Consequently, if we let ρ be the restriction of $\tilde{\pi}$ to A, then the sub-representation of ρ determined by the range of j is clearly equivalent to π, the equivalence being implemented by j. Another useful remark is that, since j is an isometry, we have that j^*j coincides with the identity on H, so

$$j^*\tilde{\pi}(a)j = j^*j\pi(a) = \pi(a),$$

for every a in A.

We may now prove our main result on dilation of partial representations.

Theorem 30.3. *Let u be a partial representation of a group G on a Hilbert space H. Then there exists a unitary group representation $\tilde{u}$ of G on a Hilbert space $\tilde{H}$, and an isometric linear operator $j : H \to \tilde{H}$, such that*

$$u_g = j^*\tilde{u}_g j, \quad \forall\, g \in G.$$

In addition, if p denotes the projection of $\tilde{H}$ onto the range of j, namely $p = jj^*$, then p commutes with $\tilde{u}_{g^{-1}} p \tilde{u}_g$ for every g in G.

Proof. We should first observe that, if we identify H with its image via j, hence thinking of j as the inclusion map, then, in line with (9.5), the formula displayed in the statement becomes

$$u_g = p\tilde{u}_g p.$$

However, in this proof we will not enforce the identification of H as a subspace of $\tilde{H}$.

Taking $\mathcal{R}$ to be the empty set of relations in (14.4) (see also (14.17) and (10.5)), we conclude that there exists a *-representation π of $C^*_{\text{par}}(G)$ on H such that

$$\pi(w_g) = u_g, \quad \forall\, g \in G,$$

where the w_g denote the canonical generators of $C^*_{\text{par}}(G)$ (please notice that these are denoted by u_g in (14.4)). In particular

$$\pi(1) = \pi(w_1) = u_1 = 1,$$

so π is a non-degenerate representation.

Recall from (14.18) that $C^*_{\text{par}}(G)$ is isomorphic to the crossed product algebra relative to the partial Bernoulli action β of G on Ω_1. By definition, this is the restriction of the global Bernoulli action η, introduced in (5.10), to Ω_1. So we conclude from (22.4) that the canonical mapping

$$\iota : C^*_{\text{par}}(G) = C(\Omega_1) \rtimes G \longrightarrow C\big(\{0,1\}^G\big) \rtimes G$$

is injective, hence we may view $C^*_{\text{par}}(G)$ as a subalgebra of $C\big(\{0,1\}^G\big) \rtimes G$.

Notice that Ω_1 is a clopen subset of $\{0,1\}^G$, whence the characteristic function of the former, which we denote by 1_{Ω_1}, lies in $C\big(\{0,1\}^G\big)$.

Given that w_g identifies with $1_g\delta_g$ by (14.18), the following identity, to be proved in the sequel, will be useful:

$$\iota(1_g\delta_g) = 1_{\Omega_1}\tilde{\delta}_g 1_{\Omega_1}, \tag{30.3.1}$$

where $\tilde{\delta}_g$ denoted the canonical unitary element of $C\big(\{0,1\}^G\big) \rtimes G$ implementing the global Bernoulli action. To see this we compute

$$1_{\Omega_1}\tilde{\delta}_g 1_{\Omega_1} = 1_{\Omega_1}1_{\eta_g(\Omega_1)}\tilde{\delta}_g = 1_{\Omega_1\cap\eta_g(\Omega_1)}\tilde{\delta}_g \overset{(5.13)}{=} 1_g\tilde{\delta}_g = \iota(1_g\delta_g),$$

proving (30.3.1).

We next choose a representation $\tilde{\pi}$ of $C\big(\{0,1\}^G\big) \rtimes G$ on a Hilbert space $\tilde{H}$, and an isometric linear operator $j : H \to \tilde{H}$, satisfying the conditions in (30.2) relative to π. In particular, the formula

$$\tilde{u}_g = \tilde{\pi}(\tilde{\delta}_g), \quad \forall\, g \in G,$$

defines a unitary representation $\tilde{u}$ of G on $\tilde{H}$, which we will prove to satisfy all of the conditions in the statement.

As a first step, insisting on the fact that w_g identifies with $1_g\delta_g$ by (14.18), we have

$$\begin{aligned} u_g &= \pi(w_g) \\ &= \pi(1_g\delta_g) \\ &= j^*\tilde{\pi}\big(\iota(1_g\tilde{\delta}_g)\big)j \\ &\overset{(30.3.1)}{=} j^*\tilde{\pi}(1_{\Omega_1}\tilde{\delta}_g 1_{\Omega_1})j \\ &= j^*\tilde{\pi}(1_{\Omega_1})\tilde{u}_g\tilde{\pi}(1_{\Omega_1})j \\ &= \ldots \end{aligned} \tag{30.3.2}$$

We would now like to get rid of the two occurrences of term $\tilde{\pi}(1_{\Omega_1})$ above, and to justify this we observe that

$$\tilde{\pi}(1_{\Omega_1})j \overset{(30.2)}{=} j\pi(1_{\Omega_1}) = j,$$

since π is a non-degenerate representation and 1_{Ω_1} represents the unit of $C\big(\Omega_1\big)\rtimes G$. Using this, and the identity obtained by taking stars on both sides, we deduce from (30.3.2) that

$$u_g = j^*\tilde{u}_g j.$$

In order to prove the last sentence in the statement, we compute

$$\begin{aligned}
\tilde{u}_g p\tilde{u}_g p &= \tilde{u}_g jj^*\tilde{u}_g jj^* \\
&= \tilde{u}_g j u_g j^* \\
&= \tilde{u}_g j\pi(1_g\delta_g)j^* \\
&\overset{(30.2)}{=} \tilde{u}_g\tilde{\pi}\big(\iota(1_g\delta_g)\big)jj^* \\
&= \tilde{\pi}\big(\tilde{\delta}_{g^{-1}}1_g\tilde{\delta}_g\big)jj^* \\
&= \tilde{\pi}\big(1_{g^{-1}}\big)jj^* \\
&= j\pi\big(1_{g^{-1}}\big)j^*.
\end{aligned}$$

This shows that the element with which we started the above computation is self-adjoint, so

$$\tilde{u}_g p\tilde{u}_g p = (\tilde{u}_g p\tilde{u}_g p)^* = p\tilde{u}_g p\tilde{u}_g,$$

proving the desired commutativity. □

Notes and remarks. We believe that Lemma (30.2) is well known among specialists, but we have not been able to find a reference for it in the literature. Theorem (30.3) was first proved in [**2**, Proposition 3.3] for the case of amenable groups.

CHAPTER 31

Semigroups of Isometries

The Toeplitz algebra, namely the C*-algebra generated by a non-unitary isometry [**25**], is among the most historically significant examples of C*-algebras. After Coburn's pioneering example, many authors have studied C*-algebras generated by sets of isometries, sometimes also including partial isometries. In many such examples, the generating isometries are parametrized by a semigroup. The present chapter is therefore dedicated to studying semigroups of isometries and their relationship to partial group representations.

Recall that the term *semigroup* in Mathematics always refers to a set P, equipped with an associative operation (often denoted multiplicatively).

Definition 31.1. Let A be a unital C*-algebra. A *semigroup of isometries* in A, based on a semigroup P, is a map $v : P \to A$, such that

(i) $v_p^* v_p = 1$,
(ii) $v_p v_q = v_{pq}$,

for every $p, q \in P$.

Notice that if P has a unit, say 1, then

$$v_1 = v_1^* v_1 v_1 = v_1^* v_1 = 1, \tag{31.2}$$

so v must necessarily send 1 to 1.

31.3. A simple example of a semigroup of isometries is obtained by taking any isometric linear operator S on a Hilbert space H, and defining

$$v : n \in \mathbb{N} \mapsto S^n \in \mathscr{L}(H).$$

31.4. To describe an example based on the semigroup $\mathbb{N}\times\mathbb{N}$, let us fix two isometric linear operators S and T on the same Hilbert space H, and let us assume that S and T commute. Then, defining

$$v : (n, m) \in \mathbb{N} \times \mathbb{N} \mapsto S^n T^m \in \mathscr{L}(H),$$

we again get a semigroup of isometries on H.

Definition 31.5. A semigroup P is said to be left-cancellative provided

$$pm = pn \quad \Rightarrow \quad m = n,$$

for every m, n and p in P.

Given a left-cancellative semigroup P, there is a somewhat canonical example of semigroup of isometries which we would now like to present. Consider the Hilbert space $\ell^2(P)$, with its usual orthonormal basis $\{e_n\}_{n\in P}$.

For each p in P, consider the bounded linear operator λ_p on $\ell^2(P)$, specified by

$$\lambda_p(e_n) = e_{pn}, \quad \forall n \in P.$$

Notice that λ_p is an isometric operator[39] because the map $n \mapsto pn$ is injective as a consequence of P being left-cancellative. It is now easy to check that λ is a semigroup of isometries in $\mathscr{L}\left(\ell^2(P)\right)$.

Definition 31.6. We shall refer to the above λ as the *regular semigroup of isometries* of P.

If P is left-cancellative, observe that $\lambda_p\lambda_p^*$ is the orthogonal projection onto

$$\lambda_p\big(\ell^2(P)\big) = \overline{\operatorname{span}}\{e_{pn} : n \in P\}. \tag{31.7}$$

In particular it is easy to see that $\lambda_p\lambda_p^*$ commutes with $\lambda_q\lambda_q^*$, for every p and q in P.

The commutativity of range projections is not always granted, even for the case of the commutative semigroup $\mathbb{N} \times \mathbb{N}$, as the following example shows. Let φ be a continuous, complex valued function on the closed unit disk

$$\mathbb{D} := \{z \in \mathbb{C} : |z| \leq 1\},$$

which is holomorphic on the interior of $\mathbb{D}$. Suppose also that φ is *unimodular*, in the sense that $|\varphi(z)| = 1$, for every

$$z \in \mathbb{T} := \{z \in \mathbb{C} : |z| = 1\}.$$

It is then easy to see that the operator

$$U_\varphi : \xi \in L^2(\mathbb{T}) \mapsto \varphi\xi \in L^2(\mathbb{T})$$

($\varphi\xi$ referring to pointwise product) is unitary.

Denoting by H^2 the *Hardy space*[40] of Classical Harmonic Analysis, one may prove that $U_\varphi(H^2) \subseteq H^2$ (this is where we need φ to be holomorphic on the interior of the disk), so the restriction of U_φ to H^2, henceforth denoted

$$T_\varphi : H^2 \to H^2,$$

is an isometry. This is often referred to in the literature as the *Toeplitz operator* with symbol φ.

Denoting by P the *Hardy projection*, namely the orthogonal projection from $L^2(\mathbb{T})$ to H^2, the invariance of H^2 under U_φ may be expressed by the formula

$$PU_\varphi P = U_\varphi P,$$

and the Toeplitz operator may be alternatively defined by

$$T_\varphi = PU_\varphi P.$$

A useful consequence of this description is the following formula for the adjoint of T_φ:

$$T_\varphi^* = PU_\varphi^* P = PU_{\bar\varphi} P. \tag{31.8}$$

Given two unimodular holomorphic functions φ and ψ, as above, it is easy to see that T_φ and T_ψ commute. So, in the context of (31.4), we get the semigroup of isometries

$$v : (n, m) \in \mathbb{N} \times \mathbb{N} \mapsto T_\varphi^n T_\psi^m \in \mathscr{L}(H^2). \tag{31.9}$$

[39]If P did not possess this property, in particular if infinitely many n's were mapped to the same element of P under left multiplication by p, there would be no *bounded* linear operator sending the e_n to e_{pn}.

[40]The Hardy space is the closed subspace of $L^2(\mathbb{T})$ spanned by the set of all functions of the form $z \mapsto z^n$, with $n \geq 0$.

In order to present our intended example illustrating that the range projections need not commute, let us fix the following two unimodular holomorphic functions on the unit disk:

$$\varphi(z) = z, \quad \text{and} \quad \psi(z) = \frac{z-a}{1-\bar{a}z}, \quad \forall z \in \mathbb{T},$$

where a is a fixed complex number with $|a| < 1$.

Incidentally, ψ is usually called a *Blaschke factor*, and it is an easy exercise to prove that it is indeed a unimodular holomorphic function on the unit disk (the denominator vanishes only for $z = 1/\bar{a}$, which lies outside the disk).

Our next goal is to analyze the commutativity of the range projections

$$Q' := v_{(1,0)}v_{(1,0)}^*, \quad \text{and} \quad Q'' := v_{(0,1)}v_{(0,1)}^*.$$

Upon identifying the Hardy space with $\ell^2(\mathbb{N})$, according to the usual orthonormal basis $\{e_n\}_{n\in\mathbb{N}}$ of H^2 given by

$$e_n(z) = z^n, \quad \forall z \in \mathbb{N},$$

we have that $v_{(1,0)}$, also known as T_φ, becomes the *unilateral shift*, in the sense that

$$T_\varphi(e_n) = e_{n+1}, \quad \forall n \in \mathbb{N}.$$

The range projection

$$Q' = T_\varphi T_\varphi^*,$$

is then the orthogonal projection onto the subspace of H^2 spanned by the e_n, with $n \in \mathbb{N} \setminus \{0\}$ and, in particular, we have that

$$Q'(e_0) = 0. \tag{31.10}$$

Assuming that Q' and Q'' commute, we have for all $j > 0$ that

$$\begin{aligned}\langle T_\psi T_\psi^*(e_0), e_j\rangle &= \langle Q''(e_0), Q'(e_j)\rangle \\ &= \langle Q'Q''(e_0), e_j\rangle \\ &= \langle Q''Q'(e_0), e_j\rangle \\ &\overset{(31.10)}{=} 0.\end{aligned}$$

It follows that all but the zeroth coefficient of $T_\psi T_\psi^*(e_0)$ relative to the canonical basis vanish, so

$$T_\psi T_\psi^*(e_0) = ce_0,$$

for some constant $c \in \mathbb{C}$. On the other hand

$$T_\psi T_\psi^*(e_0) \overset{(31.8)}{=} PU_\psi PPU_{\bar{\psi}}P(e_0) = \psi P\bar{\psi}.$$

Notice that, since ψ is holomorphic, its Taylor series around zero coincides with its Fourier expansion on the unit circle:

$$\psi(z) \sim \sum_{n=0}^{\infty} \frac{\psi^{(n)}(0)}{n!} z^n = \sum_{n=0}^{\infty} \hat{\psi}(n) z^n.$$

In particular all of the Fourier coefficients $\widehat{(\bar{\psi})}(n)$ vanish, for $n > 0$, which implies that $P\bar{\psi} = \bar{\psi}(0)e_0$. Comparing the two expressions for $T_\psi T_\psi^*(e_0)$ obtained above we deduce that

$$ce_0 = T_\psi T_\psi^*(e_0) = \psi P\bar{\psi} = \psi\bar{\psi}(0)e_0.$$

This equation entails two possibilities: in case $c \neq 0$, one has that ψ is equal to the constant $c\bar{\psi}(0)^{-1}$, almost everywhere on $\mathbb{T}$, which is clearly not the case.

The second possibility, namely that $c = 0$, in turn implies that $\psi\bar{\psi}(0) = 0$, whence $\psi(0) = 0$. This proves the following:

Proposition 31.11. *Let φ and ψ be given by*

$$\varphi(z) = z, \quad \textit{and} \quad \psi(z) = \frac{z-a}{1-\bar{a}z}, \quad \forall\, z \in \mathbb{T},$$

where a is a fixed complex number with $|a| < 1$. Then the corresponding Toeplitz operators T_φ and T_ψ are commuting isometric linear operators on Hardy's space. Moreover their range projections $T_\varphi T_\varphi^*$ and $T_\psi T_\psi^*$ commute if and only if $a = 0$.

31.12. The above result allows for the construction of an example of the situation mentioned above: choosing ψ to be any Blaschke factor with $a \neq 0$, the semigroup of isometries presented in (31.9) will be such that the final projections of $v_{(1,0)}$ and $v_{(0,1)}$ do not commute.

This is in stark contrast with (9.8.iv), according to which the final projections of the partial isometries involved in a partial group representation always commute! We must therefore recognize that the theory of partial representations of groups is unsuitable for dealing with such badly behaved examples.

We will therefore restrict our attention to special cases of semigroup of isometries which can be effectively studied via the theory of partial representations.

If we are to find a partial representation of a group somewhere behind a semigroup of isometries, we'd better make sure there is a group around. So we will assume from now on that the semigroup P is a sub-semigroup of a group G. If we are moreover given a semigroup of isometries

$$v : P \to A,$$

where A is a unital C*-algebra, we will now discuss conditions under which one may extend v to a partial representation of G.

Assume for a while that the problem has been solved, i.e., that a partial representation u of G in A has been found such that $u_n = v_n$, for all n in P. Since each v_n is an isometry, it is clearly left-invertible. Therefore, from (9.9) it follows that, for all $m, n \in P$,

$$u_{m^{-1}n} = u_{m^{-1}}u_n = u_m^* u_n = v_m^* v_n. \tag{31.13}$$

In other words, if $g \in P^{-1}P$, then u_g may be recovered from v.

Definition 31.14. A sub-semigroup P of a group G is called an *Ore sub-semigroup*[41], provided it satisfies $G = P^{-1}P$.

Lemma 31.15. *Let G be a group and let P be an Ore sub-semigroup of G. Also let A be a unital C*-algebra and $v : P \to A$ be a semigroup of isometries. Then, for every $m, n, p, q \in P$, one has that*

$$m^{-1}n = p^{-1}q \quad\Rightarrow\quad v_m^* v_n = v_p^* v_q.$$

PROOF. Given that $m^{-1}n = p^{-1}q$, we have that $pm^{-1} = qn^{-1}$. By hypothesis we may choose r and s in G, such that

$$pm^{-1} = qn^{-1} = r^{-1}s.$$

[41] An abstract semigroup P may be embedded in a group G, such that $G = P^{-1}P$, if and only if P is cancellative and for every m and n in P, there exist x and y in P, such that $xa = yb$. This is the content of Ore's Theorem. In view of this, some authors prefer to say that P is an *Ore semigroup*, rather than an *Ore sub-semigroup of G*.

Therefore $rp = sm$ and $rq = sn$, so

$$v_m = v_s^* v_s v_m = v_s^* v_{sm} = v_s^* v_{rp} = v_s^* v_r v_p,$$

whence

$$v_m^* v_n = v_p^* v_r^* v_s v_n = v_p^* v_r^* v_{sn} = v_p^* v_r^* v_{rq} = v_p^* v_r^* v_r v_q = v_p^* v_q.$$

□

We thus see that the condition that $G = P^{-1}P$ gives a canonical way to extend a semigroup of isometries v on P to a map defined on G. Simply take any g in G, decompose it as $g = m^{-1}n$, and define $u_g = v_m^* v_n$. By the above result, u_g does not depend on the chosen decomposition of g.

As already mentioned, we are interested in extending v to a partial representation of G, but it is by no means clear that the map u thus described will be a partial representation. In particular, if the range projections $v_n v_n^*$ do not commute with each other, then (9.8.iv) implies that u cannot be a partial representation.

Theorem 31.16. *Let G be a group and let P be an Ore sub-semigroup of G. Also let A be a unital C^*-algebra and $v : P \to A$ be a semigroup of isometries. Then the following are equivalent:*

(i) $v_m v_m^*$ commutes with $v_n v_n^*$, for all n, m in P,
(ii) there exists a *-partial representation u of G in A extending v.

In this case, the partial representation u referred to in (ii) is unique.

PROOF. The implication (ii)⇒(i) is an easy consequence of (9.8.iv), so let us focus on the converse. Given g in G, write $g = m^{-1}n$, with $m, n \in P$, and set $u_g = v_m^* v_n$. By (31.15) we see that u is a well defined function on G, and we will next prove it to be a *-partial representation.

Upon replacing P with $P \cup \{1\}$, we may assume that $1 \in P$, and as seen in (31.2), we have that $v_1 = 1$, so (9.1.i) is verified. The proof of (9.1.iv) is also elementary, so let us prove (9.1.ii). Given $g, h \in G$, write

$$g = m^{-1}n, \quad \text{and} \quad h = p^{-1}q,$$

with $m, n, p, q \in P$. Using the hypothesis, pick r and s in P such that

$$np^{-1} = r^{-1}s,$$

so $rn = sp$. Letting

$$m' = rm, \quad n' = rn, \quad p' = sp, \quad \text{and} \quad q' = sq,$$

notice that

$$g = m'^{-1}n', \quad \text{and} \quad h = p'^{-1}q',$$

but now $n' = p'$. This says that, in our initial choice of m, n, p and q, we could have taken $n = p$. With this extra assumption (and removing all diacritics) we then have that $gh = m^{-1}q$, and

$$\begin{aligned} u_g u_h u_{h^{-1}} &= v_m^* \; v_p v_p^* \; v_q v_q^* \; v_p \\ &= v_m^* \; v_q v_q^* \; v_p v_p^* \; v_p \\ &= v_m^* v_q v_q^* v_p \\ &= u_{gh} u_{h^{-1}}. \end{aligned}$$

This proves (9.1.ii) and, as already noticed in (9.2), axiom (9.1.iii) must also hold.

The uniqueness of u now follows immediately from (9.9), as observed in (31.13). □

Motivated by the result above, we will now study semigroups of isometries which are not necessarily based on Ore sub-semigroups, but which may be extended to a partial representation. So let us introduce the following terminology.

Definition 31.17. Let G be a group and let P be a sub-semigroup of G. Also let A be a unital C*-algebra and $v : P \to A$ be a semigroup of isometries. We will say that v is *extendable*, if there exists a *-partial representation u of G, such that $u_n = v_n$, for every $n \in P$.

In the case of Ore semigroups, as we have seen in (31.16), a necessary and sufficient condition for extendability is the commutativity of range projections. Also, in the general case it easy to see that this condition is still necessary but we have unfortunately not been able to determine an elegant set of sufficient conditions for extendability.

In fact the concept of extendability elicits many questions that we will not even attempt to answer. For example, it is not clear whether the extended partial representation is unique, even under the assumption that G is generated by P.

An example of a non-extendable semigroup of isometries is clearly presented by (31.9), provided φ and ψ are as in (31.11), with $a \neq 0$.

The following is a dilation result for extendable semigroups of isometries:

Theorem 31.18. *Let G be a group and let P be a sub-semigroup of G. Also let*

$$v : P \to \mathscr{L}(H)$$

be an extendable semigroup of isometries, where H is a Hilbert space. Then there exists a Hilbert space $\tilde{H}$ containing H, and a unitary group representation $\tilde{u}$ of G on $\tilde{H}$ such that, for every n in P, one has that:

(i) $\tilde{u}_n(H) \subseteq H$,

(ii) v_n *coincides with the restriction of* $\tilde{u}_n$ *to* H,

(iii) denoting by p the orthogonal projection from $\tilde{H}$ to H, one has that p commutes with $\tilde{u}_{g^{-1}}p\tilde{u}_g$, for every g in G.

Proof. Since v is extendable, we may pick a *-partial representation u of G on H extending v. Let $\tilde{u}$, $\tilde{H}$ and j be obtained by applying (30.3) to u. Viewing H as a subspace of $\tilde{H}$ via j, we then have that (iii) is verified. Given n in P, we have by (30.3) that

$$j^*\tilde{u}_n j = u_n = v_n,$$

and, taking into account that v_n is an isometry, we conclude that

$$1 = v_n^* v_n = (j^*\tilde{u}_n j)^* j^*\tilde{u}_n j = j^*\tilde{u}_{n^{-1}} j j^*\tilde{u}_n j.$$

Multiplying this on the left by j and on the right by j^*, we get

$$p = jj^* = j1j^* = jj^*\tilde{u}_{n^{-1}} jj^*\tilde{u}_n jj^* = p\tilde{u}_{n^{-1}} p\tilde{u}_n p = \ldots$$

By the already proved point (iii), the above equals

$$\ldots = \tilde{u}_{n^{-1}} p\tilde{u}_n pp = \tilde{u}_{n^{-1}} p\tilde{u}_n p.$$

This proves that $p = \tilde{u}_{n^{-1}} p\tilde{u}_n p$, and if we now multiply this on the left by $\tilde{u}_n$ we arrive at

$$\tilde{u}_n p = p\tilde{u}_n p,$$

which says that the range of p, namely H, is invariant under $\tilde{u}_n$, hence proving (i). As for (ii), we have for all n in P that

$$v_n = u_n = j^* \tilde{u}_n j = \tilde{u}_n|_H,$$

where the last equality is justified by our identification of H as a subspace of $\tilde{H}$, and the invariance of the former under $\tilde{u}_n$. □

Definition 31.19. Let P be a semigroup (not necessarily contained in a group) and let

$$v : P \to \mathscr{L}(H)$$

be a semigroup of isometries, where H is a Hilbert space. By a *unitary dilation* of v we mean a semigroup of isometries

$$w : P \to \mathscr{L}(\tilde{H})$$

on a Hilbert space $\tilde{H}$ containing H, such that, for every n in P, w_n is a *unitary operator*, leaving H invariant, and whose restriction to H coincides with v_n.

As a consequence of (31.18), we see that every extendable semigroup of isometries admits a dilation. In particular, every semigroup of isometries based on an Ore semigroup whose final projections commute, is extendable by (31.16), and hence admits a dilation. This should be compared with [**76**, Theorem 1.4], where a similar result has been proven without the hypothesis of the commutativity of final projections, but also without conclusion (31.18.iii).

Notes and remarks. The study of isometries and semigroups thereof dates back at least to the 1960s and has involved numerous authors, such as Arveson [**10**], Burdak [**20**], Coburn [**25**], Douglas [**39**], [**40**], Horák and Müller [**70**], Laca [**76**], Nica [**85**], Phillips and Raeburn [**90**], Popovici [**91**] and Słociński [**103**] among others.

CHAPTER 32

Quasi-Lattice Ordered Groups

Besides the case of Ore semigroups, we will now present another important example of extendable semigroup of isometries.

Let us begin by noticing that if P is a sub-semigroup of a group G, such that $P \cap P^{-1} = \{1\}$, then one may define a left-invariant partial order relation on G by

$$g \leq h \Leftrightarrow g^{-1}h \in P,$$

for all g and h in G. Conversely, if "$\leq$" is a left-invariant partial order relation on G, then

$$P := \{g \in G : g \geq 1\}$$

is a sub-semigroup with $P \cap P^{-1} = \{1\}$. So there is a one-to-one correspondence between such sub-semigroups and left-invariant partial order relations on G.

Definition 32.1. Let G be a group with a distinguished sub-semigroup P, such that $P \cap P^{-1} = \{1\}$, and denote by "$\leq$" the corresponding order relation on G. Given a nonempty subset $A \subseteq G$,

(i) we say that an element $k \in G$ is an *upper bound* for A and, if $k \geq g$, for all g in A,

(ii) we say that an element $m \in G$ is a *least upper bound* for A, if m is an upper bound for A, and $m \leq k$, whenever k is an upper bound for A,

(iii) we say that an element $\underline{m \in P}$ is a *least upper bound* $\underline{\textit{in } P}$ for A, if m is an upper bound for A, and $m \leq k$, whenever k is an upper bound for A, with $\underline{k \in P}$,

(iv) we say that (G, P) is a *quasi-lattice*, or that G is *quasi-lattice ordered*, if every nonempty finite subset $A \subseteq G$, admitting an upper bound in P, necessarily admits a least upper bound in P,

(v) we say that (G, P) is a *weak quasi-lattice*, or that G is *weakly quasi-lattice ordered*, if every nonempty finite subset $A \subseteq P$, admitting an upper bound[42], necessarily admits a least upper bound.

The first two definitions above are, of course, part of the elementary theory of ordered sets. They are listed just to provide a backdrop for the others. Moreover, given the subtleties involved in the last two definitions, it is perhaps worth repeating them in more technical terms. Given a nonempty subset $A \subseteq G$, let us denote by

$$A^{\uparrow} = \{k \in G : k \text{ is an upper bound for } A\} \tag{32.2}$$

Then

(a) (G, P) is a quasi-lattice if and only if, for every nonempty finite subset $A \subseteq G$, such that $A^{\uparrow} \cap P \neq \emptyset$, there exists a smallest element in $A^{\uparrow} \cap P$.

[42]Notice that, if $A \subseteq P$, then any upper bound of A is necessarily in P.

(b) (G,P) is a weak quasi-lattice if and only if, for every nonempty finite subset $A \subseteq P$, such that $A^{\uparrow} \neq \emptyset$, there exists a smallest element in $A^{\uparrow}$.

Notice that the notion of weak quasi-lattice may be defined using only the algebraic structure of P, while checking that a pair (G,P) is a quasi-lattice requires a knowledge of the whole group G. In other words, the notion of weak quasi-lattice easily generalizes for semigroups which are not necessarily contained in a group, while the notion of quasi-lattice must necessarily refer to a pair (G,P), as opposed to a single semigroup P.

It is clear that every subset $A \subseteq P$ which admits a least upper bound, also admits a least upper bound in P. Therefore every quasi-lattice is necessarily a weak quasi-lattice.

If a set A admits a least upper bound m, then m is clearly unique and we denote it by

$$m = \vee A.$$

In addition, if A is a two element set, say $A = \{g,h\}$, we also denote by

$$g \vee h = \vee\{g,h\}.$$

The notion of quasi-lattice ordered group was introduced by Nica in [**85**], where it was noticed that every weak quasi-lattice is a quasi-lattice, provided condition (32.1.iv) holds for sets A with a single element. Nica denoted this special case of (32.1.iv) as (QL1), which he phrased as follows:

(QL1) Any g in PP^{-1} (these are precisely the elements having an upper bound in P) has a least upper bound in P.

Ten years after the publication of Nica's paper, Crisp and Laca [**27**, Lemma 7] found what might well be called "Columbus' egg" of quasi-lattices, realizing that in fact (QL1) alone is a sufficient condition for (G,P) to be a quasi-lattice (see below).

In hindsight, the following simple characterizations should hopefully clean up our act. We suggest that the reader take this as alternative definitions for the concepts involved, should the above perhaps too extensive discussion have blurred the big picture:

Proposition 32.3. *Let G be a group and let $P \subseteq G$ be a sub-semigroup such that $P \cap P^{-1} = \{1\}$. Then*

(i) (G,P) is a weak quasi-lattice, if and only if $m \vee n$ exists for every m and n in P admitting a common upper bound,

(ii) (G,P) is a quasi-lattice, if and only if any g in PP^{-1} has a least upper bound in P.

PROOF. Point (i) is easily proven by induction.

As for (ii), assume that (G,P) is a quasi-lattice and let $g \in PP^{-1}$. Writing $g = mn^{-1}$, with n and m in P, it is clear that $g \leq m$, so $\{g\}$ admits an upper bound in P, hence also a least upper bound in P.

Conversely, suppose that any g in PP^{-1} has a least upper bound in P. We will first prove that (G,P) is a weak quasi-lattice by verifying condition (i). So suppose that n and m are in P, and that p is an upper bound of n and m. Then

$$1 = n^{-1}n \leq n^{-1}p, \quad \text{and}$$

$$n^{-1}m \leq n^{-1}p,$$

so $n^{-1}m$ admits an upper bound in P, namely $n^{-1}p$. By hypothesis there exists a least upper bound in P for $n^{-1}m$, say q. Then q is a least upper bound for the set $\{1, n^{-1}m\}$, and by left-invariance of the order relation, nq is a least upper bound for $\{n, m\}$, as desired.

We will now prove that (G, P) is a quasi-lattice. So let A be a nonempty finite subset of G admitting an upper bound in P, say p. By hypothesis we know that, for each g in A, there exists the least upper bound of g in P, which we denote by n_g. Then it is clear that $n_g \leq p$, for all g in A, so p is also an upper bound for $A' = \{n_g : g \in A\}$. Since $A' \subseteq P$, the weak quasi-lattice property of (G, P) may be used to ensure that the least upper bound in P of A' exists, but this is clearly also the least upper bound in P of A. □

One of the main examples of quasi-lattice ordered groups is obtained by taking $\mathbb{F}_n$ to be the free group on an arbitrary (finite or infinite) number n of generators, and taking $\mathbb{P}_n$ to be the smallest sub-semigroup of $\mathbb{F}_n$ containing 1 and its free generators.

An example of a weak quasi-lattice which is not a quasi-lattice was recently obtained by Scarparo [**99**], who also generalized some important results on quasi-lattice ordered groups to the context of weak quasi-lattices. Scarparo's example is as follows: take $\mathbb{F}_2$ to be the free group on a set of two generators, say $\{a, b\}$, and let

$$\mathbb{P}' := b\mathbb{P}_2 \cup \{1\},$$

where $\mathbb{P}_2$ was defined above. Then $\mathbb{P}'$ is clearly a sub-semigroup of $\mathbb{F}_2$ and the pair $(\mathbb{F}_2, \mathbb{P}')$ is not a quasi-lattice although it is a weak quasi-lattice.

The reason why it is not a quasi-lattice is that the singleton $\{ba^{-1}b^{-1}\}$ is bounded in $\mathbb{P}'$ by b and ba, but it does not admit a least upper bound in $\mathbb{P}'$, as the reader may easily verify.

On the other hand $(\mathbb{F}_2, \mathbb{P}')$ is a weak quasi-lattice because $\mathbb{P}'$ is isomorphic to $\mathbb{P}_\infty$, as a semigroup, via an isomorphism defined using the set

$$\{ba^n : n \in \mathbb{N}\},$$

as a set of free generators for $\mathbb{P}'$. Since $(\mathbb{F}_\infty, \mathbb{P}_\infty)$ is a quasi-lattice, as seen above, it is necessarily also a weak quasi-lattice.

Another interesting consequence of Scarparo's example is that a semigroup P may be seen as a sub-semigroup of two non-isomorphic groups G_1 and G_2, in such a way that both (G_1, P) and (G_2, P) are weak quasi-lattices. Namely, take

$$(G_1, P) = (\mathbb{F}_2, \mathbb{P}'), \quad \text{and} \quad (G_2, P) = (\mathbb{F}_\infty, \mathbb{P}_\infty),$$

were we are identifying $\mathbb{P}'$ and $\mathbb{P}_\infty$, as already mentioned.

Whenever possible we will strive to work with weak quasi-lattices, not only because it provides for stronger results, but also due to the fact that weak quasi-lattices may be defined intrinsically, independently of the group containing them.

32.4. Nica's theory of semigroups of isometries based on quasi-lattices is rooted on the following simple idea. Let (G, P) be a weak quasi-lattice and consider the regular semigroup of isometries

$$\lambda : P \to \mathscr{L}(\ell^2(P))$$

defined in (31.6). Reinterpreting (31.7), we may say that $\lambda_p\lambda_p^*$ is the orthogonal projection onto

$$\overline{\text{span}}\{e_n : n \geq p\}. \tag{32.5}$$

Thus, given p and q in P, the product of the commuting projections $\lambda_p\lambda_p^*$ and $\lambda_q\lambda_q^*$ coincides with the orthogonal projection onto the closed linear span of the set

$$\{e_n : n \geq p\} \cap \{e_n : n \geq q\} = \{e_n : n \ \geq \ p \vee q\},$$

assuming of course that $p \vee q$ exists. When $p \vee q$ does not exist, then by hypothesis p and q have no upper bound whatsoever, so the above intersection of sets is empty. This motivates the following:

Definition 32.6. Let (G, P) be a weak quasi-lattice and A be a C*-algebra. We shall say that a given semigroup of isometries $v : P \to A$ satisfies *Nica's covariance condition*, *NCC* for short, if for every m and n in P, one has that

$$v_m v_m^* \, v_n v_n^* = \begin{cases} v_{m\vee n} v_{m\vee n}^*, & \text{if } m \vee n \text{ exists,} \\ 0, & \text{otherwise.} \end{cases}$$

As briefly indicated above, in the case of a weak quasi-lattice, the regular semigroup of isometries satisfies NCC.

Lemma 32.7. *Let (G, P) be a weak quasi-lattice, A be a C*-algebra, and $v : P \to A$ be a semigroup of isometries satisfying NCC. Given $m, n \in P$, one has that:*

(i) *if m and n have no common upper bound, then $v_m^* v_n = 0$,*
(ii) *if m and n have a common upper bound, then there are x and y in P such that $mx = ny$, and $v_m^* v_n = v_x v_y^*$.*

PROOF. Under condition (i) we have

$$v_m^* v_n = v_m^* \underbrace{v_m v_m^* v_n v_n^*}_{(32.6)} v_n = 0.$$

On the other hand, if m and n have a common upper bound, then the least upper bound exists, and there are x and y in P such that

$$m \vee n = mx = ny.$$

So

$$v_m^* v_n = v_m^* \underbrace{v_m v_m^* v_n v_n^*}_{(32.6)} v_n = v_m^* v_{m\vee n} v_{m\vee n}^* v_n = v_m^* v_{mx} v_{ny}^* v_n = v_x v_y^*.$$

□

The main goal of this chapter is to prove extendability of a given semigroups of isometries v satisfying NCC. After this is achieved, by (12.12) we will have that the range of v is a tame set. However the strategy we will adopt will require knowing in advance that the range of v is tame, so we need to prove this as an intermediate step.

Proposition 32.8. *Let (G, P) be a weak quasi-lattice, A be a C*-algebra, and $v : P \to A$ be a semigroup of isometries satisfying NCC. Then,*

(i) *the set $\{v_m v_n^* : m, n \in P\} \cup \{0\}$ is a multiplicative semigroup,*
(ii) *the range of v is a tame set of partial isometries.*

PROOF. Given m, n, p and q in P, let us suppose that n and p have no common upper bound. Then

$$v_m v_n^* v_p v_q^* \overset{(32.7.i)}{=} 0.$$

On the other hand, if n and p have a common upper bound, then by (32.7.ii) we may write $v_n^* v_p = v_x v_y^*$, for suitable x and y, so

$$v_m v_n^* v_p v_q^* = v_m v_x v_y^* v_q^* = v_{mx} v_{qy}^*.$$

This proves (i).

In order to prove (ii), let S be the semigroup referred to in (i). Since P contains 1 by hypothesis, and since $v_1 = 1$ by (31.2), we see that both the range of v and its adjoint are contained in S. Therefore we see that S coincides with the multiplicative semigroup generated by the range of v and its adjoint. To prove that the range of v is tame, it therefore suffices to verify that every element in S is a partial isometry. Given m and n in P, we have that

$$(v_m v_n^*)(v_m v_n^*)^* = v_m v_n^* v_n v_m^* = v_m v_m^*,$$

which is clearly a projection, and hence $v_m v_n^*$ is a partial isometry by (12.4). This concludes the proof of (ii). □

We will now present a crucial technical result designed to help proving the extendability of semigroups of isometries satisfying NCC. It should be compared to (31.15).

Lemma 32.9. *Let (G, P) be a weak quasi-lattice, A be a unital C^*-algebra and let*

$$v : P \to A,$$

be a semigroup of isometries satisfying NCC. Given m, n, p and q in P such that $mn^{-1} = pq^{-1}$, one has that $v_m v_n^*$ and $v_p v_q^*$ are compatible partial isometries, as defined in (12.20).

PROOF. We will soon see that the question of upper bounds for the set $\{n, q\}$ is a crucial one, so let us begin by analyzing it. Suppose first that n and q admit an upper bound k. So there are x and y in P such that

$$k = nx = qy.$$

Then

$$(mx)(nx)^{-1} = mn^{-1} = pq^{-1} = (py)(qy)^{-1},$$

so $mx = py$. Defining

$$h = mx = py,$$

it is then clear that h is an upper bound for $\{m, p\}$. Denoting by $g = mn^{-1} = pq^{-1}$, notice that

$$gk = mn^{-1}nx = mx = h,$$

so this provides for a well defined function

$$k \in \{n, q\}^{\uparrow} \mapsto gk \in \{m, p\}^{\uparrow},$$

where we are using the notation introduced in (32.2) for the set of upper bounds. Given the symmetric roles played by m, n, p and q, it is clear that

$$h \in \{m, p\}^{\uparrow} \mapsto g^{-1}h \in \{n, q\}^{\uparrow}$$

is also well defined. Therefore $\{n, q\}$ admits an upper bound if and only if $\{m, p\}$ does, and in this case the left-invariance of our order relation implies that

$$g(n \vee q) = m \vee p. \tag{32.9.1}$$

Returning to our semigroup of isometries, and letting

$$s = v_m v_n^*,$$

we have

$$ss^* = v_m v_n^* v_n v_m^* = v_m v_m^*,$$

so we see that s is indeed a partial isometry, sharing final projection with v_m. It is also easy to see that the initial projection of s coincides with the final projection of v_n. Likewise, if we let

$$t = v_p v_q^*,$$

then the initial and final projections of t are $v_q v_q^*$ and $v_p v_p^*$, respectively.

Recalling that our task is to prove that

$$st^*t = ts^*s, \quad \text{and} \quad tt^*s = ss^*t, \tag{32.9.2}$$

let us assume first that $\{n, q\}$ admits no upper bound. In this case, NCC implies that $v_n v_n^*$ is orthogonal to $v_q v_q^*$, whence $v_n^* v_q = 0$. So,

$$st^*t = v_m v_n^* v_q v_q^* = 0.$$

On the other hand,

$$ts^*s = v_p v_q^* v_n v_n^* = v_p (v_n^* v_q)^* v_n^* = 0,$$

proving the first equation in (32.9.2). In order to prove the second one, observe that, by the reasoning at the beginning of this proof, $\{m, p\}$ admits no upper bound either, so $v_p^* v_m = 0$, whence

$$tt^*s = v_p v_p^* v_m v_n^* = 0 = v_m v_m^* v_p v_q^* = ss^*t,$$

completing the proof of (32.9.2) under the assumption that $\{n, q\}$ admits no upper bound. Let us therefore assume that $\{n, q\}$ admits an upper bound, and hence by hypothesis also a least upper bound. We may then find x and y in P such that

$$n \vee q = nx = qy.$$

Therefore

$$\begin{aligned} st^*t &= v_m v_n^* v_q v_q^* \\ &= v_m v_n^* v_n v_n^* v_q v_q^* \\ &= v_m v_n^* v_{n\vee q} v_{n\vee q}^* \\ &= v_m v_n^* v_{nx} v_{nx}^* \\ &= v_m v_n^* v_n v_x v_{nx}^* \\ &= v_m v_x v_{nx}^* \\ &= v_{mx} v_{nx}^*. \end{aligned}$$

By (32.9.1) we have that $m \vee p$ exists and

$$\begin{aligned} m \vee p = g(n \vee q) &= mn^{-1}nx = mx \\ &= pq^{-1}qy = py. \end{aligned}$$

Therefore

$$\begin{aligned} ts^*s &= v_p v_q^* v_n v_n^* \\ &= v_p v_q^* v_q v_q^* v_n v_n^* \\ &= v_p v_q^* v_{q\vee n} v_{q\vee n}^* \\ &= v_p v_q^* v_{qy} v_{nx}^* \\ &= v_p v_q^* v_q v_y v_{nx}^* \\ &= v_{py} v_{nx}^* \\ &= v_{mx} v_{nx}^*, \end{aligned}$$

thus proving that $st^*t = ts^*s$, which is the first equation in (32.9.2). The second one may now be verified similarly, or by applying what has already been proved to

$$s' := s^* = v_n v_m^*, \quad \text{and} \quad t' := t^* = v_q v_p^*.$$

This concludes the proof. □

The following is the main result of this chapter. It was first proved in [**94**] for the case of quasi-lattices. Our plan to prove it for weak quasi-lattices will require that the range of our semigroup of isometries lies in a von Neumann algebra.

Theorem 32.10. *Let (G, P) be a weak quasi-lattice, and let A be a unital C^*-algebra. Suppose moreover that either*

(i) A is a von Neumann algebra, or
(ii) (G, P) is a quasi-lattice.

Then every semigroup of isometries $v : P \to A$ which satisfies NCC is extendable.

PROOF. Given g in $G \setminus PP^{-1}$, we define

$$u_g = 0.$$

On the other hand, given g in PP^{-1}, consider the collection of partial isometries

$$T_g = \{v_m v_n^* : m, n \in P,\ mn^{-1} = g\}.$$

By (32.9) we have that the elements of T_g are mutually compatible and we would now like to argue that $\vee T_g$ exists. Under the hypothesis that A is a von Neumann algebra we may simply use (12.26), so let us prove the existence of $\vee T_g$ under (ii). In this case we will actually prove that T_g contains a maximum element.

Observing that g may be written as $g = pq^{-1}$, for some $p, q \in P$, notice that we then have that $g \preceq p$, so we see that g admits an upper bound in P. Using the quasi-lattice property, the least upper bound in P of $\{g\}$, here denoted by $\sigma(g)$, exists. In particular $g \preceq \sigma(g)$, so

$$\tau(g) := g^{-1}\sigma(g) \in P.$$

We may then write

$$g = \sigma(g)\tau(g)^{-1}, \tag{32.10.1}$$

which might be thought of as the *most efficient way* of writing g. Evidently $v_{\sigma(g)} v_{\tau(g)}^*$ lies in T_g, and we claim that this element dominates every other element of T_g. In fact, given any $m, n \in P$ such that $g = mn^{-1}$, we have that $g \preceq m$.

Since $\sigma(g)$ is the least upper bound in P of g, it follows that $\sigma(g) \preceq m$ as well, so there exists x in P such that $m = \sigma(g)x$. Consequently

$$n = g^{-1}m = \tau(g)\sigma(g)^{-1}\sigma(g)x = \tau(g)x.$$

To prove our claim that $v_{\sigma(g)}v^*_{\tau(g)}$ is the biggest element in T_g, we must prove that $v_m v_n^* \preceq v_{\sigma(g)}v^*_{\tau(g)}$, so we compute

$$\begin{aligned} v_{\sigma(g)}v^*_{\tau(g)}(v_m v_n^*)^* v_m v_n^* &= v_{\sigma(g)}v^*_{\tau(g)} v_n v_m^* v_m v_n^* \\ &= v_{\sigma(g)}v^*_{\tau(g)} v_n v_n^* \\ &= v_{\sigma(g)}v^*_{\tau(g)} v_{\tau(g)} v_x v_n^* \\ &= v_{\sigma(g)x} v_n^* \\ &= v_m v_n^*. \end{aligned}$$

This proves that $v_{\sigma(g)}v^*_{\tau(g)}$ is the biggest element in T_g, so it is obvious that $\vee T_g$ exists.

Having proven the existence of $\vee T_g$ under either (i) or (ii), we may define

$$u_g = \vee T_g,$$

and we will now set out to prove that u is a partial representation, the strategy being to apply (9.6). We first notice that, thanks to (32.8), the range of v is a tame set, so

$$S := \langle v(P) \cup v(P)^* \rangle,$$

meaning the multiplicative sub-semigroup of A generated by $v(P) \cup v(P)^*$, is an inverse semigroup by (12.11).

Notice that by construction, under hypothesis (ii), the values of u lie in S itself, while under (i), the values of u lie in the $\vee$-closure of S, which is an inverse semigroup by (12.30).

In either case we may view u as a map from G to some inverse semigroup, which is a necessary condition for the application of (9.6). In fact we need u to take values in a *unital* inverse semigroup, so we actually have to throw in the unit of A.

We will now check conditions (9.6.i)–(9.6.iii), but since the verification of (9.6.i) & (9.6.iii) is trivial, our task is simply to show that

$$u_g u_h \preceq u_{gh}, \tag{32.10.2}$$

for all g and h in G. If either g or h lie outside PP^{-1}, the left-hand-side of (32.10.2) vanishes, so there is nothing to do. We therefore assume that both g and h lie in PP^{-1}, and in this case we claim that u_{gh} dominates $T_g T_h$. To see this, choose elements in T_g and T_h, respectively of the form $v_m v_n^*$ and $v_p v_q^*$, where $g = mn^{-1}$ and $h = pq^{-1}$. We must then prove that

$$v_m v_n^* v_p v_q^* \preceq u_{gh}. \tag{32.10.3}$$

If $\{n, p\}$ admits no upper bound then (32.7.i) implies that $v_n^* v_q = 0$, so the left-hand-side of (32.10.3) vanishes and, again, there is nothing to do. On the other hand, if $\{n, p\}$ admits an upper bound, we have by (32.7.ii) that there are x and y in P such that $nx = py$, and $v_n^* v_p = v_x v_y^*$, whence

$$v_m v_n^* v_p v_q^* = v_m v_x v_y^* v_q^* = v_{mx} v_{qy}^*. \tag{32.10.4}$$

Observe that

$$mx(qy)^{-1} = mxy^{-1}q^{-1} = mn^{-1}pq^{-1} = gh,$$

so $v_{mx}v_{qy}^* \in T_{gh}$, and we then deduce from (32.10.4) that

$$v_m v_n^* v_p v_q^* \preceq \vee T_{gh} = u_{gh}.$$

This proves (32.10.3), and hence that u_{gh} indeed dominates $T_g T_h$. Consequently

$$u_g u_h = (\vee T_g)(\vee T_h) \overset{(12.28)}{=} \vee(T_g T_h) \preceq u_{gh},$$

taking care of (32.10.2), whence allowing us to apply (9.6), proving that u is a partial representation, as desired.

Let us now prove that u extends v. For this observe that if p is in P, then

$$u_p = \vee T_p = \vee\{v_m v_n^* : m, n \in P,\ mn^{-1} = p\}.$$

Given m and n in P such that $mn^{-1} = p$, we have that $m = pn$, and we claim that

$$v_m v_n^* \preceq v_p.$$

In fact,

$$v_p(v_m v_n^*)^*(v_m v_n^*) = v_p v_n v_m^* v_m v_n^* = v_{pn} v_n^* = v_m v_n^*.$$

This proves the claim, so we see that v_p indeed dominates T_p. Since v_p moreover belongs to T_p, we deduce that

$$v_p = \vee T_p = u_p,$$

proving that v indeed extends u, which is to say that v is extendable. This concludes the proof. □

In the above proof, we did not worry about the uniqueness of the extended partial representation u. However this will later become relevant, so let us discuss it now. In doing so we will be content with the quasi-lattice case, leaving the study of uniqueness in the case of weak quasi-lattices as an outstanding problem.

Again referring to the strategy adopted in the above proof, notice that our first step was to set $u_g = 0$, whenever g is not in PP^{-1}. Evidently this is equivalent to saying that

$$e_g = u_g u_{g^{-1}} = 0.$$

On the other hand, in the quasi-lattice case, we saw that every g in PP^{-1} has a *most efficient* decomposition, namely (32.10.1), in which case recall that we have defined

$$u_g = \vee T_g = v_{\sigma(g)} v_{\tau(g)}^*,$$

whence

$$\begin{aligned} e_g &= u_g u_{g^{-1}} \\ &= v_{\sigma(g)} v_{\tau(g)}^* v_{\tau(g)} v_{\sigma(g)}^* \\ &= v_{\sigma(g)} v_{\sigma(g)}^* \\ &= u_{\sigma(g)} u_{\sigma(g)^{-1}} \\ &= e_{\sigma(g)} \\ &= e_{g \vee 1}. \end{aligned}$$

Proposition 32.11. *Let (G, P) be a quasi-lattice, and let A be a unital C^*-algebra. Then every semigroup of isometries $v : P \to A$ satisfying NCC admits a* unique *extension to a partial representation u of G in A, satisfying*

$$e_g = \begin{cases} e_{g\vee 1}, & \text{if } g \in PP^{-1}, \\ 0, & \text{otherwise}, \end{cases} \tag{32.11.1}$$

for all g in G, where $e_g = u_g u_{g^{-1}}$, as usual.

PROOF. The existence follows from (32.10) and our discussion above. Regarding uniqueness, let u be an extension of v satisfying the above properties. Then, for all g in $G \setminus PP^{-1}$, one has that

$$u_g = u_g u_{g^{-1}} u_g = e_g u_g = 0 u_g = 0.$$

On the other hand, if g is in PP^{-1}, then g admits an upper bound in P, so

$$m := g \vee 1$$

exists by hypothesis, so we may write $g = mn^{-1}$, where $n = g^{-1}m \in P$. By hypothesis we then have that $e_g = e_m$, so,

$$u_g = u_g u_{g^{-1}} u_g = e_g u_g = e_m u_{mn^{-1}} = u_m u_{m^{-1}} u_{mn^{-1}} = u_m u_{n^{-1}} = v_m v_n^*.$$

This proves that u is unique. □

Although apparently innocuous, condition (32.11.1) has very interesting consequences for our purposes:

Proposition 32.12. *Let (G, P) be a quasi-lattice, and let A be a unital C^*-algebra. Given any partial representation $u : G \to A$ satisfying* (32.11.1), *the restriction of u to P is a semigroup of isometries satisfying NCC.*

PROOF. Given any n in P, notice that $n^{-1} \leq 1$, so $n^{-1} \vee 1 = 1$. Consequently

$$1 = e_1 = e_{n^{-1}\vee 1} = e_{n^{-1}} = u_n^* u_n,$$

from where we conclude that u_n is an isometry. In particular each u_n is left-invertible, so by (9.9.iii) we have that

$$u_m u_n = u_{mn}, \quad \forall m, n \in P.$$

This shows that $u|_P$ is a semigroup of isometries.

To conclude, let us now prove NCC. For this, suppose first that m and n are given elements of P not possessing a common upper bound. Then we claim that $m^{-1}n \notin PP^{-1}$. To see this, assume by contradiction that

$$m^{-1}n = pq^{-1},$$

where $p, q \in P$. Then

$$m, n \leq nq = mp,$$

contradicting our assumption. By hypothesis we then have $e_{m^{-1}n} = 0$, so also $u_{m^{-1}n} = 0$. Focusing on (32.6) we then compute

$$u_m u_m^* \, u_n u_n^* = u_m u_{m^{-1}} u_n u_{n^{-1}} \overset{(9.1.\text{ii})}{=} u_m u_{m^{-1}n} u_{n^{-1}} = 0.$$

Suppose now that m and n are elements of P possessing a common upper bound. Then $m \vee n$ exists and there are x and y in P such that

$$m \vee n = mx = ny.$$

Moreover, given the left invariance of the order relation on G, we have

$$1 \vee (m^{-1}n) = m^{-1}(m \vee n) = x,$$

whence, by hypothesis we have $e_{m^{-1}n} = e_x$, and then

$$\begin{aligned} u_m^* \, u_n &\overset{(9.9.\text{iii})}{=} u_{m^{-1}n} \\ &= e_{m^{-1}n} u_{m^{-1}n} \\ &= e_x u_{m^{-1}n} \\ &= u_x u_{x^{-1}} u_{m^{-1}n} \\ &= u_x u_{x^{-1}m^{-1}n} \\ &= u_x u_{y^{-1}}. \end{aligned}$$

Therefore,

$$\begin{aligned} u_m u_m^* \, u_n u_n^* &= u_m u_x u_{y^{-1}} u_{n^{-1}} \\ &= u_{mx} u_{(ny)^{-1}} \\ &= u_{m \vee n} u_{m \vee n}^*, \end{aligned}$$

as desired. □

CHAPTER 33

C*-Algebras Generated by Semigroups of Isometries

Given a semigroup P, we may consider the universal C*-algebra $C^*(P)$ generated by a set

$$\{w_n : n \in P\},$$

subject to the requirement that the correspondence

$$n \mapsto w_n$$

be a semigroup of isometries. The universal property of $C^*(P)$ may then be phrased as follows: for every unital C*-algebra A, and every semigroup of isometries $v : P \to A$, there exists a unique *-homomorphism $\varphi : C^*(P) \to A$, such that the diagram

$$\begin{array}{ccc} P & \xrightarrow{\ v\ } & A \\ \Big\downarrow{\scriptstyle w} & \nearrow{\scriptstyle \varphi} & \\ C^*(P) & & \end{array}$$

commutes. So, to study the representation theory of $C^*(P)$ is equivalent to studying all semigroups of isometries based on P.

In the case of $\mathbb{N}\times\mathbb{N}$, observe that given any semigroup of isometries $v : \mathbb{N}\times\mathbb{N} \to A$, one has that $v_{(1,0)}$ and $v_{(0,1)}$ are commuting isometries. Conversely, given any pair of commuting isometries, we obtain by (31.4) a semigroup of isometries based on $\mathbb{N}\times\mathbb{N}$. Thus one may argue that, to fully understand $C^*(\mathbb{N}\times\mathbb{N})$ is to understand all commuting pairs of isometries, and in fact many authors have dedicated a significant amount of energy in this endeavor.

In order to give the reader a feeling for the difficulty of the problem at hand, we point out that many results in the literature on this subject have the following format: given any pair of commuting isometries S and T on a Hilbert space H, there is an orthogonal decomposition

$$H = H_1 \oplus \cdots \oplus H_n$$

in invariant subspaces, in such a way that the restrictions of S and T to the H_i may be described in more or less concrete terms. For example,

(i) S and T are unitary on H_1,
(ii) $S|_{H_2}$ is unitary and $T|_{H_2}$ is a multiple of the unilateral shift,
(iii) $S|_{H_3}$ is a multiple of the unilateral shift and $T|_{H_3}$ is unitary, etc.

However, as one approaches the last space H_n, the available descriptions often become more and more technical and sometimes include an *evanescent* space, about which little can be said. We refer the reader to [**103**] and [**70**] for some of the earliest results, and to [**91**] and [**20**] for what we believe represents the state of the art in this hard subject.

The difficulties presented by the theory of semigroups of isometries based on the otherwise nicely behaved semigroup $\mathbb{N}\times\mathbb{N}$ is perhaps a sign that insurmountable obstacles lie ahead of the corresponding theory for more general semigroups. We will therefore restrict our study to a nicer class of semigroups of isometries, namely the extendable ones.

Given a group G and a sub-semigroup $P \subseteq G$, recall from (31.17) that a semigroup of isometries

$$v : P \to A$$

is extendable if there exists a partial representation u of G in A such that $u_n = v_n$, for every n in P. Thus, to study extendable semigroups of isometries based on P is the same as studying partial representations of G which ascribe isometries to the elements of P.

Definition 33.1. Let G be a group and let $P \subseteq G$ be a sub-semigroup. We will denote[43] by $C^*(G, P)$ the universal unital C*-algebra generated by a set $\{u_g : g \in G\}$ subject to relations (9.1.i)–(9.1.iv), in addition to

$$u_n^* u_n = 1, \quad \forall n \in P.$$

Observe that this is a special case of (14.2), in the sense that $C^*(G, P)$ coincides with $C^*_{\text{par}}(G, \mathscr{P})$, where $\mathscr{P}$ consists of the above set of relations stating that u_n is an isometry for every n in P.

As a consequence of (14.16), we therefore have that $C^*(G, P)$ may be described as the partial crossed product algebra $C(\Omega_{\mathscr{P}}) \rtimes G$, and we will now give a precise description of the space $\Omega_{\mathscr{P}}$.

Our relations, if written in the form (14.1), become

$$e_{n^{-1}} - 1 = 0, \quad \forall n \in P,$$

so the set $\mathcal{F}_{\mathcal{R}}$ mentioned in (14.8) is formed by the functions

$$\omega \in \{0, 1\}^G \mapsto [n^{-1} \in \omega] - 1 \in \mathbb{C}.$$

Still according to (14.8), we have that $\Omega_{\mathscr{P}}$ consists of all $\omega \in \Omega_1$ such that

$$[n^{-1} \in g^{-1}\omega] - 1 = 0, \quad \forall n \in P, \quad \forall g \in \omega.$$

The above condition for an element $\omega \in \Omega_1$ to be in $\Omega_{\mathscr{P}}$ may be interpreted as

$$g \in \omega \ \Rightarrow \ gn^{-1} \in \omega, \quad \forall n \in P. \tag{33.2}$$

Since $h \leq g$ if and only if $h = gn^{-1}$, for some n in P, we see that (33.2) precisely expresses that ω is hereditary[44]. The above analysis and (14.16) therefore give us the following:

Theorem 33.3. *Let G be a group and let $P \subseteq G$ be a sub-semigroup. Consider the closed subspace $\Omega_{\mathscr{P}}$ of Ω_1 formed by the hereditary elements, equipped with the partial action of G obtained by restricting the partial Bernoulli action. Then there is a natural *-isomorphism between $C^*(G, P)$ and the partial crossed product*

---

[43]We should warn the reader that our choice of notation here is not standard. In particular it conflicts with the notation adopted in [**85**] for a C*-algebra which we will later study under the notation $N(P)$.

[44]Recall that a subset S of an ordered set X is said to be hereditary if, whenever x and y are elements of X, such that $x \leq y \in S$, one has that $x \in S$.

$C(\Omega_{\mathscr{P}})\rtimes G$. Under this isomorphism, each u_g corresponds to $1_g\delta_g$, where 1_g denotes the characteristic function of

$$D_g^{\mathscr{P}} = \{\omega \in \Omega_{\mathscr{P}} : g \in \omega\}.$$

One may argue that the above result is not fully satisfactory since the definition of $C^*(G, P)$ puts too much emphasis on the group G, while one might actually only be interested in the semigroup P. Further research is therefore needed in order to answer a few outstanding questions such as:

Question 33.4. Can extendability of a semigroup of isometries v be characterized via algebraic relations involving only the isometries v_n, for n in P?

Recall that (31.16) provides an answer to this question in the case of Ore semigroups.

We have already mentioned that the commutativity of range projections is a necessary condition for extendability of a semigroup of isometries, so any answer to (33.4) is likely to include the commutativity of range projections. This also motivates the following:

Definition 33.5. Given a semigroup P, we denote by $C^*_{ab}(P)$ the universal unital C*-algebra generated by a set $\{w_n : n \in P\}$, subject to the relations

(i) $w_n^*w_n = 1$,
(ii) $w_mw_n = w_{mn}$,
(iii) $w_mw_m^*$ commutes with $w_nw_n^*$,

for all n, m in P.

Another relevant open question is as follows:

Question 33.6. Given a sub-semigroup P of a group G, can the subalgebra of $C^*(G, P)$ generated by $\{u_p : p \in P\}$ be concretely described?

In the above context, observe that, given a group element $g \in P^{-1}P$, and writing $g = m^{-1}n$, with $m, n \in P$, one has by (9.9) that

$$u_g = u_{m^{-1}n} = u_{m^{-1}}u_n,$$

since u_n is an isometry, and hence left-invertible. The image of $P^{-1}P$ under u is therefore contained in the subalgebra referred to in (33.6). If G coincides with $P^{-1}P$, that is, if P is an Ore sub-semigroup of G, the answer to question (33.6) is therefore that the subalgebra mentioned there coincides with the whole of $C^*(G, P)$.

Being able to satisfactorily answer the above questions for Ore sub-semigroups, we may give a nice description of $C^*_{ab}(P)$, as follows:

Theorem 33.7. *Let P be an Ore sub-semigroup of a group G. Then $C^*_{ab}(P)$ is naturally isomorphic to $C^*(G, P)$ and consequently there is a *-isomorphism*

$$\varphi : C^*_{ab}(P) \to C(\Omega_{\mathscr{P}}) \rtimes G,$$

such that

$$\varphi(w_n) = 1_n\delta_n, \quad \forall n \in P,$$

where $\Omega_{\mathscr{P}}$ and 1_n are as in (33.3).

PROOF. We will prove that there is an isomorphism

$$\psi : C^*_{ab}(P) \to C^*(G, P), \tag{33.7.1}$$

such that

$$\psi(w_n) = u_n, \quad \forall p \in P. \tag{33.7.2}$$

The conclusion will then follow immediately from (33.3). As a first step, observe that the correspondence

$$n \in P \mapsto u_n \in C^*(G, P)$$

is a semigroup of isometries, therefore evidently satisfying (33.5.i) & (33.5.ii), and which also satisfies (33.5.iii) thanks to (9.8.iv). The universal property of $C^*_{ab}(P)$ may therefore be invoked to prove the existence of a *-homomorphism ψ, as in (33.7.1), satisfying (33.7.2), and we need only prove that ψ is an isomorphism.

Noticing that the correspondence

$$n \in P \mapsto w_n \in C^*_{ab}(P)$$

is a semigroup of isometries satisfying (31.16.i), we have by (31.16) that there exists a partial representation

$$\tilde{w} : G \to C^*_{ab}(P),$$

such that $\tilde{w}_n = w_n$, for every n in P. Since the $\tilde{w}_n$ are isometries, the universal property of $C^*(G, P)$ implies that there exists a *-homomorphism

$$\gamma : C^*(G, P) \to C^*_{ab}(P),$$

such that $\gamma(u_n) = \tilde{w}_n$, for all n in P. We then conclude that, for all $n \in P$, one has that

$$\gamma\big(\psi(w_n)\big) \overset{(33.7.2)}{=} \gamma(u_n) = \tilde{w}_n = w_n,$$

from where we see that $\gamma \circ \psi$ is the identity on $C^*_{ab}(P)$. In particular this shows that ψ is one-to-one. In order to prove that ψ is onto $C^*(G, P)$, it is enough to show that the standard generating set $\{u_g : g \in G\}$ of $C^*(G, P)$ lies in the range of ψ. For this, given g in G, write $g = m^{-1}n$, with m and n in P. Then

$$\begin{aligned} u_g &= u_{m^{-1}n} \\ &\overset{(9.9)}{=} u_{m^{-1}} u_n \\ &= (u_m)^* u_n \\ &= \psi(w_m)^* \psi(w_n) \\ &= \psi(w_m^* w_n) \\ &\in \psi\big(C^*_{ab}(P)\big). \end{aligned}$$

This proves that ψ is an isomorphism, as desired. □

Since the range of a partial representation is always a tame set by (12.12), the above result implies that the isometries canonically generating $C^*_{ab}(P)$ form a tame set in case P is an Ore sub-semigroup. However the following seems to be open:

Question 33.8. Given a semigroup P, is the subset $\{w_n : n \in P\}$ of $C^*_{ab}(P)$ tame? If not, what is the smallest set of relations one can add to the definition of $C^*_{ab}(P)$ to make the answer affirmative?

CHAPTER 34

Wiener-Hopf C*-Algebras

In this chapter we continue the study initiated above of C*-algebras associated to semigroups of isometries. One of the most important among these is the Wiener-Hopf C*-algebra associated to a quasi-lattice ordered group (G, P), so we dedicate the present chapter, in its entirety, to the study of this example.

We will initially concentrate on the study of a C*-algebra introduced by Nica in [**85**], which we shall denote by $N(P)$, and which should be seen as the *full* version of the actual Wiener-Hopf algebra to be defined later.

▶ We will now fix, for the entire duration of this chapter, a group G, and a sub-semigroup $P \subseteq G$, such that (G, P) is a quasi-lattice.

Definition 34.1. We will denote by $N(P)$ the universal unital C*-algebra for semigroups of isometries based on P satisfying NCC. More precisely, $N(P)$ is the universal unital C*-algebra generated by a set $\{v_n : n \in P\}$, subject to the relations below for all n and m in P:

(i) $v_n^* v_n = 1$,

(ii) $v_m v_n = v_{mn}$,

(iii) $v_m v_m^* \; v_n v_n^* = \begin{cases} v_{m\vee n} v_{m\vee n}^*, & \text{if } m \vee n \text{ exists,} \\ 0, & \text{otherwise.} \end{cases}$

Some authors denote the algebra introduced above by $C^*(G, P)$, but we have chosen $N(P)$ in order to avoid conflict with the notation introduced in (33.1). Our notation is also intended to emphasize that $N(P)$ depends only on the algebraic structure of P, rather than on G.

As we have seen in (32.11) and (32.12), there exists a one-to-one correspondence between semigroup of isometries satisfying NCC and partial representations of G satisfying (32.11.1).

Proposition 34.2. *Let $\mathcal{N}$ be the set of relations* (32.11.1), *for all g in G, and consider the universal C*-algebra $C^*_{par}(G, \mathcal{N})$ for partial representations of G satisfying $\mathcal{N}$, as defined in* (14.2). *Then there is a *-isomorphism*

$$\varphi : N(P) \to C^*_{par}(G, \mathcal{N}),$$

such that $\varphi(v_n) = u_n$, for all n in P, where we denote the canonical partial representation of G in $C^*_{par}(G, \mathcal{N})$ by u.

PROOF. Since u satisfies (32.11.1) by construction, (32.12) implies that the restriction of u to P is a semigroup of isometries satisfying NCC. The universal property of $N(P)$ then provides for a unital *-homomorphism

$$\varphi : N(P) \to C^*_{\text{par}}(G, \mathcal{N}),$$

such that

$$\varphi(v_n) = u_n, \quad \forall n \in P. \tag{34.2.1}$$

On the other hand, by (32.11), the universal semigroup of isometries $v : P \to N(P)$ extends to a partial representation $\tilde{v} : G \to N(P)$ satisfying (32.11.1). So, again by universality, there is a unital *-homomorphism

$$\psi : C^*_{\text{par}}(G, \mathcal{N}) \to N(P),$$

such that

$$\psi(u_g) = \tilde{v}_g, \quad \forall g \in G.$$

Given any $n \in P$, we have that

$$\psi\big(\varphi(v_n)\big) = \psi\big(u_n\big) = \tilde{v}_n = v_n,$$

from where we see that $\psi \circ \varphi$ is the identity mapping on $N(P)$.

Notice that the correspondence

$$n \in P \mapsto \varphi(v_n) \in C^*_{\text{par}}(G, \mathcal{N})$$

is a semigroup of isometries satisfying NCC, since this is obtained by composing v with a *-homomorphism. We could now apply (32.11) to prove that it extends to a partial representation of G, except that we already know two extensions of it, the first one being

$$g \in G \mapsto \varphi(\tilde{v}_g) \in C^*_{\text{par}}(G, \mathcal{N}),$$

and the second one being simply u, as one may easily deduce from (34.2.1). Clearly both of these extensions satisfy (32.11.1), so the uniqueness part of (32.11) implies that

$$\varphi(\tilde{v}_g) = u_g, \quad \forall g \in G.$$

Consequently we have

$$\varphi\big(\psi(u_g)\big) = \varphi(\tilde{v}_g) = u_g,$$

so $\varphi \circ \psi$ coincides with the identity mapping of $C^*_{\text{par}}(G, \mathcal{N})$ on a generating set, whence $\varphi \circ \psi$ is the identity. This proves that φ is indeed an isomorphism, as desired. □

Using (14.16) we may then describe $C^*_{\text{par}}(G, \mathcal{N})$, and hence also $N(P)$, as a partial crossed product. But, in order to state a more concrete result, let us first give a description of the spectrum of $\mathcal{N}$ which is more geometrical than the one given in (14.8).

Proposition 34.3. *The spectrum of the set of relations $\mathcal{N}$ is given by*

$$\Omega_{\mathcal{N}} = \big\{\omega \in \Omega_1 : \omega \textit{ is hereditary and directed}^{45}\big\}.$$

PROOF. We will first prove "$\subseteq$", so we pick any $\omega \in \Omega_{\mathcal{N}}$. For every n in P, we have that $n^{-1} \vee 1 = 1$, so $\mathcal{N}$ includes the relation "$e_{n^{-1}} = e_1$". Consequently, for every $g \in \omega$, we have by the definition of $\Omega_{\mathcal{N}}$, that

$$[n^{-1} \in g^{-1}\omega] = [1 \in g^{-1}\omega] = [g \in \omega] = 1,$$

which is to say that $gn^{-1} \in \omega$, thus proving that ω is hereditary.

[45] Recall that a subset S of an ordered set X is said to be directed if, for all $x, y \in S$, there exists z in S, with $z \geq x, y$.

We next claim that

$$\omega \subseteq PP^{-1}. \tag{34.3.1}$$

To see this let $g \in G \setminus PP^{-1}$. Then $\mathcal{N}$ includes the relation "$e_g = 0$" and, again by the definition of $\Omega_{\mathcal{N}}$, we have that $[g \in \omega] = 0$, so $g \notin \omega$, proving the claim.

We will next prove that ω is directed. For this let $g, h \in \omega$, so by (34.3.1) we may write $g = mn^{-1}$ and $h = pq^{-1}$. Assuming, without loss of generality, that $m = g \vee 1$ and $p = h \vee 1$, we have that

$$1 = [g \in \omega] = [g \vee 1 \in \omega] = [m \in \omega],$$

so $m \in \omega$, and similarly $p \in \omega$. Observing that

$$g = mn^{-1} \leq m, \quad \text{and} \quad h = pq^{-1} \leq p,$$

we see that it is enough to find a common upper bound for m and p in ω, since this will automatically be a common upper bound for g and h.

Recalling that $\Omega_{\mathcal{N}}$ is invariant under the partial Bernoulli action, and since $m \in \omega$, we have that $m^{-1}\omega \in \Omega_{\mathcal{N}}$. So the conclusion reached in (34.3.1) also applies to $m^{-1}\omega$, whence

$$m^{-1}p \in m^{-1}\omega \subseteq PP^{-1},$$

and therefore $(m^{-1}p) \vee 1$ exists. The relation "$e_{m^{-1}p} = e_{(m^{-1}p)\vee 1}$" is then among the relations in $\mathcal{N}$, so for all k in ω one has

$$[m^{-1}p \in k^{-1}\omega] = [(m^{-1}p)\vee 1 \in k^{-1}\omega].$$

Plugging in $k = m$, left-hand-side above evaluates to 1, so the right-hand-side does too, meaning that

$$(m^{-1}p) \vee 1 \in m^{-1}\omega,$$

whence

$$\omega \ni m\big((m^{-1}p) \vee 1\big) = p \vee m,$$

where the last step above, including the existence of $p \vee m$, is a consequence of the left-invariance of the order relation on G. This proves that m and p have a common upper bound in ω, thus proving that ω is directed.

Let us now prove the inclusion "$\supseteq$" between the sets mentioned in the statement, so we pick an hereditary and directed $\omega \in \Omega_1$. We will first prove that ω is contained in PP^{-1}. For this pick $g \in \omega$ and observe that, since 1 is also in ω, there exists an upper bound m for $\{g, 1\}$ in ω. Thus m is in P and, since $g \leq m$, we deduce that $n := g^{-1}m$ also lies in P. So

$$g = mn^{-1} \in PP^{-1}.$$

This verifies our claim that $\omega \subseteq PP^{-1}$.

In order to prove that $\omega \in \Omega_{\mathcal{N}}$, we must show that $f(h^{-1}\omega) = 0$, for all h in ω, and for all f in $\mathcal{F}_{\mathcal{N}}$ (see (14.8) for the definition of $\mathcal{F}_{\mathcal{N}}$).

According to (32.11.1) we need to consider two cases. Firstly, when g is not in PP^{-1}, the corresponding relation "$e_g = 0$" in $\mathcal{N}$ leads to the function

$$f(x) = [g \in x], \quad \forall\, x \in \Omega_1,$$

so we must prove that

$$g \notin h^{-1}\omega, \quad \forall\, h \in \omega. \tag{34.3.2}$$

By the left-invariance of the order relation on G, for every h in ω, we have that $h^{-1}\omega$ is also hereditary and directed. Moreover, if $h \in \omega$, then $h^{-1}\omega$ also lies in Ω_1, so (34.3.1) holds for $h^{-1}\omega$, just like it does for ω. So

$$h^{-1}\omega \subseteq PP^{-1},$$

from where (34.3.2) follows immediately.

The second case to be considered is when $g \in PP^{-1}$, in which case $g \vee 1$ exists and we are led to the function f in $\mathcal{F}_{\mathcal{N}}$ given by

$$f(\xi) = [g \in \xi] - [g\vee 1 \in \xi], \quad \forall \xi \in \Omega_1.$$

Given h in ω we must then show that $f(h^{-1}\omega) = 0$, which translates into

$$[g \in h^{-1}\omega] = [g\vee 1 \in h^{-1}\omega],$$

or, equivalently

$$\begin{array}{ccc} hg \in \omega \;\Leftrightarrow & h(g \vee 1) & \in \omega, \\ & \| & \\ & (hg) \vee h & \end{array}$$

where the vertical equal sign is a consequence of left-invariance. Using that ω is hereditary and directed, and that h is in ω, this may now be easily verified. □

Interpreting (14.16) in the present case, and using (34.2), we have the following concrete description of $N(P)$ as a partial crossed product.

Theorem 34.4. *Given a quasi-lattice (G, P), consider the subset $\Omega_{\mathcal{N}}$ of 2^G formed by the hereditary directed subsets of G containing 1. Then $\Omega_{\mathcal{N}}$ is compact and invariant under the partial Bernoulli action. In addition there is a *-isomorphism*

$$\psi : N(P) \to C(\Omega_{\mathcal{N}}) \rtimes G,$$

such that $\psi(v_n) = 1_n\delta_n$, where 1_n denotes the characteristic function of the set $\{\omega \in \Omega_{\mathcal{N}} : n \in \omega\}$.

Observe that if ω is a hereditary and directed subset of G, then $\omega \cap P$ is a hereditary and directed subset of P. We will next explore this correspondence in order to obtain another description of $\Omega_{\mathcal{N}}$.

Proposition 34.5. *The map*

$$\omega \in \Omega_{\mathcal{N}} \mapsto \omega \cap P \in 2^P$$

is a homeomorphism from $\Omega_{\mathcal{N}}$ onto the subset of 2^P formed by all nonempty, hereditary, directed subsets of P.

PROOF. We leave it for the reader to verify the continuity of our map.

Given any ω in $\Omega_{\mathcal{N}}$, let $\xi = \omega \cap P$. Using that ω is hereditary and directed by (34.3), it is easy to see that

$$\omega = \xi P^{-1} = \{g \in G : g \le n, \text{ for some } n \in \xi\},$$

from where it follows that the map in the statement is injective.

It is obvious that $\omega \cap P$ is a hereditary directed subset of P for every ω in $\Omega_{\mathcal{N}}$, and conversely, for every nonempty, hereditary, directed subset $\xi \subseteq P$, one has that

$$\omega_\xi := \xi P^{-1}$$

lies in $\Omega_{\mathcal{N}}$, and $\xi = \omega_\xi \cap P$. This proves that the range of our map is as described in the statement. Since $\Omega_{\mathcal{N}}$ is compact and 2^P is Hausdorff, it follows that our map is a homeomorphism onto its range. □

The above result can easily be used to transfer the partial Bernoulli action on $\Omega_{\mathcal{N}}$ to a partial action of G on the set of all nonempty, hereditary, directed subsets of P, thus providing yet another model for the partial dynamical system leading up to $N(P)$ in (34.4).

Given any m in P, notice that the set

$$mP^{-1} = \{mn^{-1} : n \in P\} = \{g \in G : g \leq m\} \tag{34.6}$$

is a hereditary and directed subset of G containing 1, hence mP^{-1} is an element of $\Omega_{\mathcal{N}}$.

Given any $\omega \in \Omega_{\mathcal{N}}$, we may then consider the net

$$\{mP^{-1}\}_{m \in \omega \cap P}, \tag{34.7}$$

where $\omega \cap P$, carrying the order induced from G, is a directed set by (34.5), hence suitable for playing the role of the index set for a net.

Proposition 34.8. *For every ω in $\Omega_{\mathcal{N}}$, the net described in* (34.7) *converges to ω.*

PROOF. Given ω in $\Omega_{\mathcal{N}}$ and a neighborhood V of ω, by definition of the product topology there are

$$g_1, g_2, \ldots, g_r;\ h_1, h_2, \ldots, h_s \in G,$$

such that the *basic neighborhood*

$$U := \{\eta \in \Omega_{\mathcal{N}} : g_i \in \eta,\ h_j \notin \eta,\ \forall i \leq r,\ \forall j \leq s\}$$

satisfies

$$\omega \in U \subseteq V.$$

In particular the g_i lie in ω, so we may use the fact that ω is directed to find some m_0 in ω such that $m_0 \geq g_i$, for all $i \leq r$. Upon assuming without loss of generality that 1 is among the g_i, we have that $m_0 \in \omega \cap P$.

We then claim that, for all m in $\omega \cap P$ with $m \geq m_0$, one has that $mP^{-1} \in U$, which is to say that

(a) $g_i \in mP^{-1}$, for all $i \leq r$, and
(b) $h_j \notin mP^{-1}$, for all $j \leq s$.

Point (a) is an obvious consequence of the fact that each

$$g_i \leq m_0 \leq m.$$

With respect to (b), assume by way of contradiction, that some $h_j \in mP^{-1}$. Then $h_j \leq m$, whence $h_j \in \omega$, because $m \in \omega$, and ω is hereditary. This is a contradiction with the fact that ω belongs to U, hence verifying (b). This concludes the proof. □

Recall from (32.4) that the regular representation of P on $\ell^2(P)$ satisfies NCC. Thus, by universality, there is a *-representation of $N(P)$ on $\ell^2(P)$ consistent with the regular representation. Our next goal will be to study the range of such a representation and, in particular, to show that it is isomorphic to the *reduced* crossed product relative to the same partial dynamical system whose associated *full* crossed product was shown to coincide with $N(P)$ in (34.4).

Definition 34.9. The closed *-algebra of bounded operators on $\ell^2(P)$ generated by the range of λ is called the *Wiener-Hopf* algebra of P, and it is denoted by $\mathcal{W}(P)$.

In [**85**], Nica denotes the C*-algebra defined above by $\mathcal{W}(G,P)$, but since the above definition is given only in terms of the semigroup P, we prefer not to explicitly mention the ambient group G.

We will next describe $\mathcal{W}(P)$ as a reduced partial crossed product.

Theorem 34.10. *There is a* **-isomorphism*

$$\chi : \mathcal{W}(P) \to C(\Omega_{\mathcal{N}}) \rtimes_{red} G,$$

such that $\chi(\lambda_n) = 1_n\delta_n$, *where* 1_n *is as in* (34.4).

Proof. The universal property of $N(P)$ implies that there exists a *-homomorphism

$$\sigma : N(P) \to \mathcal{W}(P),$$

such that $\sigma(v_n) = \lambda_n$, for all n in P, which is clearly onto.

Considering the *-isomorphism ψ given in (34.4) we may define a surjective *-homomorphism $\rho = \sigma \circ \psi^{-1}$, as indicated in the diagram:

$$\begin{array}{ccc} C(\Omega_{\mathcal{N}}) \rtimes G & \xrightarrow{\rho} & \mathcal{W}(P) \\ \psi \nwarrow & & \nearrow \sigma \\ & N(P) & \end{array}$$

We will now prove that the kernel of ρ coincides with the kernel of the regular representation

$$\Lambda : C(\Omega_{\mathcal{N}}) \rtimes G \to C(\Omega_{\mathcal{N}}) \rtimes_{\text{red}} G,$$

from where the statement will follow.

Consider the conditional expectation F from $\mathscr{L}\left(\ell^2(P)\right)$ onto the subalgebra formed by the diagonal operators relative to the standard orthonormal basis of $\ell^2(P)$. To be precise,

$$F(T) = \sum_{n \in P} e_{n,n} T e_{n,n}, \quad \forall T \in \mathscr{L}\left(\ell^2(P)\right),$$

where the sum is interpreted in the strong operator topology, and each $e_{n,n}$ is the orthogonal projection onto the one-dimensional subspace generated by the corresponding basis vector. It is a well known fact that F is a faithful conditional expectation onto the algebra of diagonal operators.

We will also consider the conditional expectation E, given by the composition

$$\overbrace{C(\Omega_{\mathcal{N}}) \rtimes G \xrightarrow{\Lambda} C(\Omega_{\mathcal{N}}) \rtimes_{\text{red}} G \xrightarrow{E_1}}^{E} C(\Omega_{\mathcal{N}}),$$

where E_1 is the faithful conditional expectation given by (17.8). We then claim that the diagram

$$\begin{array}{ccc} C(\Omega_{\mathcal{N}}) \rtimes G & \xrightarrow{\rho} & \mathcal{W}(P) \\ \downarrow E & & \downarrow F \\ C(\Omega_{\mathcal{N}}) & \xrightarrow{\rho} & \mathscr{L}\left(\ell^2(P)\right) \end{array}$$

commutes, where the bottom arrow is to be interpreted as the restriction of ρ to the canonical copy of $C(\Omega_{\mathcal{N}})$ within the crossed product.

By (32.8.i) we have that $N(P)$ is the closed linear span of the set of elements of the form $v_m v_n^*$, as m and n range in P. Using (34.4) we therefore have that $C(\Omega_{\mathcal{N}}) \rtimes G$ is spanned by the elements y of the form

$$\begin{aligned} y &= \psi(v_m v_n^*) \\ &= (1_m \delta_m)(1_n \delta_n)^* \\ &= \beta_m(1_{m^{-1}} 1_{n^{-1}}) \delta_{mn^{-1}} \\ &= 1_m 1_{mn^{-1}} \delta_{mn^{-1}}, \end{aligned} \tag{34.10.1}$$

and the commutativity of the above diagram will follow once we check that $F\rho(y) = \rho E(y)$, for every y of the above form.

Assuming that $m \neq n$, one has that $mn^{-1} \neq 1$, and $E(y) = 0$. On the other hand,

$$\rho(y) = \sigma \circ \psi^{-1} \circ \psi(v_m v_n^*) = \sigma(v_m v_n^*) = \lambda_m \lambda_n^*.$$

Because $m \neq n$, one has that the matrix of $\lambda_m \lambda_n^*$ has no nonzero diagonal entry, which is to say that

$$F\big(\rho(y)\big) = 0 = \rho\big(E(y)\big),$$

as desired.

Assuming now that $m = n$, we have that y lies in $C(\Omega_{\mathcal{N}})$ (which we identify with $C(\Omega_{\mathcal{N}})\delta_1$ as usual), so $E(y) = y$. On the other hand

$$\rho(y) = \lambda_m \lambda_m^*,$$

which is a diagonal operator by (31.7), whence

$$F\big(\rho(y)\big) = \rho(y) = \rho\big(E(y)\big),$$

proving our diagram to be commutative, as claimed.

We will eventually like to show that ρ is one-to-one on $C(\Omega_{\mathcal{N}})$. With this goal in mind we next claim that, for every k in P, and every f in $C(\Omega_{\mathcal{N}})$, one has that the kth diagonal entry of $\rho(f)$, here denoted by $\rho(f)_{k,k}$, is given by

$$\rho(f)_{k,k} = f(kP^{-1}), \tag{34.10.2}$$

where we are seeing kP^{-1} as an element of $\Omega_{\mathcal{N}}$, as we already did in (34.7).

Let us first assume that $f = 1_m$, where m is in P. In this case, identifying 1_m with $1_m\delta_1$, as usual, we have that

$$1_m \delta_1 \overset{(34.10.1)}{=} \psi(v_m v_m^*),$$

so $\rho(1_m) = \lambda_m \lambda_m^*$, whence for every k in P, the corresponding diagonal entry of $\rho(1_m)$ is given by

$$\rho(1_m)_{k,k} = (\lambda_m \lambda_m^*)_{k,k} \overset{(32.5)}{=} [k \geq m] = [m \in kP^{-1}] = 1_m(kP^{-1}),$$

proving (34.10.2) for the particular case of $f = 1_m$. To prove the general case of this identity, it is enough to verify that the 1_m generate $C(\Omega_{\mathcal{N}})$ as a C*-algebra, which we do as follows: by Stone-Weierstrass we have that

$$\{1_g : g \in G\} \tag{34.10.3}$$

generates $C(\Omega_{\mathcal{N}})$ as a C*-algebra. On the other hand, since each ω in $\Omega_{\mathcal{N}}$ is hereditary and directed, and since $1 \in \omega$, we have that

$$g \in \omega \Leftrightarrow g \vee 1 \in \omega, \quad \forall g \in G.$$

Consequently $1_g = 1_{g\vee 1}$, when $g \vee 1$ exists, and $1_g = 0$, otherwise. Except for the identically zero function we then have that (34.10.3) coincides with

$$\{1_n : n \in P\},$$

which is therefore also a generating set for $C(\Omega_{\mathcal{N}})$, concluding the proof of (34.10.2).

Still aiming at the proof that the kernel of ρ coincides with the kernel of the regular representation of $C(\Omega_{\mathcal{N}}) \rtimes G$, we will next prove that ρ is injective on $C(\Omega_{\mathcal{N}})$.

For this, suppose that $\rho(f) = 0$, for some f in $C(\Omega_{\mathcal{N}})$. From (34.10.2) we conclude that f vanishes on every kP^{-1} in $\Omega_{\mathcal{N}}$. However, the set formed by these is dense in $\Omega_{\mathcal{N}}$ by (34.8), so $f = 0$.

With this we may now describe the null space of ρ, as follows: given any y in $C(\Omega_{\mathcal{N}}) \rtimes G$, and using that F and E_1 are faithful and ρ is injective on $C(\Omega_{\mathcal{N}})$, we have that

$$\begin{aligned}
\rho(y) = 0 &\Leftrightarrow F\big(\rho(y)^*\rho(y)\big) = 0 \\
&\Leftrightarrow \rho\big(E(y^*y)\big) = 0 \\
&\Leftrightarrow E(y^*y) = 0 \\
&\Leftrightarrow E_1\big(\Lambda(y^*y)\big) = 0 \\
&\Leftrightarrow \Lambda(y) = 0.
\end{aligned}$$

This shows that ρ and Λ share kernels, as claimed, so ρ factors through Λ, producing a *-isomorphism

$$C(\Omega_{\mathcal{N}}) \rtimes_{\text{red}} G \xrightarrow{\tilde{\rho}} \mathcal{W}(P),$$

whose inverse satisfies all of the required conditions. □

We may now use the results on amenability of Fell bundles to obtain conditions under which $N(P)$ coincides with the Wiener-Hopf algebra:

Theorem 34.11. *Let (G, P) be a quasi-lattice, where G is an amenable group. Then $N(P)$ is naturally isomorphic to $\mathcal{W}(P)$.*

Proof. Follows immediately from (34.4), (34.10), and (20.7). □

We dedicate the remainder of this chapter to prove a useful characterization of faithful representations of the Wiener-Hopf algebra.

Proposition 34.12. *Let (G, P) be a quasi-lattice. Then the partial Bernoulli action restricted to $\Omega_{\mathcal{N}}$ is topologically free.*

Proof. Let $g \in G \setminus \{1\}$ and assume by contradiction that there is an open subset V of the domain of β_g, where β refers to the partial Bernoulli action, formed by fixed points for β_g.

By (34.8) the collection of all mP^{-1}, as m range in P, forms a dense subset of $\Omega_{\mathcal{N}}$, so there is some m in P such that $mP^{-1} \in V$. Therefore

$$mP^{-1} = \beta_g(mP^{-1}) = gmP^{-1}.$$

Since m is the maximum element in mP^{-1}, and gm is the maximum element in gmP^{-1}, we deduce that $gm = m$, and hence that $g = 1$, a contradiction. This concludes the proof. □

In view of (34.10) and the result above, Corollary (29.6) applies to give a characterization of faithful representations of the Wiener-Hopf algebra as those whose restriction to $C(\Omega_{\mathcal{N}})$ are faithful. However, a more careful analysis will lead us to an even more precise result. But before that we need the following auxiliary result.

Lemma 34.13. *Let (G, P) be a quasi-lattice and let W be a subset of $\Omega_{\mathcal{N}}$ which is nonempty, open and invariant. Then there are $p_1, p_2, \ldots, p_k \in P \setminus \{1\}$, such that W contains the subset*

$$\{\omega \in \Omega_{\mathcal{N}} : p_i \notin \omega, \text{ for all } i = 1, \ldots, k\}.$$

PROOF. As before, we denote the partial Bernoulli action of G on $\Omega_{\mathcal{N}}$ by β, and for each g in G, we denote the range of β_g by $D_g^{\mathcal{N}}$.

Regarding the elements mP^{-1} of $\Omega_{\mathcal{N}}$ described in (34.6), we claim that $1P^{-1}$ lies in W. To see this observe that, for every g in G, one has

$$1P^{-1} \in D_g^{\mathcal{N}} \overset{(5.14)}{\Leftrightarrow} g \in 1P^{-1} \Leftrightarrow g^{-1} \in P.$$

The orbit of $1P^{-1}$ under β is therefore given by

$$\{\beta_n(1P^{-1}) : n \in P\} = \{nP^{-1} : n \in P\},$$

which is a dense subset of $\Omega_{\mathcal{N}}$ by (34.8). Since W is nonempty, it follows that W contains nP^{-1} for some $n \in P$. So, using the invariance of W, we deduce that

$$1P^{-1} = \beta_{n^{-1}}(nP^{-1}) \in \beta_{n^{-1}}\big(W \cap D_n^{\mathcal{N}}\big) \subseteq W,$$

proving the claim. We may therefore pick

$$g_1, g_2, \ldots, g_r;\ h_1, h_2, \ldots, h_s \in G,$$

such that the basic neighborhood

$$U := \{\omega \in \Omega_{\mathcal{N}} : g_i \in \omega,\ h_j \notin \omega,\ \forall i \leq r,\ \forall j \leq s\}$$

satisfies

$$1P^{-1} \in U \subseteq W.$$

Consequently, for each $i \leq r$, one has that $g_i \in 1P^{-1}$, so g_i has the form $g_i = n_i^{-1}$, for some n_i in P. However notice that $n_i^{-1} \in \omega$, for every single ω in $\Omega_{\mathcal{N}}$, because

$$n_i^{-1} \leq 1 \overset{(5.14)}{\in} \omega,$$

and ω is hereditary by (34.3). So the condition "$g_i \in \omega$", appearing in the definition of U above, is innocuous and hence may be omitted, meaning that

$$U = \{\omega \in \Omega_{\mathcal{N}} : h_j \notin \omega,\ \forall j \leq s\}.$$

Since every ω in $\Omega_{\mathcal{N}}$ is directed by (34.3), one may easily see that $\omega \subseteq PP^{-1}$ (this was in fact explicitly proved in (34.3.1)). So, if any given h_j is not in PP^{-1}, the condition "$h_j \notin \omega$" is true for any ω in $\Omega_{\mathcal{N}}$, and hence may also be eliminated from the definition of U. We may therefore assume, without loss of generality, that the h_j all lie in PP^{-1}.

Recall that such elements are precisely the ones which admit an upper bound in P, and hence $h_j \vee 1$ exist, for every j. Again because every ω in $\Omega_{\mathcal{N}}$ is hereditary and directed, it is easy to see that

$$h_j \in \omega \Leftrightarrow h_j \vee 1 \in \omega, \quad \forall \omega \in \Omega_{\mathcal{N}}.$$

We then conclude that the condition "$h_j \notin \omega$" in the definition of U may be replaced by "$h_j \vee 1 \notin \omega$", allowing us to assume without loss of generality that $h_j \in P$. Moreover we must have $h_j \neq 1$, for every j, since otherwise $U = \emptyset$. This concludes the proof. □

The following is the characterization of faithful representations of the Wiener-Hopf algebra announced earlier.

Theorem 34.14. *Let (G, P) be a quasi-lattice and let π be a representation of $\mathcal{W}(P)$ on a Hilbert space. Then π is faithful if and only if, given any $p_1, p_2, \ldots, p_k \in P \setminus \{1\}$, one has that*

$$(1 - V_1V_1^*)(1 - V_2V_2^*) \cdots (1 - V_kV_k^*) \neq 0,$$

where each $V_i = \pi(\lambda_{p_i})$.

PROOF. Recall from (34.9) that $\mathcal{W}(P)$ is the closed *-algebra of operators on $\ell^2(P)$ generated by the range of the regular semigroup of isometries of P, namely

$$\{\lambda_p : p \in P\}.$$

It is also interesting to point out that, apart from λ_1, which is the identity operator, all of the other λ_p's are *proper* isometries, in the sense that

$$1 - \lambda_p\lambda_p^* \neq 0, \quad \forall p \in P \setminus \{1\}.$$

Denoting by e_1 the first element of the canonical basis of $\ell^2(P)$, observe that, for each p in $P \setminus \{1\}$, one has that e_1 is orthogonal to the range of $\lambda_p\lambda_p^*$ by (32.5), so that in fact we have

$$(1 - \lambda_p\lambda_p^*)(e_1) = e_1. \tag{34.14.1}$$

Therefore, given $p_1, p_2, \ldots, p_k \in P \setminus \{1\}$, as above, we have that

$$(1 - \lambda_{p_1}\lambda_{p_1}^*)(1 - \lambda_{p_2}\lambda_{p_2}^*) \cdots (1 - \lambda_{p_k}\lambda_{p_k}^*) \neq 0,$$

because this operator sends e_1 to itself, as one may easily verify by successive applications of (34.14.1).

If π is a faithful representation of $\mathcal{W}(P)$, it therefore does not vanish on the above operator, hence proving the "only if" part of the statement.

In order to prove the converse, let us assume by contradiction that π satisfies the condition in the statement and yet it is not faithful. Using the isomorphism

$$\chi : \mathcal{W}(P) \to C(\Omega_{\mathcal{N}}) \rtimes_{\text{red}} G$$

provided by (34.10), we have that

$$\rho := \pi \circ \chi^{-1}$$

is a representation of the above reduced crossed product, which is likewise non-faithful.

Denoting by J the null space of ρ, we then have that J is a nontrivial ideal of $C(\Omega_{\mathcal{N}}) \rtimes_{\text{red}} G$. So, putting together (34.12) and (29.5), we deduce that

$$K := J \cap C(\Omega_{\mathcal{N}}) \neq \{0\},$$

and moreover K is invariant by (23.11). Writing $K = C_0(W)$, where W is an open subset of $\Omega_{\mathcal{N}}$, we then have that W is nonempty and invariant.

We may then invoke (34.13) to produce $p_1, p_2, \ldots, p_k$ in $P \setminus \{1\}$ such that

$$U := \{\omega \in \Omega_{\mathcal{N}} : p_i \notin \omega, \text{ for all } i = 1, \ldots, k\} \subseteq W.$$

It follows that any function in $C(\Omega_{\mathcal{N}})$, whose support is contained in U, necessarily lies in the null space of ρ.

Denoting, as usual, the characteristic function of the set

$$\{\omega \in \Omega_{\mathcal{N}} : p \in \omega\}$$

by 1_p, observe that the characteristic function of U, here denoted by 1_U, is given by

$$1_U = (1 - 1_{p_1})(1 - 1_{p_2}) \ldots (1 - 1_{p_k}).$$

Therefore

$$\begin{aligned} 0 &= \rho(1_U) \\ &= \prod_{i=1}^{k} \rho(1 - 1_{p_i}) \\ &\overset{(8.14.f)}{=} \prod_{i=1}^{k} \rho\big(1 - (1_{p_i}\delta_{p_i})(1_{p_i}\delta_{p_i})^*\big) \\ &\overset{(34.10)}{=} \prod_{i=1}^{k} \rho\big(1 - \chi(\lambda_{p_i})\chi(\lambda_{p_i})^*\big) \\ &= \prod_{i=1}^{k} \big(1 - \pi(\lambda_{p_i})\pi(\lambda_{p_i})^*\big) \\ &= \prod_{i=1}^{k} \big(1 - V_i V_i^*\big), \end{aligned}$$

where the V_i are as in the statement. This is a contradiction, and hence the proof is concluded. □

Notes and remarks. Wiener-Hopf operators were first introduced by Norbert Wiener and Eberhard Hopf in [**69**], who studied them from the point of view of integral equations. In the context of quasi-lattice ordered groups, the Wiener-Hopf C*-algebra $\mathcal{W}(P)$ and the algebra $N(P)$ (with a different notation) was first studied by Nica in [**85**], where he observed that both $\mathcal{W}(P)$ and $N(P)$ have a "crossed product type structure", emphasizing the role of the abelian subalgebra $\mathcal{D} = C(\Omega_{\mathcal{N}})$ [**85**, Section 6.2]. We believe the above description of these algebras as reduced and full partial crossed products vindicates Nica's suspicion in a telling way. Theorem (34.14) is due to Laca and Raeburn [**77**, Proposition 2.3(3)].

CHAPTER 35

The Toeplitz C*-Algebra of a Graph

As we have already mentioned, C*-algebras generated by partial isometries have played a very important role in the theory. Besides the C*-algebras associated to semigroups of isometries treated in the previous chapter, some of the most prevalent examples are the Cuntz-Krieger [**29**], Exel-Laca [**57**] and the graph C*-algebras [**95**].

The class of Exel-Laca algebras includes the Cuntz-Krieger algebras, but the relationship between the former and graph algebras is not as straightforward. When a certain matrix which parametrizes Exel-Laca algebras have the property that two distinct rows are either equal or orthogonal (a well known property of the incidence matrix of a graph), Exel-Laca algebras are easily seen to produce all graph algebras, except for graphs containing vertices which are not the range of any edge (known as sources). However, up to Morita-Rieffel-equivalence, the class of Exel-Laca algebras coincides with the class of graph algebras, as recently proved by Katsura, Muhly, Sims and Tomforde [**73**].

Both Exel-Laca and graph C*-algebras may be described as partial crossed products, and in fact these algebras have been among the motivating examples for the development of the theory of partial actions. Since the partial crossed product description of graph C*-algebras is a bit easier than that of Exel-Laca algebras, we will dedicate the remaining chapters of this book to studying graph C*-algebras from the point of view of partial actions.

▶ From now on we will fix a *graph*

$$E = (E^0, E^1, r, d),$$

where E^0 is the set of *vertices*, E^1 is the set of *edges*, and

$$r, d : E^1 \to E^0$$

are the *range* and *domain* (or *source*) maps, respectively.

Definition 35.1. The *Toeplitz C*-algebra* of E, denoted $\mathcal{T}_E$, is the universal C*-algebra generated by a set of mutually orthogonal projections

$$\mathscr{G}_0 = \{p_v : v \in E^0\},$$

and a set of partial isometries

$$\mathscr{G}_1 = \{s_a : a \in E^1\},$$

subject to the relations:

(i) $s_a^* s_b = [a = b]\, p_{d(a)}$, where the brackets correspond to Boolean value,
(ii) $s_a s_a^* \leq p_{r(a)}$,

for all a and b in E^1. The *graph C*-algebra of* E, denoted $C^*(E)$, is likewise defined, where in addition to the above relations, one has

(iii) $p_v = \sum_{r(a)=v} s_a s_a^*$,

for all $v \in E^0$, such that $r^{-1}(v)$ is finite and nonempty.

The first requirement for a C*-algebra generated by partial isometries to be tractable with our methods is that the partial isometries involved form a tame set, so our first medium term goal will be to prove this fact. Since $C^*(E)$ is derived from the Toeplitz algebra, we will initially concentrate our attention on $\mathcal{T}_E$.

As usual, a (finite) *path* in E is defined to be a sequence $\alpha = \alpha_1 \ldots \alpha_n$, where $n \geq 1$, and the α_i are edges in E, such that $d(\alpha_i) = r(\alpha_{i+1})$, for all $i = 1, \ldots, n-1$. This is the usual convention when treating graphs from a categorical point of view, in which functions compose from right to left.

The length of α, denoted $|\alpha|$, is the number n of edges in it. As a special case we will also consider each vertex of E as a *path of length zero*.

The *source* of a path α of length $n > 0$ is defined by $d(\alpha) = d(\alpha_n)$, and its *range* is defined by $r(\alpha) = r(\alpha_1)$. In the special case that α is a path of length zero, hence consisting of a single vertex, the range and source of α are both defined to be that vertex.

The set of all paths of length n will be denoted by E^n (this being consistent with the notations for E^0 and E^1 already in use), and the set of all (finite) paths will be denoted by E^*.

By an *infinite path* in E we shall mean an infinite sequence

$$\alpha = \alpha_1\alpha_2 \ldots$$

where each α_i is an edge in E, and $d(\alpha_i) = r(\alpha_{i+1})$, for all $i \geq 1$. The set of all infinite paths will be denoted by E^∞. We define the range of an infinite path α to be $r(\alpha) = r(\alpha_1)$, and the length of such a path to be $|\alpha| = \infty$. There is no sensible notion of domain for an infinite path.

The set formed by all finite or infinite paths in E will be denoted by $E^\sharp$, namely

$$E^\sharp = E^* \cup E^\infty.$$

Definition 35.2. Given paths $\alpha \in E^*$ and $\beta \in E^\sharp$, with $d(\alpha) = r(\beta)$, we will denote by $\alpha\beta$ the path defined as follows:

(i) if $|\alpha| > 0$, and $|\beta| > 0$, then $\alpha\beta$ is simply the concatenation of α and β,
(ii) if $|\alpha| = 0$, and $|\beta| > 0$, then $\alpha\beta = \beta$,
(iii) if $|\alpha| > 0$, and $|\beta| = 0$, then $\alpha\beta = \alpha$,
(iv) if $|\alpha| = |\beta| = 0$, in which case α necessarily coincides with β (because $\alpha = d(\alpha) = r(\beta) = \beta$), then $\alpha\beta$ is defined to be either α or β.

Thus we see that paths of length zero get absorbed at either end of path multiplication except, of course, when both paths being multiplied have length zero, when only one of them disappears. One should note, however, that even though paths of length zero do not show up in the resulting product, they play an important role in determining whether or not the multiplication is defined. In particular, if α is a path of length zero consisting of a single vertex v, then $\alpha\beta$ is not defined, hence forbidden, unless $r(\beta) = v$.

This should be compared with the operation of composition of morphisms in a category, where vertices play the role of identities.

Definition 35.3. Given two paths $\alpha \in E^*$, and $\beta \in E^\sharp$, we will say that α is a *prefix* of β if there exists a path $\gamma \in E^\sharp$, such that $d(\alpha) = r(\gamma)$, and $\beta = \alpha\gamma$.

As an example, notice that for every path α, one has that $r(\alpha)\alpha = \alpha$, so $r(\alpha)$ is a prefix of α. The following is an alternate way to describe this concept:

Proposition 35.4. *Given two paths $\alpha \in E^*$, and $\beta \in E^\sharp$, one has that α is a prefix of β if and only if*

(*i*) $r(\alpha) = r(\beta)$,
(*ii*) $|\alpha| \leq |\beta|$,
(*iii*) $\alpha_i = \beta_i$, *for all* $i = 1, \ldots, |\alpha|$.

Notice that when α and β are paths of nonzero length, in order to check that α is a prefix of β it is enough to verify (35.4.ii)–(35.4.iii), since (35.4.i), follows from (35.4.iii), with $i = 1$.

On the other hand, if $|\alpha| = 0$, then α is a prefix of β if and only if (35.4.i) holds.

Given a finite path $\alpha = \alpha_1 \ldots \alpha_n$ of length $n > 0$, we will denote by s_α the element of $\mathcal{T}_E$ given by

$$s_\alpha = s_{\alpha_1} \ldots s_{\alpha_n}.$$

In the special case that v is a vertex in E^0, and α is the path of length zero given by $\alpha = v$, we will let

$$s_\alpha = p_v.$$

Lemma 35.5. *For every $\alpha \in E^*$, one has*

(*i*) $s_\alpha = p_{r(\alpha)} s_\alpha$,
(*ii*) $s_\alpha = s_\alpha p_{d(\alpha)}$,
(*iii*) $s_\alpha s_\alpha^* \leq p_{r(\alpha)}$.

PROOF. Observing that the case $|\alpha| = 0$ is trivial, we assume that $|\alpha| \geq 1$. Notice that, for every edge $a \in E^1$, one has

$$p_{r(a)} s_a = p_{r(a)} s_a s_a^* s_a \overset{(35.1.ii)}{=} s_a s_a^* s_a = s_a.$$

Applying this to the leading edge of α, we deduce (i). On the other hand,

$$s_a = s_a s_a^* s_a \overset{(35.1.i)}{=} s_a p_{d(a)},$$

which, applied to the trailing edge of α, provides (ii).

Regarding (iii), we have

$$(\dagger) \qquad s_\alpha s_\alpha^* \overset{(i)}{=} p_{r(\alpha)} s_\alpha s_\alpha^* p_{r(\alpha)} \leq \|s_\alpha\|^2 p_{r(\alpha)} \leq p_{r(\alpha)}.$$

In the first inequality above we have used the identity

$$abb^*a^* \leq \|b\|^2 aa^*,$$

known to hold in any C*-algebra, while the second inequality in (†) holds because s_α is a product of partial isometries, each of which has norm no bigger than 1, hence $\|s_\alpha\| \leq 1$. ☐

Remark 35.6. If $\alpha = \alpha_1 \ldots \alpha_n$ is a random sequence of edges, not necessarily forming a path, we may still define s_α, as above. However, unless α is a path, we will have $s_\alpha = 0$. The reason is as follows: if there exists i with $d(\alpha_i) \neq r(\alpha_{i+1})$, then

$$s_{\alpha_i} s_{\alpha_{i+1}} = s_{\alpha_i} p_{d(\alpha_i)} p_{r(\alpha_{i+1})} s_{\alpha_{i+1}} = 0,$$

because the vertex projections p_v are pairwise orthogonal by construction.

Lemma 35.7. *If α is a finite path in E, then $s_\alpha^* s_\alpha = p_{d(\alpha)}$.*

PROOF. This is evident if $|\alpha| = 0$, while (35.1.i) gives the case $|\alpha| = 1$. If $|\alpha| \geq 2$, then

$$(\star) \qquad s_{\alpha_1}^* s_{\alpha_1} s_{\alpha_2} = p_{d(\alpha_1)} s_{\alpha_2} = p_{r(\alpha_2)} s_{\alpha_2} \overset{(35.5.i)}{=} s_{\alpha_2},$$

so

$$s_\alpha^* s_\alpha = s_{\alpha_n}^* \ldots s_{\alpha_2}^* \underbrace{s_{\alpha_1}^* s_{\alpha_1} s_{\alpha_2}}_{(\star)} \ldots s_{\alpha_n} = s_{\alpha_n}^* \ldots s_{\alpha_2}^* s_{\alpha_2} \ldots s_{\alpha_n},$$

and the result follows by induction. □

Lemma 35.8. *Given α and β in E^* such that $s_\alpha^* s_\beta \neq 0$, then either α is a prefix of β, or vice versa.*

PROOF. Since $s_\beta^* s_\alpha = (s_\alpha^* s_\beta)^* \neq 0$, we see that α and β play symmetric roles, so we may assume, without loss of generality, that $|\alpha| \leq |\beta|$.

Observe that

$$0 \neq s_\alpha^* s_\beta \overset{(35.5.i)}{=} s_\alpha^* p_{r(\alpha)} p_{r(\beta)} s_\beta,$$

which yields $p_{r(\alpha)} p_{r(\beta)} \neq 0$, and consequently $r(\alpha) = r(\beta)$. If $|\alpha| = 0$, the proof is thus concluded, so we suppose that $|\alpha| > 0$, writing $\alpha = \alpha_1 \alpha_2 \ldots \alpha_n$, and $\beta = \beta_1 \beta_2 \ldots \beta_m$.

Notice that $\alpha_1 = \beta_1$, since otherwise we would have $s_{\alpha_1}^* s_{\beta_1} = 0$ by (35.1.i), which would imply $s_\alpha^* s_\beta = 0$. Writing $v = d(\beta_1) = r(\beta_2)$, we then have

$$\begin{aligned} 0 \neq s_\alpha^* s_\beta &= s_{\alpha_n}^* \ldots s_{\alpha_2}^* s_{\alpha_1}^* s_{\beta_1} s_{\beta_2} \ldots s_{\beta_m} \\ &\overset{(35.1.i)}{=} s_{\alpha_n}^* \ldots s_{\alpha_2}^* p_v s_{\beta_2} \ldots s_{\beta_m} \\ &\overset{(35.5.i)}{=} s_{\alpha_n}^* \ldots s_{\alpha_2}^* s_{\beta_2} \ldots s_{\beta_m}, \end{aligned}$$

and the proof again follows by induction. □

Our goal of proving the generating set of $\mathcal{T}_E$ to be a tame set of partial isometries depends on an understanding of the semigroup they generate. With the above preparations we are now able to describe this semigroup.

Proposition 35.9. *Let $\mathscr{G}$ be the subset of $\mathcal{T}_E$ given by*

$$\mathscr{G} = \mathscr{G}_0 \cup \mathscr{G}_1 = \{p_v : v \in E^0\} \cup \{s_a : a \in E^1\}.$$

Then the multiplicative sub-semigroup of $\mathcal{T}_E$ generated by $\mathscr{G} \cup \mathscr{G}^* \cup \{0\}$ coincides with $\{s_\alpha s_\beta^* : \alpha, \beta \in E^*\} \cup \{0\}$.

PROOF. We will initially prove that the set M defined by

$$M = \{s_\alpha s_\beta^* : \alpha, \beta \in E^*\} \cup \{0\},$$

is closed under multiplication. So pick α, β, μ and ν in E^*, and let us show that

$$s_\alpha s_\beta^* s_\mu s_\nu^* \in M. \tag{35.9.1}$$

If $s_\beta^* s_\mu = 0$, then our claim follows trivially. Otherwise $s_\beta^* s_\mu \neq 0$, and (35.8) implies that β is a prefix of μ, or vice versa. Assuming without loss of generality that β is a prefix of μ, we have that $s_\mu = s_\beta s_\xi$, for some path ξ, whence

$$s_\alpha s_\beta^* s_\mu s_\nu^* = s_\alpha s_\beta^* s_\beta s_\xi s_\nu^* \overset{(35.7)}{=} s_\alpha p_{d(\beta)} s_\xi s_\nu^* \overset{(35.5.\text{i})}{=} s_\alpha p_{d(\beta)} p_{r(\xi)} s_\xi s_\nu^* = \cdots$$

If $d(\beta) \neq r(\xi)$, the above vanishes, in which case (35.9.1) is proved, or else $d(\beta) = r(\xi)$, whence $p_{d(\beta)} p_{r(\xi)} = p_{r(\xi)}$, and the above equals

$$\cdots = s_\alpha p_{r(\xi)} s_\xi s_\nu^* = s_\alpha s_\xi s_\nu^* = s_{\alpha\xi} s_\nu^* \in M.$$

Strictly speaking, $\alpha\xi$ might not be a path, but in this case $s_{\alpha\xi} = 0$, by (35.6), so $s_{\alpha\xi} s_\nu^*$ lies in M, as well.

This concludes the proof that M is a semigroup, as claimed. Using angle brackets "$\langle\ \rangle$" to denote generated semigroup, we clearly have

$$\mathscr{G} \cup \mathscr{G}^* \cup \{0\} \subseteq M \subseteq \big\langle \mathscr{G} \cup \mathscr{G}^* \cup \{0\} \big\rangle,$$

from where the proof follows easily. □

Speaking of the elements $s_\alpha s_\beta^*$ forming the above semigroup M, notice that when $d(\alpha) \neq d(\beta)$, one has

$$s_\alpha s_\beta^* \overset{(35.5.\text{ii})}{=} s_\alpha p_{d(\alpha)} p_{d(\beta)} s_\beta^* = 0.$$

So, we may alternatively describe M as

$$M = \{s_\alpha s_\beta^* : \alpha, \beta \in E^*,\ d(\alpha) \neq d(\beta)\} \cup \{0\}.$$

Having a concrete description of our semigroup, we may now face our first main objective:

Theorem 35.10. *For every graph E, the set of standard generators of $\mathcal{T}_E$, namely $\mathscr{G} = \mathscr{G}_0 \cup \mathscr{G}_1$, is a tame set of partial isometries.*

PROOF. By (35.9), it suffices to prove that $s_\alpha s_\beta^*$ is a partial isometry for every $\alpha, \beta \in E^*$. We have

$$\begin{aligned}(s_\alpha s_\beta^*)(s_\alpha s_\beta^*)^*(s_\alpha s_\beta^*) &= s_\alpha s_\beta^* s_\beta s_\alpha^* s_\alpha s_\beta^* \\ &\overset{(35.7)}{=} s_\alpha p_{d(\beta)} p_{d(\alpha)} s_\beta^* \\ &= s_\alpha p_{d(\alpha)} p_{d(\beta)} s_\beta^* \\ &\overset{(35.5.\text{ii})}{=} s_\alpha s_\beta^*.\end{aligned}$$

This shows that $s_\alpha s_\beta^*$ is a partial isometry, and hence that $\mathscr{G}$ is tame. □

It follows from the above that the set $\{s_a : a \in E^1\}$ is also tame. So, letting $\mathbb{F}$ be the free group on the set E^1, we may invoke (12.13) to conclude that there exists a unique semi-saturated *-partial representation

$$u : \mathbb{F} \to \widetilde{\mathcal{T}}_E, \tag{35.11}$$

such that $u_a = s_a$, for all $a \in E^1$, where $\widetilde{\mathcal{T}_E}$ is the algebra obtained from $\mathcal{T}_E$ by adding a unit to it, even if it already has one[46].

As usual let us denote by $e_g = u_g u_{g^{-1}}$.

Proposition 35.12. *The partial representation described in* (35.11) *satisfies, for every* a *and* b *in* E^1,

(*i*) $d(a) = d(b) \Rightarrow e_{a^{-1}} = e_{b^{-1}}$,
(*ii*) $d(a) \neq d(b) \Rightarrow e_{a^{-1}} e_{b^{-1}} = 0$,
(*iii*) $r(a) = d(b) \Rightarrow e_a \leq e_{b^{-1}}$,
(*iv*) $a \neq b \Rightarrow e_a e_b = 0$.

PROOF. Noticing that

$$e_{a^{-1}} = u_{a^{-1}} u_a = s_a^* s_a = p_{d(a)},$$

points (i) and (ii) follow at once. As for (iii) we have

$$e_a = u_a u_{a^{-1}} = s_a s_a^* \overset{(35.1.\text{ii})}{\leq} p_{r(a)} = p_{d(b)} = e_{b^{-1}}.$$

Finally, if $a \neq b$, then

$$e_a e_b = u_a u_{a^{-1}} u_b u_{b^{-1}} = s_a s_a^* s_b s_b^* \overset{(35.1.\text{i})}{=} 0.$$

□

In order to avoid technical details, we will now restrict ourselves to studying graphs for which $\mathcal{T}_E$ is generated by the s_a alone, a property that follows from the absence of sinks, as defined below:

Definition 35.13. A vertex $v \in E^0$ is said to be a:

(i) *sink*, if $d^{-1}(v)$ is the empty set,
(ii) *source*, if $r^{-1}(v)$ is the empty set,
(iii) *regular vertex*, if $r^{-1}(v)$ is finite and nonempty.

If the vertex v is not a sink, then there exists some edge a in E^1 such that $d(a) = v$. Therefore (35.1.i) implies that

$$s_a^* s_a = p_{d(a)} = p_v,$$

whence p_v lies in the algebra generated by the s_a.

► From now on we will concentrate on graphs without sinks, so we suppose throughout that E has no sinks.

In this case, as seen above, all of the p_v may be expressed in terms of the s_a, which is to say that

$$\mathscr{G}_1 = \{s_a : a \in E^1\}$$

generates $\mathcal{T}_E$, as a C*-algebra.

Theorem 35.14. *Let* E *be a graph without sinks, and let* $\mathbb{F}$ *be the free group on the set* E^1, *equipped with the usual word-length function. Also let*

(*i*) $\mathcal{R}'$ *be the set consisting of relations* (9.1.i)–(9.1.iv),
(*ii*) $\mathcal{R}_{sat}$ *be the set of relations described in* (14.20), *and*

[46]It is easy to see that $\mathcal{T}_E$ is unital if and only if E^0 is finite, in which case the unit of $\mathcal{T}_E$ is given by the sum of all the p_v, with v ranging in E^0.

(*iii*) $\mathcal{R}_E$ *be the set consisting of relations* (35.12.i)–(35.12.iv), *for each* a *and* b *in* E^1. *(By this we mean that we take the relation on the right-hand-side of "$\Rightarrow$", whenever the condition in the left-hand-side holds).*

Then $\widetilde{\mathcal{T}}_E$ *is* **-isomorphic to the universal unital C*-algebra generated by a set* $\{u_g : g \in \mathbb{F}\}$, *subject to the set of relations* $\mathcal{R}' \cup \mathcal{R}_{sat} \cup \mathcal{R}_E$.

PROOF. It clearly suffices to prove that $\widetilde{\mathcal{T}}_E$ possesses the corresponding universal property. As a first step, observe that the elements in the range of the partial representation u given by (35.11) satisfy $\mathcal{R}'$ for obvious reasons, satisfy $\mathcal{R}_{\text{sat}}$ because u is semi-saturated, and satisfy $\mathcal{R}_E$ by (35.12).

Since $s_a = u_a$, for every a in E^1, we see that the range of u contains $\mathscr{G}_1$. Moreover, for every $v \in E^0$, one may pick $a \in E^1$ such that $d(a) = v$, because E has no sinks, whence

$$p_v = s_a^* s_a = u_{a^{-1}} u_a,$$

and so we see that the range of u also contains $\mathscr{G}_0$, hence generating $\widetilde{\mathcal{T}}_E$ as a C*-algebra.

Let us now assume we are given a unital C*-algebra B which contains a set $\{\hat{u}_g : g \in \mathbb{F}\}$, satisfying relations $\mathcal{R}' \cup \mathcal{R}_{\text{sat}} \cup \mathcal{R}_E$. As usual, for each g in $\mathbb{F}$, we will let $\hat{e}_g = \hat{u}_g \hat{u}_{g^{-1}}$.

For each a in E^1, define, $\hat{s}_a = \hat{u}_a$, and for every $v \in E^0$, choose an edge $b \in E^1$, with $d(b) = v$, and put $\hat{p}_v = \hat{e}_{b^{-1}}$. Notice that such an edge exists because v is not a sink. Moreover, if a is another edge with $d(a) = v$, then $\hat{e}_{b^{-1}} = \hat{e}_{a^{-1}}$ by (35.12.i), a relation which is part of $\mathcal{R}_E$, and is therefore satisfied by the $\hat{u}_g$.

We now claim that the sets

$$\widehat{\mathscr{G}_0} = \{\hat{p}_v : v \in E^0\}, \quad \text{and} \quad \widehat{\mathscr{G}_1} = \{\hat{s}_a : a \in E^1\},$$

satisfy conditions (35.1.i) & (35.1.ii). In fact, since $\hat{u}$ is clearly a partial representation, we have that each $\hat{s}_a$ is a partial isometry by (12.2), and hence the $\hat{p}_v$ are projections.

If $v_1, v_2 \in E^0$ are two distinct vertices, choose $a_1, a_2 \in E^1$, with $d(a_i) = v_i$. Then

$$\hat{p}_{v_1} \hat{p}_{v_2} = \hat{e}_{a_1^{-1}} \hat{e}_{a_2^{-1}} = 0,$$

by (35.12.ii), so the p_v are pairwise orthogonal, as required.

In order to check (35.1.i), let $a, b \in E^1$. Then

$$\hat{s}_a^* \hat{s}_b = \hat{u}_{a^{-1}} \hat{u}_b = \hat{u}_{a^{-1}} \hat{u}_a \hat{u}_{a^{-1}} \hat{u}_b \hat{u}_{b^{-1}} \hat{u}_b = \hat{u}_{a^{-1}} \hat{e}_a \hat{e}_b \hat{u}_b.$$

If $a \neq b$, the above vanishes by (35.12.iv). Otherwise, we have

$$\hat{s}_a^* \hat{s}_b = \hat{u}_{a^{-1}} \hat{u}_a = \hat{e}_{a^{-1}} = \hat{p}_{d(a)}.$$

Let us now prove (35.1.ii), so pick $a \in E^1$. Since E has no sinks, there exists some $b \in E^1$, such that $d(b) = r(a)$. Then

$$\hat{s}_a \hat{s}_a^* = \hat{u}_a \hat{u}_{a^{-1}} = \hat{e}_a \overset{(35.12.\text{iii})}{\leq} \hat{e}_{b^{-1}} = \hat{p}_{d(b)} = \hat{p}_{r(a)}.$$

Therefore, by the universal property of $\mathcal{T}_E$, we conclude that there exists a *-homomorphism $\varphi : \mathcal{T}_E \to B$, such that

$$\varphi(p_v) = \hat{p}_v, \quad \text{and} \quad \varphi(s_a) = \hat{s}_a, \tag{35.14.1}$$

for all $v \in E^0$, and all $a \in E^1$. By mapping identity to identity, we may extend φ to a unital map from $\widetilde{\mathcal{T}}_E$ to B.

The proof will then be concluded once we prove that $\varphi(u_g) = \hat{u}_g$, for all g in $\mathbb{F}$, which we do by induction on $|g|$. The case $|g| = 0$, follows since φ preserve identities, while the case $|g| = 1$, follows from (35.14.1).

If $|g| \geq 2$, write $g = g_1g_2$, with $|g| = |g_1| + |g_2|$, and $|g_1|, |g_2| < |g|$. Then, since both u and $\hat{u}$ are semi-saturated, we have by induction that

$$\varphi(u_g) = \varphi(u_{g_1g_2}) = \varphi(u_{g_1}u_{g_2}) = \varphi(u_{g_1})\varphi(u_{g_2}) = \hat{u}_{g_1}\hat{u}_{g_2} = \hat{u}_{g_1g_2} = \hat{u}_g.$$

This concludes the proof. □

According to (14.2), our last result may then be interpreted as saying that

$$\widetilde{\mathcal{T}}_E \simeq C^*_{\text{par}}(\mathbb{F}, \mathcal{R}_{\text{sat}} \cup \mathcal{R}_E),$$

so we may apply (14.16) in order to describe $\widetilde{\mathcal{T}}_E$ as a partial crossed product.

Corollary 35.15. *Given a graph E without sinks, denote by Ω_E the spectrum of the set of relations $\mathcal{R}_{sat} \cup \mathcal{R}_E$, and consider the partial action of $\mathbb{F}$ on Ω_E obtained by restricting the partial Bernoulli action. Then there is a *-isomorphism*

$$\varphi : \widetilde{\mathcal{T}}_E \to C(\Omega_E) \rtimes \mathbb{F},$$

such that

$$\varphi(s_a) = 1_a\delta_a, \quad \forall\, a \in E^1.$$

In what follows we will denote the already mentioned restriction of the partial Bernoulli action to Ω_E by

$$\theta_E = \big(\{D^E_g\}_{g\in G}, \{\theta^E_g\}_{g\in G}\big). \tag{35.16}$$

By direct inspection, using (14.8), it is easy to see that

$$\omega_0 = \{1\}, \tag{35.17}$$

is an element of Ω_E. Since ω_0 does not lie in any D^E_g (see (5.14)), for $g \neq 1$, we have that $\{\omega_0\}$ is an invariant subset of Ω_E, whence its complement is an open invariant set.

This moreover implies that $C_0\big(\Omega_E \setminus \{\omega_0\}\big)$ is an invariant ideal of $C(\Omega_E)$, whence $C_0\big(\Omega_E \setminus \{\omega_0\}\big) \rtimes \mathbb{F}$ is an ideal in $C(\Omega_E) \rtimes \mathbb{F}$ by (22.9).

Theorem 35.18. *In the context of* (35.15), *the restriction of φ to $\mathcal{T}_E$ is an isomorphism from $\mathcal{T}_E$ onto $C_0\big(\Omega_E \setminus \{\omega_0\}\big) \rtimes \mathbb{F}$.*

Proof. As already seen, we have that $\omega_0 \notin D^E_a$, for every edge a, whence $D^E_a \subseteq \Omega_E \setminus \{\omega_0\}$. Consequently

$$1_a\delta_a \in C_0\big(\Omega_E \setminus \{\omega_0\}\big) \rtimes \mathbb{F},$$

showing that $\varphi(\mathcal{T}_E)$ is contained in $C_0\big(\Omega_E \setminus \{\omega_0\}\big) \rtimes \mathbb{F}$. By (22.9) we have that

$$\frac{C(\Omega_E) \rtimes \mathbb{F}}{C_0\big(\Omega_E \setminus \{\omega_0\}\big) \rtimes \mathbb{F}} \simeq C(\{\omega_0\}) \rtimes \mathbb{F},$$

which is a one-dimensional algebra given that the intersection of the D^E_g with $\{\omega_0\}$ is empty, except for when $g = 1$. Thus, the co-dimension of $C_0\big(\Omega_E \setminus \{\omega_0\}\big) \rtimes \mathbb{F}$ in $C(\Omega_E) \rtimes \mathbb{F}$ is the same as the co-dimension of $\mathcal{T}_E$ in $\widetilde{\mathcal{T}}_E$, both being equal to 1. Since φ is an isomorphism, one then sees that $\varphi(\mathcal{T}_E)$ must coincide with in $C_0\big(\Omega_E \setminus \{\omega_0\}\big) \rtimes \mathbb{F}$. □

Notes and remarks. In 1980, Enomoto and Watatani [**41**] realized that certain properties of the recently introduced Cuntz-Krieger C*-algebras [**29**] could be described by notions from graph theory. Seventeen years later, Kumjian, Pask, Raeburn and Renault [**75**] began studying C*-algebras for certain infinite graphs, in the wake of which many authors established an intense area of research going by the name of "graph C*-algebras". Definition (35.1) first appeared in [**64**]. It was inspired by ideas from [**75**] and [**57**]. The reader will find a long list of references for the theory of graph C*-algebras in [**95**].

CHAPTER 36

Path Spaces

The definition of the spectrum of a set of relations given in (14.8) is a one-size-fits-all description, having many applications to concrete examples, such as the various C*-algebras associated to semigroups of isometries and the Toeplitz C*-algebra of a graph studied above. However experience shows that, once we have focused on a specific example, the description of $\Omega_{\mathcal{R}}$ given in its formal definition may often be greatly simplified, leading to a much more intuitive and geometrically meaningful picture of this space.

The purpose of this chapter is thus to develop a careful analysis of the spectrum Ω_E of the relations leading up to $\widetilde{\mathcal{T}}_E$, as described in (35.15).

▶ As above, we will continue working with a fixed graph E without sinks. Since Ω_E is a subset of $\{0,1\}^{\mathbb{F}} = \mathscr{P}(\mathbb{F})$, each $\omega \in \Omega_E$ may be viewed as a subset of $\mathbb{F}$. A useful mental picture of an element ω in Ω_E is thus to imagine the vertices of the Cayley graph of $\mathbb{F}$ painted black whenever they correspond to an element in ω.

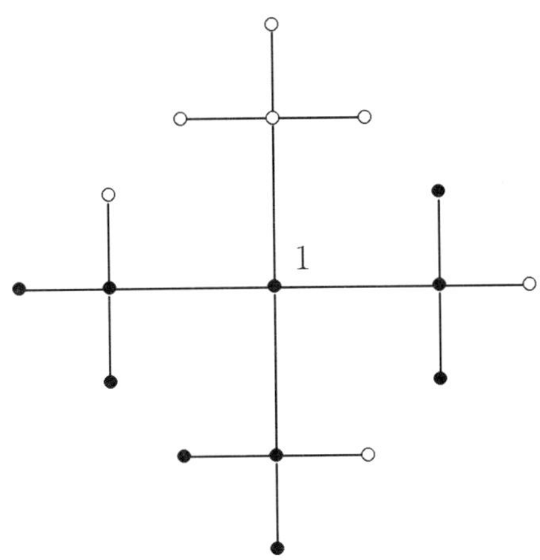

Observing that Ω_E is a subspace of the space Ω_1 defined in (5.11), every $\omega \in \Omega_E$ is a subset of $\mathbb{F}$ containing the unit group element, so the vertex of the Cayley graph corresponding to 1 will always be painted black. Moreover, the presence of relations $\mathcal{R}_{\rm sat}$ imply that every ω is convex, according to (14.20), and one may easily show that the abstract notion of convexity given in (14.19.b) coincides with the corresponding geometric notion.

Our alternative description of the elements ω belonging to Ω_E will be aided by the following concept:

Definition 36.1. Let $\omega \in \Omega_1$, and let $g \in \omega$. The *local configuration* of ω at g is the set

$$\mathrm{loc}_g(\omega) = \{h \in \mathbb{F} : |h| = 1,\ gh \in \omega\}.$$

Since the vertices in the Cayley graph adjacent to a given g are precisely those of the form gh, with $|h| = 1$, the local configuration of ω at g determines which of the adjacent vertices correspond to elements belonging to ω. Also notice that the condition $|h| = 1$ implies that h is either an edge or the inverse of an edge. Therefore $\mathrm{loc}_g(\omega) \subseteq E^1 \cup (E^1)^{-1}$, where the inverse is evidently taken with respect to the group structure of $\mathbb{F}$.

Proposition 36.2. *A given $\omega \in \Omega_1$ lies in Ω_E if and only if ω is convex, and for all g in ω, the local configuration of ω at g is of one of the following mutually exclusive types:*

(a) there is an edge $a \in E^1$, such that

$$\mathrm{loc}_g(\omega) = \{a\} \cup \{b^{-1} : b \in E^1,\ d(b) = r(a)\},$$

(b) there is a vertex $v \in E^0$, such that

$$\mathrm{loc}_g(\omega) = \{b^{-1} : b \in E^1,\ d(b) = v\},$$

(c) $\mathrm{loc}_g(\omega)$ is the empty set.

PROOF. Let us first prove the only if part, so let us suppose that $\omega \in \Omega_E$. Since the relations involved in the definition of Ω_E (see (35.15)) include $\mathcal{R}_{\mathrm{sat}}$, we have that $\Omega_E \subseteq \Omega_{\mathcal{R}_{\mathrm{sat}}}$, so ω is in $\Omega_{\mathcal{R}_{\mathrm{sat}}}$ and then (14.20) implies that ω is convex.

Let us now suppose that we are given g in G, and there exists an edge $a \in \mathrm{loc}_g(\omega)$. If so, we claim that the local configuration of ω at g is of type (a). In fact, suppose by contradiction that some other edge b lies in $\mathrm{loc}_g(\omega)$. Then, taking relation (35.12.iv) into account, one sees that the function

$$f = \varepsilon_a \varepsilon_b$$

(see (14.8)) lies in $\mathcal{F}_{\mathcal{R}_E}$, whence

$$0 = f(g^{-1}\omega) = [a \in g^{-1}\omega][b \in g^{-1}\omega] = [ga \in \omega][gb \in \omega],$$

and since $ga \in \omega$, it follows that $gb \notin \omega$, whence $b \notin \mathrm{loc}_g(\omega)$. This proves that a is the only edge (as opposed to the inverse of an edge) lying in $\mathrm{loc}_g(\omega)$.

Let us now prove that $b^{-1} \in \mathrm{loc}_g(\omega)$, whenever the edge b is such that $d(b) = r(a)$. Under these conditions relation (35.12.iii) is in $\mathcal{R}_E$, so the function

$$f = \varepsilon_a - \varepsilon_a \varepsilon_{b^{-1}} = \varepsilon_a(1 - \varepsilon_{b^{-1}})$$

lies in $\mathcal{F}_{\mathcal{R}_E}$, whence

$$\begin{aligned} 0 &= f(g^{-1}\omega) \\ &= [a \in g^{-1}\omega]\big(1 - [b^{-1} \in g^{-1}\omega]\big) \\ &= [ga \in \omega]\big(1 - [gb^{-1} \in \omega]\big). \end{aligned}$$

Since $ga \in \omega$, it follows that $gb^{-1} \in \omega$, so $b^{-1} \in \mathrm{loc}_g(\omega)$, as desired.

To conclude the proof of the claim that the local configuration of ω at g is of type (a), it now suffices to prove that $b^{-1} \notin \text{loc}_g(\omega)$, whenever $d(b) \neq r(a)$. Using that $r(a)$ is not a sink, choose $c \in E^1$ such that $d(c) = r(a)$, so $c^{-1} \in \text{loc}_g(\omega)$ by the previous paragraph. We then have that $d(c) \neq d(b)$, so relation (35.12.ii) applies and letting

$$f = \varepsilon_{b^{-1}}\varepsilon_{c^{-1}},$$

we have

$$0 = f(g^{-1}\omega) = [b^{-1} \in g^{-1}\omega][c^{-1} \in g^{-1}\omega] = [gb^{-1} \in \omega][gc^{-1} \in \omega].$$

Since $gc^{-1} \in \omega$, as seen above, we have that $gb^{-1} \notin \omega$, and consequently $b^{-1} \notin \text{loc}_g(\omega)$.

This proves the claim under the assumption that $\text{loc}_g(\omega)$ contains an edge, so let us suppose the contrary. If $\text{loc}_g(\omega)$ is empty, the proof is over, so we are left with the case in which b^{-1} lies in $\text{loc}_g(\omega)$ for some edge b. Denoting by $v = d(b)$, we have for an arbitrary edge c that

$$[gc^{-1} \in \omega] = [gc^{-1} \in \omega][gb^{-1} \in \omega] = \varepsilon_{c^{-1}}(g^{-1}\omega)\varepsilon_{b^{-1}}(g^{-1}\omega) = \cdots$$

Considering the appropriate function in $\mathcal{F}_{\mathcal{R}_E}$ related to (35.12.i)–(35.12.ii), according to whether or not $d(c) = d(b)$, the above equals

$$\cdots = [d(c) = d(b)]\,\varepsilon_{b^{-1}}(g^{-1}\omega) = [d(c) = v],$$

which says that $c \in \text{loc}_g(\omega)$ if and only if $d(c) = v$. This proves that $\text{loc}_g(\omega)$ is of type (b), and hence the proof of the only if part is concluded.

The proof of the converse essentially consists in reversing the arguments above and is left to the reader. □

As already observed in (35.17), the element $\omega_0 = \{1\}$ is a member of Ω_E. By convexity, this is the only element displaying a local configuration of type (36.2.c).

By inspection of the possible local configuration types above one may easily prove the following:

Proposition 36.3. *Let ω be in Ω_E, and let g be in ω.*

(*i*) *If $\text{loc}_g(\omega)$ contains an edge a, then*

$$b^{-1} \in \text{loc}_g(\omega) \ \Leftrightarrow\ d(b) = r(a), \quad \forall\, b \in E^1.$$

(*ii*) *If $\text{loc}_g(\omega)$ contains c^{-1}, for some edge c, then*

$$b^{-1} \in \text{loc}_g(\omega) \ \Leftrightarrow\ d(b) = d(c), \quad \forall\, b \in E^1.$$

We now begin to work towards giving a description of the elements of Ω_E based on local configurations.

Lemma 36.4. *Let ω be in Ω_E, and let g be in ω. Given a finite path ν, with $m := |\nu| \geq 1$, and $\nu_m^{-1} \in \text{loc}_g(\omega)$, then $g\nu^{-1} \in \omega$.*

PROOF. We prove by (backwards) induction that $\mu\nu_m^{-1}\nu_{m-1}^{-1}\dots\nu_k^{-1} \in \omega$, for every k. By the definition of local configurations it is immediate that $g\nu_m^{-1} \in \omega$. Now supposing that $k_0 \leq m$, and

$$\mu\nu_m^{-1}\nu_{m-1}^{-1}\dots\nu_k^{-1} \in \omega, \quad \forall\, k \geq k_0,$$

let $h = \mu\nu_m^{-1}\nu_{m-1}^{-1}\dots\nu_{k_0}^{-1}$. Then $\nu_{k_0} \in \mathrm{loc}_h(\omega)$, and $d(\nu_{k_0-1}) = r(\nu_{k_0})$, so $\nu_{k_0-1}^{-1}$ is in $\mathrm{loc}_h(\omega)$ by (36.3.i), meaning that

$$\omega \ni h\nu_{k_0-1}^{-1} = \mu\nu_m^{-1}\nu_{m-1}^{-1}\dots\nu_{k_0}^{-1}\nu_{k_0-1}^{-1},$$

as desired. □

In our next main result we will use the above local description of the elements in Ω_E to give a parametrization of Ω_E by the set of all (finite and infinite) paths. In order to do this we must first fine tune our interpretation of finite paths as elements of $\mathbb{F}$: if α is a finite path of positive length, then we may write $\alpha = \alpha_1\alpha_2\dots\alpha_n$, and there is really only one way to see α as an element in $\mathbb{F}$, namely as the product of the α_i. However, when α has length zero then it consists of a single vertex and we will adopt the perhaps not so obvious convention that α represents the unit element of $\mathbb{F}$. This is not to say that we are identifying all paths of length zero with one another, the convention only applying when paths are seen as elements of $\mathbb{F}$.

The correspondence, about to defined, of elements of $\mathbb{F}$ with paths in E, will send each path α to the subset ω_α introduced below:

Definition 36.5. Given $\alpha \in E^\sharp$, let ω_α be the subset of $\mathbb{F}$ defined by

$$\omega_\alpha = \{\mu\nu^{-1} : \mu, \nu \in E^*,\ d(\mu) = d(\nu),\ \mu \text{ is a prefix of } \alpha\}.$$

Taking μ to be the range of α, considered as a path of length zero, as well as a prefix of α, one sees that ω_α contains every element of the form ν^{-1}, where ν is a path with $d(\nu) = r(\alpha)$. Likewise, if μ is any prefix of α, then, taking ν to be the path of length zero consisting of the vertex $d(\mu)$, we see that μ is in ω_α.

Let us now discuss the reduced form of elements in ω_α.

Proposition 36.6. *For any $\alpha \in E^\sharp$, and for every $g \in \omega_\alpha$, there are $\mu, \nu \in E^*$ such that*

(*i*) $g = \mu\nu^{-1}$,
(*ii*) $\ell(g) = |\mu| + |\nu|$,
(*iii*) $d(\mu) = d(\nu)$, *and*
(*iv*) μ *is a prefix of* α.

Proof. Given $g = \mu\nu^{-1} \in \omega_\alpha$, as in (36.5), there is nothing to do in case $|\mu|$ or $|\nu|$ vanish. So we may assume that $|\mu|$ and $|\nu|$ are both at least equal to 1, and we may then write $\mu = \mu_1\mu_2\dots\mu_n$, and $\nu = \nu_1\nu_2\dots\nu_m$, with the μ_i and the ν_j in E^1, whence

$$g = \mu_1\mu_2\dots\mu_{n-1}\mu_n\nu_m^{-1}\nu_{m-1}^{-1}\dots\nu_2^{-1}\nu_1^{-1}.$$

If $\mu_n \neq \nu_m$, then g is in reduced form, so $\ell(g) = |\mu| + |\nu|$, as needed.

Otherwise let us use induction on $|\mu|+|\nu|$. So, supposing that $\mu_n = \nu_m$, we may clearly cancel out the term "$\mu_n\nu_m^{-1}$" above, and we are then left with $g = \mu'\nu'^{-1}$, where $\mu' = \mu_1\mu_2\dots\mu_{n-1}$, and $\nu' = \nu_1\nu_2\dots\nu_{m-1}$. In this case, notice that

$$d(\mu') = d(\mu_{n-1}) = r(\mu_n) = r(\nu_m) = d(\nu_{m-1}) = d(\nu'),$$

so μ' and ν' satisfy the conditions required of μ and ν in the definition of ω_α, and the proof follows by induction.

In fact we still need to take into account the case in which m or n coincide with 1, since e.g. when $m = 1$, there is no ν_{m-1}. Under this situation we may still cancel out the term "$\mu_n\nu_m^{-1}$", so that $g = \mu'\nu'^{-1}$, where μ' is as above and ν' is the path of length zero consisting of the vertex $d(\mu_{n-1})$.

If, instead, one had $n = 1$, we could take $\mu' = d(\nu_{m-1})$ and ν', as above, observing that

$$\mu' = d(\nu_{m-1}) = r(\nu_m) = r(\mu_1) = r(\mu) = r(\alpha),$$

so μ' is still a prefix of α and we could proceed as above. Of course this still leaves out the case in which $n = m = 1$, but then $g = 1$ and it is enough to take $\mu = \nu = r(\alpha)$. □

As we will soon find out, the elements of $\mathbb{F}$ of the form described in (36.6) have a special importance to us. We shall thus introduce a new concept to highlight these elements.

Definition 36.7. Given g in $\mathbb{F}$ of the form $g = \mu\nu^{-1}$, where $\mu, \nu \in E^*$, we will say that g is in *standard form* if conditions (36.6.ii)–(36.6.iii) are satisfied.

The following is the main technical result leading up to our concrete description of Ω_E.

Theorem 36.8. *For each finite or infinite path α, one has that ω_α belongs to Ω_E. In addition, the correspondence $\alpha \mapsto \omega_\alpha$ is a one-to-one mapping from $E^\sharp$ onto $\Omega_E \setminus \{\omega_0\}$, where ω_0 is defined in* (35.17).

PROOF. By checking convexity and analyzing local configurations it is easy to use (36.2) in order to prove that ω_α is in $\Omega_E \setminus \{\omega_0\}$, for every α in $E^\sharp$. It is also evident that the mapping referred to in the statement is injective.

In order to prove surjectivity, let $\omega \in \Omega_E \setminus \{\omega_0\}$, and define

$$\omega_+ = \omega \cap \mathbb{F}_+,$$

where $\mathbb{F}_+$ is the positive cone[47] of $\mathbb{F}$. We claim that any

$$\alpha = \alpha_1\alpha_2 \ldots \alpha_n \in \omega_+,$$

where each α_i is an edge, and $n \geq 1$, is necessarily a path. In fact, for every $i = 1, \ldots, n-1$, notice that, by convexity, $g := \alpha_1\alpha_2 \ldots \alpha_i \in \omega_+$, and

$$\alpha_i^{-1}, \alpha_{i+1} \in \operatorname{loc}_g(\omega).$$

The local configuration of ω at g is therefore of type (36.2.a), since it contains an edge, whence $d(\alpha_i) = r(\alpha_{i+1})$, proving that α is indeed a path.

We next claim that, given any two finite paths $\alpha, \beta \in \omega_+$, with $1 \leq |\alpha| \leq |\beta|$, then necessarily α is a prefix of β. Otherwise let i be the smallest integer such that $\alpha_i \neq \beta_i$. Again by convexity we have that

$$g := \alpha_1\alpha_2 \ldots \alpha_{i-1} = \beta_1\beta_2 \ldots \beta_{i-1} \in \omega_+,$$

(in case $i = 0$, then g is to be interpreted as 1), and $\alpha_i, \beta_i \in \operatorname{loc}_g(\omega)$.

Including two different edges, $\operatorname{loc}_g(\omega)$ fails to be of any of the types described in (36.2), thus bringing about a contradiction. This proves that α is a prefix of β, as desired.

We will now build a path α which will later be proven to satisfy $\omega = \omega_\alpha$.

(i) Assuming that the set of finite paths of positive length in ω_+ is finite and nonempty, we let α be the longest such path.

---

[47] $\mathbb{F}_+$ is defined as the sub-semigroup of $\mathbb{F}$ generated by E^1. We assume that $\mathbb{F}_+$ also contains the unit of $\mathbb{F}$.

(ii) Should there exist arbitrarily long paths in ω_+, we let α be the infinite path whose prefixes are the finite paths in ω.

(iii) In case ω contains no path of positive length, then the local configuration of ω at 1 is necessarily of type (36.2.b) (it cannot be of type (36.2.c) since ω is convex and $\omega \neq \omega_0$). In this case we put $\alpha = v$, where v is the vertex referred to in (36.2.b).

As already indicated, we will prove that $\omega_\alpha = \omega$. By construction it is clear that $\omega_\alpha \cap \mathbb{F}_+ = \omega \cap \mathbb{F}_+$. Therefore, for any g in $\mathbb{F}_+$, we have that $g \in \omega$, if and only if g is a prefix of α.

In order to prove that $\omega_\alpha \subseteq \omega$, pick any $g \in \omega_\alpha$, and write $g = \mu\nu^{-1}$, with μ a prefix of α and ν a path with $d(\mu) = d(\nu)$.

If $|\nu| = 0$, then g is a prefix of α, so g is in ω. We therefore suppose that

$$\nu = \nu_1\nu_2 \ldots \nu_m,$$

where $m \geq 1$.

CASE 1. Assuming that $n := |\mu| \geq 1$, we have that μ is in ω. Moreover $\mu_n^{-1} \in \operatorname{loc}_\mu(\omega)$, and $d(\mu_n) = d(\mu) = d(\nu) = d(\nu_m)$. Therefore $\nu_m^{-1} \in \operatorname{loc}_g(\omega)$ by (36.3.ii). Employing (36.4) we then have that $g = \mu\nu^{-1} \in \omega$.

CASE 2. Assuming that $|\mu| = 0$, we have that $\mu = v$, for some vertex v. The fact that μ is a prefix of α in this case means that $v = r(\alpha)$, therefore

$$d(\nu_m) = d(\nu) = d(\mu) = v = r(\alpha).$$

Temporarily assuming that $|\alpha| \geq 1$, we then have that $\alpha_1 \in \operatorname{loc}_1(\omega)$, and $d(\nu_m) = r(\alpha_1)$, so $\nu_m^{-1} \in \operatorname{loc}_1(\omega)$, by (36.3.i). Another application of (36.4) then gives

$$\omega \ni 1\nu^{-1} = \mu\nu^{-1} = g.$$

Still under case (2), but now assuming that $|\alpha| = 0$, write $\alpha = v$, for some vertex v. This implies that we are under situation (iii) above, whence the local configuration of ω at 1 is of type (36.2.b). We then have by definition that $\operatorname{loc}_1(\omega) = \{b^{-1} : b \in E^1,\ d(b) = v\}$. Observe that

$$d(\nu_m) = d(\nu) = d(\mu) = d(\alpha) = v,$$

so $\nu_m^{-1} \in \operatorname{loc}_1(\omega)$, and again by (36.4) we conclude that $\omega \ni 1\nu^{-1} = \mu\nu^{-1} = g$.

This concludes the proof that $\omega_\alpha \subseteq \omega$, so we are left with the task of proving the reverse inclusion. Thus, let $g \in \omega$, and write

$$g = x_1x_2 \ldots x_n,$$

in reduced form. We claim that there is no i for which $x_i \in (E^1)^{-1}$ and $x_{i+1} \in E^1$. Otherwise, write $x_i = a^{-1}$, and $x_{i+1} = b$, with a and b in E^1, and let $g = x_1x_2 \ldots x_i$. Then, by convexity,

$$g,\ ga,\ gb \in \omega,$$

whence $a, b \in \operatorname{loc}_g(\omega)$. This results in a local configuration with two distinct edges, contradicting (36.2). Therefore we see that the factors of the reduced form of g lying in E^1 must be to the left of those lying in $(E^1)^{-1}$. In other words

$$g = \mu\nu^{-1},$$

where μ and ν are in $\mathbb{F}_+$. One may now easily employ (36.3) to prove that μ and ν are paths, and also that $d(\mu) = d(\nu)$. By convexity we see that μ is in ω, hence

μ is a prefix of α. Consequently $g \in \omega_\alpha$, thus verifying that $\omega \subseteq \omega_\alpha$, and hence concluding the proof. □

In order to obtain a model for Ω_E in terms of paths which accounts for the exceptional element ω_0, we make the following:

Definition 36.9. (a) We fix any element in the universe not belonging to $E^\sharp$, denote it by $\varnothing$, and call it the *empty path*[48].

(b) No path in $E^\sharp$ is considered to be a prefix of $\varnothing$.

(c) The *full path space* of E is the set $\widetilde{E}^\sharp = E^\sharp \cup \{\varnothing\}$.

(d) Given any $\omega \in \Omega_E \setminus \{\omega_0\}$, the *stem* of ω, denoted $\sigma(\omega)$, is the unique element α in $E^\sharp$, such that $\omega_\alpha = \omega$, according to (36.8).

(e) The stem of ω_0 is defined to be the empty path $\varnothing$.

The stem may then be seen as a function

$$\sigma : \Omega_E \to \widetilde{E}^\sharp,$$

which is bijective thanks to (36.8).

The restriction of the partial Bernoulli action to Ω_E, which gives rise to $\widetilde{\mathcal{T}}_E$ according to (35.15), may therefore be transferred via the stem function over to $\widetilde{E}^\sharp$. It is our next goal to give a concrete description of this partial action. We begin with a technical result.

Lemma 36.10. *Let g be any element of $\mathbb{F} \setminus \{1\}$. Then:*

(i) If D_g^E is nonempty, then g admits a standard form $g = \mu\nu^{-1}$ (see Definition (36.7)),

(ii) If $g = \mu\nu^{-1}$ is in standard form then

$$\sigma(D_g^E) = X_\mu := \{\alpha \in E^\sharp : \mu \text{ is a prefix of } \alpha\}.$$

Proof. Assuming that D_g^E is nonempty, let $\omega \in D_g^E$, and let $\alpha = \sigma(\omega)$. So

$$g \overset{(14.11)}{\in} \omega = \omega_\alpha,$$

and point (i) follows from (36.6).

In order to prove (ii), let $\omega \in D_g^E$, and let $\alpha = \sigma(\omega)$. Since $g \in \omega = \omega_\alpha$, we conclude from (36.6) that there is a prefix of α, say μ', and a finite path ν' with $d(\mu') = d(\nu')$, such that

$$g = \mu\nu^{-1} = \mu'\nu'^{-1},$$

and $\ell(g) = |\mu'| + |\nu'|$. By uniqueness of reduced forms, we conclude that $\mu = \mu'$, and $\nu = \nu'$. In particular μ is a prefix of α, so

$$\sigma(\omega) = \alpha \in X_\mu.$$

This proves that $\sigma(D_g^E) \subseteq X_\mu$. In order to prove the reverse inclusion, let $\alpha \in X_\mu$. Then μ is a prefix of α, so it is clear from the definition of ω_α that $g \in \omega_\alpha$, which in turn implies by (14.11) that $\omega_\alpha \in D_g^E$. Therefore

$$\alpha = \sigma(\omega_\alpha) \in \sigma(D_g^E).$$

□

[48]The empty path should not be confused with any of the paths of length zero we have so far been working with.

Having understood the mirror images of the D_g^E through the stem function, we will now describe the corresponding partial action on the path space.

Proposition 36.11. *Given $g = \mu\nu^{-1}$ in standard form, consider the mapping*

$$\tau_{\mu,\nu} : \nu\gamma \in X_\nu \mapsto \mu\gamma \in X_\mu.$$

Then the diagram

$$\begin{array}{ccc} D_{g^{-1}}^E & \xrightarrow{\theta_g^E} & D_g^E \\ \downarrow{\scriptstyle\sigma} & & \downarrow{\scriptstyle\sigma} \\ X_\nu & \xrightarrow{\tau_{\mu,\nu}} & X_\mu \end{array}$$

commutes.

Proof. Given any ω in $D_{g^{-1}}^E$, let

$$\alpha = \sigma(\omega), \quad \text{and} \quad \beta = \sigma(g\omega).$$

Then $\alpha \in X_\nu$, and $\beta \in X_\mu$, so we may write

$$\alpha = \nu\gamma, \quad \text{and} \quad \beta = \mu\delta,$$

for suitable paths γ and δ.

Given any finite prefix γ' of γ, notice that $\nu\gamma'$ is a prefix of $\alpha = \nu\gamma$, whence $\nu\gamma' \in \omega$, and

$$g\omega \ni g\nu\gamma' = \mu\gamma',$$

so we conclude that $\mu\gamma'$ is a prefix of β. Since γ' may be taken to be equal to γ when γ is finite, or arbitrarily large if γ is infinite, we conclude that $\mu\gamma$ is a prefix of $\beta = \mu\delta$, whence γ is a prefix of δ.

By repeating the above reasoning relative to g^{-1}, we similarly conclude that δ must be a prefix of γ, which is to say that $\gamma = \delta$. Therefore

$$\tau_{\mu,\nu}(\alpha) = \tau_{\mu,\nu}(\nu\gamma) = \mu\gamma = \mu\delta = \beta,$$

concluding the proof. □

We will now describe a topology on $\widetilde{E}^\sharp$, relative to which the stem is a homeomorphism. Before doing this, let us observe that, since Ω_E is a topological subspace of the product space $\{0,1\}^{\mathbb{F}}$, its topology is generated by the inverse images of open subsets of $\{0,1\}$, under the canonical projections,

$$\pi_g : \Omega_E \subseteq \{0,1\}^{\mathbb{F}} \to \{0,1\},$$

for $g \in \mathbb{F}$. The topology of $\{0,1\}$, in turn, is generated by the open subsets $\{0\}$ and $\{1\}$, so we may generate the topology of Ω_E using the sets

$$\pi_g^{-1}(\{0\}), \quad \text{and} \quad \pi_g^{-1}(\{1\}),$$

for all g in $\mathbb{F}$. Under our identification of $\{0,1\}^{\mathbb{F}}$ with the set $\mathscr{P}(\mathbb{F})$ of all subsets of $\mathbb{F}$, recall from (5.8) that the π_g are given by $\pi_g(\omega) = [g \in \omega]$. We then have by (14.11) that

$$\pi_g^{-1}(\{1\}) = D_g^E, \quad \text{and} \quad \pi_g^{-1}(\{0\}) = \Omega_E \setminus D_g^E.$$

So we see that the topology of Ω_E is generated by the D_g^E and their complements. However, by (36.10.i) we need only care for these when g admits a standard form.

Proposition 36.12. *Consider the topology on $\widetilde{E}^\sharp$ generated by all of the X_μ, together with their complements relative to $\widetilde{E}^\sharp$, where μ ranges in E^*. Then the stem*

$$\sigma : \Omega_E \to \widetilde{E}^\sharp,$$

is a homeomorphism, and consequently $\widetilde{E}^\sharp$ is a compact, Hausdorff, totally disconnected topological space.

PROOF. Follows immediately from (36.10.ii) and the above comments about the topology of Ω_E. □

36.13. Summarizing, let us give a detailed description of the restricted partial Bernoulli action on Ω_E, once the latter is identified with the full path space of E.

(a) The acting group is the free group $\mathbb{F}$ on the set E^1 of edges in our graph,

(b) The space $\widetilde{E}^\sharp$ consists of all finite and infinite paths on E, plus a so called empty path $\varnothing$. Thus

$$\widetilde{E}^\sharp = E^\sharp \cup \{\varnothing\} = E^* \cup E^\infty \cup \{\varnothing\}.$$

(c) For each g in $\mathbb{F}$ admitting a standard form $g = \mu\nu^{-1}$, we set

$$\widetilde{E}^\sharp_g = X_\mu = \{\alpha \in E^\sharp : \mu \text{ is a prefix of } \alpha\}.$$

(d) For each g in F not admitting a standard form, we let $\widetilde{E}^\sharp_g$ be the empty set.

(e) For each g in $\mathbb{F}$ admitting a standard form $g = \mu\nu^{-1}$, we let τ_g be the partial homeomorphism of $\widetilde{E}^\sharp$ given by

$$\begin{array}{ccccccc} \tau_g = \tau_{\mu,\nu} & : & \nu\gamma & \in & X_\nu & \longmapsto & \mu\gamma \in X_\mu. \\ & & & & \| & & \| \\ & & & & \widetilde{E}^\sharp_{g^{-1}} & & \widetilde{E}^\sharp_g \end{array}$$

(f) For each g in $\mathbb{F}$ not admitting a standard form, we let τ_g be the empty map from the empty set $\widetilde{E}^\sharp_{g^{-1}}$ to the empty set $\widetilde{E}^\sharp_g$.

We thus obtain a topological partial action

$$\tau = \big(\{\widetilde{E}^\sharp_g\}_{g\in\mathbb{F}},\ \{\tau_g\}_{g\in\mathbb{F}}\big),$$

of $\mathbb{F}$ on $\widetilde{E}^\sharp$, which is evidently topologically equivalent to the restriction of the partial Bernoulli action to Ω_E.

Since equivalent partial actions clearly give rise to isomorphic crossed products, we have the following immediate consequence of (35.15):

Theorem 36.14. *Given a graph E with no sinks, consider the partial action τ described above. Then there exists a *-isomorphism*

$$\varphi : \widetilde{\mathcal{T}}_E \to C(\widetilde{E}^\sharp) \rtimes \mathbb{F},$$

such that

$$\varphi(s_a) = 1_{X_a}\delta_a,$$

for all $a \in E^1$.

Since the stem of ω_0 coincides with $\varnothing$, we have that $\dot\sigma$ establishes a covariant homeomorphism from $\Omega_E \setminus \{\omega_0\}$ onto $\widetilde{E}^\sharp \setminus \{\varnothing\} = E^\sharp$, so we have, as in (35.18) that:

Theorem 36.15. *In the context of (36.14), the restriction of φ to $\mathcal{T}_E$ is an isomorphism from $\mathcal{T}_E$ onto $C_0(E^\sharp) \rtimes \mathbb{F}$.*

Let us take a few moments to discuss the question of compactness of the various spaces appearing above. The starting point is of course the fact the Ω_E is compact by (14.9). Being homeomorphic to Ω_E by (36.12), one also has that $\widetilde{E}^\sharp$ is compact.

Proposition 36.16. *Let E be a graph without sinks. Then the following are equivalent:*

(i) E^0 is finite,
(ii) $E^\sharp$ is compact,
(iii) ω_0 is an isolated point in Ω_E,
(iv) $\varnothing$ is an isolated point in $\widetilde{E}^\sharp$.

PROOF. (i) $\Rightarrow$ (iii): For each vertex $v \in E^0$, choose an edge $a_v \in E^1$ such that $d(a_v) = v$, and consider the compact-open subset of Ω_E given by

$$U_v = D^E_{a_v^{-1}} = \{\omega \in \Omega_E : a_v^{-1} \in \omega\}.$$

By (36.3.ii) applied to $g = 1$, we see that the definition of U_v does not depend on the choice of a_v. Using (36.2) we than conclude that

$$\Omega_E \setminus \{\omega_0\} = \bigcup_{v \in E^0} U_v.$$

Assuming that E^0 is finite, we see that the above union of sets is closed, whence its complement, namely $\{\omega_0\}$, is open, proving ω_0 to be an isolated point.

(iii) $\Leftrightarrow$ (iv): Follows immediately from the fact that the stem is a homeomorphism from Ω_E to $\widetilde{E}^\sharp$, sending ω_0 to $\varnothing$.

(iv) $\Rightarrow$ (ii): If $\varnothing$ is isolated, then

$$E^\sharp = \widetilde{E}^\sharp \setminus \{\varnothing\}$$

is closed in the compact space $\widetilde{E}^\sharp$, hence $E^\sharp$ is compact.

(ii) $\Rightarrow$ (i): Assuming (ii), the open cover $\{X_v\}_{v \in E^0}$ of $E^\sharp$ admits a finite subcover, say

$$E^\sharp = X_{v_1} \cup X_{v_2} \cup \ldots \cup X_{v_n}.$$

Given any vertex v, viewed as a path of length zero, hence an element of $E^\sharp$, there exists some k such that $v \in X_{v_k}$, so v_k is a prefix of v, which is to say that $v_k = v$. Thus $E^0 = \{v_1, v_2, \ldots, v_n\}$ is a finite set. □

Definition 36.17. We shall denote by $\mathscr{B}^E$ the semi-direct product bundle for the partial action τ given in (36.13).

As a consequence of (36.14) and (16.28), we have that $\widetilde{\mathcal{T}}_E$ is isomorphic to the full cross sectional C*-algebra of $\mathscr{B}^E$.

Recall from (35.11) that there exists a semi-saturated partial representation u of $\mathbb{F}$ in $\widetilde{\mathcal{T}}_E$, such that $u_a = s_a$, for all $a \in E^1$. Composing u with the isomorphism given in (36.14) we obtain a semi-saturated partial representation

$$v : \mathbb{F} \to C(\widetilde{E}^\sharp) \rtimes \mathbb{F}$$

such that

$$v_a = 1_{X_a}\delta_a, \quad \forall a \in E^1. \tag{36.18}$$

Proposition 36.19. *The Fell bundle $\mathscr{B}^v$ associated to the above partial representation v, in line with* (16.7), *is isomorphic to $\mathscr{B}^E$.*

PROOF. Writing $\mathscr{B}^E = \{B_g^E\}_{g\in\mathbb{F}}$, we will identify each B_g^E with the corresponding grading subspace

$$C_0(\widetilde{E}_g^\sharp)\delta_g \subseteq C(\widetilde{E}^\sharp) \rtimes \mathbb{F}.$$

With this identification, we will in fact prove that $\mathscr{B}^v$ is *equal to* $\mathscr{B}^E$.

By (36.18) we clearly have that

$$v_g \in B_g^E, \quad \forall\, g \in E^1,$$

and, by taking adjoints, this implies that the same holds for all g in $(E^1)^{-1}$.

Given a general element g in $\mathbb{F}$, write $g = x_1 \cdots x_n$ in reduced form, that is, each x_i lies in $E^1 \cup (E^1)^{-1}$, and $x_i^{-1} \neq x_{i+1}$. Then, using the fact that v is semi-saturated, we deduce that

$$v_g = v_{x_1} \cdots v_{x_n} \in B_{x_1}^E \cdots B_{x_n}^E \subseteq B_{x_1\cdots x_n}^E = B_g^E. \tag{36.19.1}$$

If we now suppose that $g = h_1 \cdots h_n$, no longer necessarily in reduced form, we have that

$$v_{h_1} \cdots v_{h_n} \in B_{h_1}^E \cdots B_{h_n}^E \subseteq B_{h_1\cdots h_n}^E = B_g^E.$$

By definition of $\mathscr{B}^v$ (see (16.7)) we then see that $B_g^v \subseteq B_g^E$, for every g in $\mathbb{F}$, and the proof will be concluded once we prove that in fact B_g^v coincides with B_g^E, which is to say that B_g^E is the closed linear span of the set of elements of the form

$$v_{h_1} \cdots v_{h_n},$$

with $g = h_1 \cdots h_n$.

By definition $\mathcal{T}_E$ is generated, as a C*-algebra, by the s_a, for $a \in E^1$, whence $\widetilde{\mathcal{T}}_E$ is generated by $\{s_a : a \in E^1\} \cup \{1\}$. Consequently $C(\widetilde{E}^\sharp) \rtimes \mathbb{F}$ is generated by $\{1_{X_a}\delta_a : a \in E^1\} \cup \{1\}$, and evidently also by the range of v. Thus, given y in any B_g^E, we may write

$$y = \lim_{n\to\infty} y_n,$$

where each y_n is a linear combination of elements of the form $v_{h_1} \cdots v_{h_n}$, with h_i in $\mathbb{F}$.

Let P_g be the composition

$$P_g : C^*\big(\mathscr{B}^E\big) \xrightarrow{\Lambda} C^*_{\text{red}}\big(\mathscr{B}^E\big) \xrightarrow{E_g} B_g^E,$$

where E_g is the Fourier coefficient operator given by (17.8). We then have that

$$y = P_g(y) = \lim_{n\to\infty} P_g(y_n).$$

By (36.19.1) it is easy to see that

$$P_g(v_{h_1} \cdots v_{h_n}) = \begin{cases} v_{h_1} \cdots v_{h_n}, & \text{if } h_1 \cdots h_n = g, \\ 0, & \text{otherwise.} \end{cases}$$

So, upon replacing each y_n by $P_g(y_n)$, we may assume that each y_n is a linear combination of elements of the form $v_{h_1} \cdots v_{h_n}$, with $h_1 \cdots h_n = g$. This implies that y_n is in B_g^v, and hence also y is in B_g^v. This shows that $B_g^v = B_g^E$, concluding the proof. ∎

A generalization of the above result could be proved for any C*-algebra of the form considered in (14.16), although we shall not pursue this here.

Let us now discuss the question of amenability for $\mathscr{B}^E$.

Theorem 36.20. *Given a graph E without sinks, one has that*

(i) $\mathscr{B}^E$ satisfies the approximation property and hence is amenable,
(ii) $C(\widetilde{E}^\sharp) \rtimes \mathbb{F}$ is naturally isomorphic to $C(\widetilde{E}^\sharp) \rtimes_{red} \mathbb{F}$,
(iii) $C(E^\sharp) \rtimes \mathbb{F}$ is naturally isomorphic to $C(E^\sharp) \rtimes_{red} \mathbb{F}$,
(iv) $\widetilde{\mathcal{T}}_E$ and $\mathcal{T}_E$ are nuclear C-algebras.*

PROOF. We have already seen that the partial representation u of (35.11) is semi-saturated. By (35.1.i) we have that u is also orthogonal, in the sense of (20.13.i). Since v is the composition of u with an isomorphism, it is clear that v is also semi-saturated and orthogonal. It therefore follows from (20.13.i) that $\mathscr{B}^v$, and hence also $\mathscr{B}^E$, satisfy the approximation property.

We then conclude from (20.6) that $\mathscr{B}^E$ is amenable, which is to say that the regular representation

$$\Lambda : C^*(\mathscr{B}^E) \to C^*_{\text{red}}(\mathscr{B}^E)$$

is an isomorphism. So, (ii) follows from the characterization of the full crossed product as a full cross sectional C*-algebra (see (16.28)), and the definition of the reduced crossed product as the reduced cross sectional C*-algebra (see (17.10)).

As already noticed, $E^\sharp$ is an invariant open subset of $\widetilde{E}^\sharp$. Thus, if $\mathscr{A}$ denotes the semi-direct product bundle for the restriction of τ to $E^\sharp$, we have that $\mathscr{A}$ is naturally isomorphic to a Fell sub-bundle of $\mathscr{B}^E$. Since the unit fiber algebra of $\mathscr{A}$, namely $C_0(E^\sharp)$, is an ideal in $C(\widetilde{E}^\sharp)$, the unit fiber algebra of $\mathscr{B}$, we deduce from (21.32) that $\mathscr{A}$ also satisfies the approximation property. Thus (iii) follows as above.

Finally, (iv) follows from (25.10), in view of the fact that commutative C*-algebras are nuclear. □

Notes and remarks. The partial crossed product description of $\mathcal{T}_E$ given in (36.15) follows the general method adopted in [**57**] to give a similar description of Cuntz-Krieger algebras for infinite matrices.

CHAPTER 37

Graph C*-Algebras

Having obtained a description of $\mathcal{T}_E$ in terms of a concrete partial dynamical system in (36.15), we will now proceed to giving a similar model for the graph C*-algebra $C^*(E)$. There are two methods for doing this, the most obvious and equally effective one being to repeat the above procedure, applying (14.16) to the set of relations defining $C^*(E)$, and then reinterpreting the spectrum of these relations as a path space.

For a change we will instead see $C^*(E)$ as a quotient of $\mathcal{T}_E$, and then we will apply the results of chapter (22).

▶ As before, let us fix a graph $E = (E^0, E^1, r, d)$, assumed to have no sinks.

Recall that $C^*(E)$ is defined in much the same way as $\mathcal{T}_E$, the only difference being that relations (35.1.iii) are required to hold in $C^*(E)$ but not in $\mathcal{T}_E$. These are the relations

$$p_v = \sum_{r(a)=v} s_a s_a^*,$$

for each regular vertex $v \in E^0$ (see Definition (35.13.iii)).

Since we are working with graphs without sinks, we may phrase these relations in terms of edges only, as follows: given any vertex v as above, choose an edge b such that $d(b) = v$, in which case $p_v = s_b^* s_b$, by (35.1.i). Therefore (35.1.iii) may be rewritten as

$$s_b^* s_b = \sum_{r(a)=d(b)} s_a s_a^*, \tag{37.1}$$

for every b in E^1 such that $d(b)$ is a regular vertex. These relations actually carry a greater similarity to the original relations studied by Cuntz and Krieger [**29**].

Since relations (35.1.i)–(35.1.ii) are already satisfied in $\mathcal{T}_E$, we clearly obtain the following:

Proposition 37.2. *Let L be the closed two-sided ideal in $\mathcal{T}_E$ generated by the elements*

$$s_b^* s_b - \sum_{r(a)=d(b)} s_a s_a^*,$$

for every b in E^1 such that $d(b)$ is a regular vertex. Then there exists a *-isomorphism

$$\chi : C^*(E) \to \mathcal{T}_E / L,$$

sending each canonical generating partial isometry of $C^*(E)$, which we henceforth denote by $\tilde{s}_a$, to $s_a + L$.

Observe that, as a consequence of (35.1.i), and of our assumption that E has no sinks, the $\tilde{s}_a$ are enough to generate $C^*(E)$, whence the isomorphism referred to above is uniquely determined by the fact that

$$\chi(\tilde{s}_a) = s_a + L.$$

We will now describe the elements of $C_0(E^\sharp) \rtimes \mathbb{F}$ corresponding to the above generators of the ideal L via the isomorphism given in (36.15). As we will see, these lie inside the canonical image of $C_0(E^\sharp)$ in the crossed product, so we will be able to characterize the quotient algebra as a partial crossed product using (22.10).

Given any edge b in E^1 such that $d(b)$ is regular, consider the complex valued function on $\widetilde{E}^\sharp$ defined by

$$f_b(\alpha) = [d(b) \text{ is a prefix of } \alpha] - \sum_{r(a)=d(b)} [a \text{ is a prefix of } \alpha], \quad \forall \alpha \in \widetilde{E}^\sharp,$$

where brackets correspond to Boolean value. Clearly f_b is continuous. Since $\varnothing$ has no prefixes, we see that f_b vanishes on $\varnothing$, whence $f_b \in C_0(E^\sharp)$.

Recalling from (36.10.ii) that X_μ is the set of all finite and infinite paths admitting the given finite path μ as a prefix, notice that

$$[\mu \text{ is a prefix of } \alpha] = 1_{X_\mu}(\alpha).$$

Consequently the function f_b defined above may be alternatively described as

$$f_b = 1_{X_{d(b)}} - \sum_{r(a)=d(b)} 1_{X_a}.$$

For each edge a in E^1, notice that the isomorphism φ of (36.14) satisfies

$$\varphi(s_a s_a^*) = (1_{X_a}\delta_a)(1_{X_a}\delta_a)^* \overset{(8.14)}{=} 1_{X_a}\delta_1,$$

and similarly,

$$\varphi(s_b^* s_b) = (1_{X_b}\delta_b)^*(1_{X_b}\delta_b) = \tau_{b^{-1}}(1_{X_b})\delta_1.$$

The standard form of b^{-1} is clearly $d(b)b^{-1}$, from where we see that $\tau_{b^{-1}}(1_{X_b}) = 1_{X_{d(b)}}$, so we deduce from the above that

$$\varphi(s_b^* s_b) = 1_{X_{d(b)}}\delta_1.$$

Interpreting f_b within $C_0(E^\sharp) \rtimes \mathbb{F}$ via the map ι introduced in (11.13) then produces

$$\begin{aligned}\iota(f_b) &= f_b\delta_1\\ &= 1_{X_{d(b)}}\delta_1 - \sum_{r(a)=d(b)} 1_{X_a}\delta_1\\ &= \varphi(s_b^* s_b) - \sum_{r(a)=d(b)} \varphi(s_a s_a^*)\\ &= \varphi\Big(s_b^* s_b - \sum_{r(a)=d(b)} s_a s_a^*\Big),\end{aligned}$$

which should be compared with (37.1). As an immediate consequence of (36.15) and (37.2), we therefore conclude that:

Proposition 37.3. *Let W be the subset of $C_0(E^\sharp)$ formed by the functions*

$$f_b = 1_{X_{d(b)}} - \sum_{r(a)=d(b)} 1_{X_a},$$

for each $b \in E^1$ such that $d(b)$ is regular. Also let K be the closed two-sided ideal of $C_0(E^\sharp) \rtimes \mathbb{F}$ generated by $\iota(W)$. Then the isomorphism φ given in (36.15) sends the ideal L described in (37.2) onto K, and consequently there exists a *-isomorphism

$$\psi : C^*(E) \longrightarrow \frac{C_0(E^\sharp) \rtimes \mathbb{F}}{K},$$

such that

$$\psi(\tilde{s}_a) = 1_{X_a}\delta_a + K,$$

where we are again denoting by $\tilde{s}_a$ the standard generating partial isometries of $C^*(E)$.

PROOF. It is enough to take ψ to be the composition

$$C^*(E) \quad \xrightarrow{(37.2)} \quad \frac{\mathcal{T}_E}{L} \quad \xrightarrow{(36.15)} \quad \frac{C_0(E^\sharp) \rtimes \mathbb{F}}{K}.$$

□

We are now precisely under the hypotheses of (22.10), but before invoking it we will give a concrete description of the ideal J mentioned there.

Proposition 37.4. *Let $E^\flat$ be the subset of $E^\sharp$ consisting of all paths α which satisfy any one of the following conditions:*

(*i*) *α is infinite,*
(*ii*) *α is finite and $r^{-1}(d(\alpha))$ is empty,*
(*iii*) *α is finite and $r^{-1}(d(\alpha))$ is infinite.*

Then $E^\flat$ is closed and τ-invariant. Moreover, denoting by U the complement of $E^\flat$ in $E^\sharp$, we have that $C_0(U)$ is the smallest τ-invariant ideal of $C_0(E^\sharp)$ containing the set W referred to in (37.3).

PROOF. We claim that every α in U is an isolated point of $E^\sharp$. In fact, if α is not in $E^\flat$, then α is necessarily a finite path and $r^{-1}(d(\alpha))$ is a nonempty finite set, so $d(\alpha)$ is a regular vertex and we may write

$$r^{-1}(d(\alpha)) = \{a_1, \ldots, a_n\}.$$

Observe that the set

$$X_\alpha \cap \bigcap_{i=1}^{n} E^\sharp \backslash X_{\alpha a_i}$$

is open by (36.12). It consists of all paths admitting α as a prefix, but not admitting as a prefix any one of the paths αa_i. The unique such path is evidently α, so the above set coincides with $\{\alpha\}$, thus proving that α is an isolated point. Consequently U is open, whence $E^\flat$ is closed relative to $E^\sharp$.

Although this is not relevant to us at the moment, notice that any path α satisfying (ii) above is also an isolated point, since $X_\alpha = \{\alpha\}$.

Noticing that the conditions (i–iii) above are related to the "right end" of α, while τ affects its "left end", one may easily show that $E^\flat$ is τ-invariant.

We next claim that every f_b in W vanishes on $E^\flat$. In fact, given any α in $E^\flat$, and given any edge b such that $d(b)$ is regular, we have

$$f_b(\alpha) = [d(b) \text{ is a prefix of } \alpha] - \sum_{r(a)=d(b)} [a \text{ is a prefix of } \alpha].$$

If $d(b)$ is not a prefix of α, meaning that $r(\alpha) \neq d(b)$, then evidently $r(\alpha) \neq r(a)$, for all edges a considered in the above sum. This implies that none of these a's are prefixes of α, given that (35.4.i) fails. Therefore all terms making up $f_b(\alpha)$ vanish, whence $f_b(\alpha)$ itself vanishes.

The remaining case to be treated is when $d(b)$ is a prefix of α, that is, when $r(\alpha) = d(b)$. Should α be a path of length zero, necessarily consisting of the vertex $d(b)$, then $d(\alpha) = d(b)$, so $d(\alpha)$ is regular by assumption, and then α will not satisfy any one of conditions (i–iii) above, contradicting the fact that α was taken in $E^\flat$. This said we see that α must not have length zero, so we may write

$$\alpha = \alpha_1\alpha_2\ldots$$

Given that $r(\alpha_1) = r(\alpha) = d(b)$, we see that α_1 is one of the edges a considered in the sum appearing in the definition of f_b above, and for this edge, the value of "a is a prefix of α" is evidently 1. This shows that $f_b(\alpha) = 0$, showing our claim that f_b vanishes on $E^\flat$.

We will now prove that $E^\flat$ is the biggest invariant subset of $E^\sharp$ where the f_b vanish. For this, suppose that Λ is an invariant set properly containing $E^\flat$. Choosing any α in $\Lambda \setminus E^\flat$, we see that α is a finite path, so we may let $v = d(\alpha)$, noticing that v is regular. The standard form of $g := \alpha^{-1}$ is given by $g = d(\alpha)\alpha^{-1}$ so, by invariance of Λ, we have

$$\Lambda \ni \tau_g(\alpha) = d(\alpha) = v.$$

Choosing any edge b with $d(b) = v$, we have that f_b lies in W, and clearly $f_b(v) = 1$, because v is a prefix of itself, while v can have no prefix of positive length. This implies that f_b does not vanish on Λ, proving that indeed $E^\flat$ is the biggest invariant subset of $E^\sharp$ where the f_b vanish. This also shows that U is the smallest invariant subset outside of which the f_b vanish, thus concluding the proof. □

The above classification of paths in types (i), (ii) and (iii) has important consequences in what follows and in particular the first two kinds play a special role, justifying the introduction of the following terminology:

Definition 37.5. A path α in $\widetilde{E}^\sharp$ is said to be *maximal* if it satisfies (37.4.i) or (37.4.ii).

This terminology is justified because a maximal path α cannot be enlarged, either because $d(\alpha)$ is a source or because α is already infinite.

Let us take a few moments to study the topology of $E^\flat$. Recall from (36.12) that the topology of $\widetilde{E}^\sharp$ is generated by all the X_μ plus their complements. Therefore, the sets of the form

$$\bigcap_{i=1}^{n} X_{\mu_i} \cap \bigcap_{j=1}^{m} \widetilde{E}^\sharp \setminus X_{\nu_j}$$

where $\mu_1, \ldots, \mu_n$ and $\nu_1, \ldots, \nu_m$ are finite paths, constitute a basis for the topology of $\widetilde{E}^\sharp$. Consequently the intersections of these with $E^\flat$ form a basis of open sets for the latter. However, maximal paths have a neighborhood basis of a simpler nature, as we will now see.

Proposition 37.6. *Given a graph E without sinks, and given a maximal path α in $E^\flat$, the collection of sets*

$$\{X_\mu \cap E^\flat : \mu \in E^*,\ \mu \textit{ is a prefix of } \alpha\}$$

is a neighborhood basis for α in $E^\flat$.

PROOF. We first assume that α satisfies (37.4.ii), that is, α cannot be extended any further due to the fact that $d(\alpha)$ is a source. Then the only path admitting α as a prefix is α itself, meaning that

$$X_\alpha = \{\alpha\}, \tag{37.6.1}$$

so the result follows trivially.

Assume now that α satisfies (37.4.i), that is, α is infinite. Given any neighborhood U of α, there are finite paths $\mu_1, \ldots, \mu_n$ and $\nu_1, \ldots, \nu_m$ such that

$$\alpha \in \bigcap_{i=1}^n X_{\mu_i} \cap \bigcap_{j=1}^m \widetilde{E}^\sharp \setminus X_{\nu_j} \subseteq U.$$

This implies that every μ_i is a prefix of α, while no ν_j has this property.

Let k be any integer larger that the length of every μ_i and every ν_j, and let μ be the finite path formed by the first k edges of α. It is therefore evident that each μ_i is a prefix of μ, so one has that

$$\alpha \in X_\mu \subseteq \bigcap_{i=1}^n X_{\mu_i}. \tag{37.6.2}$$

Observe that X_μ is disjoint from X_{ν_j}, for every j because, otherwise there is a path $\gamma \in X_\mu \cap X_{\nu_j}$, and hence both μ and ν_j are prefixes of γ. But since $|\mu| \geq |\nu_j|$, we see that ν_j is a prefix of μ, which in turn is a prefix of α. It follows that ν_j is a prefix of α, whence $\alpha \in X_{\nu_j}$, a contradiction. So $X_\mu \subseteq \widetilde{E}^\sharp \setminus X_{\nu_j}$ and, building on top of (37.6.2), we conclude that

$$\alpha \in X_\mu \subseteq \bigcap_{i=1}^n X_{\mu_i} \cap \bigcap_{j=1}^m \widetilde{E}^\sharp \setminus X_{\nu_j} \subseteq U.$$

This concludes the proof. □

Another useful fact about the topology of $E^\flat$ is as follows:

Proposition 37.7. *Given a graph E without sinks, the subset of $E^\flat$ formed by the maximal paths is dense in $E^\flat$.*

PROOF. Before starting the proof, observe that, by (37.6.1), the paths satisfying (37.4.ii) are isolated points, so they cannot be left out of any dense subset!

In order to prove the statement, it is enough to prove that any path α satisfying (37.4.iii) is an accumulation point of maximal paths.

Given an arbitrary neighborhood U of α, pick finite paths $\mu_1, \ldots, \mu_n$ and $\nu_1, \ldots, \nu_m$ such that

$$\alpha \in \bigcap_{i=1}^n X_{\mu_i} \cap \bigcap_{j=1}^m \widetilde{E}^\sharp \setminus X_{\nu_j} \subseteq U$$

(notice that we are unfortunately unable to use the simplification provided by (37.6) here).

Thus every μ_i is a prefix of α, hence we may assume without loss of generality that the μ_i all coincide with α, so in fact

$$\alpha \in X_\alpha \cap \bigcap_{j=1}^m \widetilde{E}^\sharp \setminus X_{\nu_j} \subseteq U. \tag{37.7.1}$$

By assumption we have that $r^{-1}\big(d(\alpha)\big)$ is infinite, so we may pick an edge a such that αa is a path which is not a prefix of any ν_j.

We next extend αa as far as possible, obtaining a path of the form

$$\beta = \alpha a \gamma,$$

which is either infinite or cannot be extended any further, and hence is maximal. We then claim that

$$\beta \in X_\alpha \cap \bigcap_{j=1}^{m} \widetilde{E}^\sharp \setminus X_{\nu_j}. \tag{37.7.2}$$

Since it is obvious that α is a prefix of β, it suffices to check that $\beta \notin X_{\nu_j}$, for all j. Arguing by contradiction, suppose that some ν_j is a prefix of β. If $|\nu_j| \leq |\alpha|$, then ν_j is a prefix of α, contradicting (37.7.1). Thus $|\nu_j| > |\alpha|$, and then necessarily αa is a prefix of ν_j, which is again a contradiction, by the choice of a. This verifies (37.7.2), so in particular $\beta \in U$, concluding the proof. □

Let us now give a description of $C^*(E)$ as a partial crossed product.

Theorem 37.8. *Let E be a graph with no sinks, and let $E^\flat$ be the τ-invariant subset of $E^\sharp$ described in (37.4), equipped with the restricted partial action. Then there is a *-isomorphism*

$$\rho : C^*(E) \to C_0(E^\flat) \rtimes \mathbb{F},$$

such that

$$\rho(\tilde{s}_a) = 1_a \delta_a, \quad \forall\, a \in E^1,$$

where 1_a is the characteristic function of $X_a \cap E^\flat$.

Proof. Letting $C_0(U)$ be the ideal referred to in (37.4), we have

$$C^*(E) \overset{(37.3)}{\simeq} \frac{C_0(E^\sharp) \rtimes \mathbb{F}}{K} \overset{(22.10)}{\simeq} \left(\frac{C_0(E^\sharp)}{C_0(U)} \right) \rtimes \mathbb{F} \simeq C_0(E^\flat) \rtimes \mathbb{F}.$$

That the isomorphism resulting from the composition of the above isomorphisms does satisfy the last assertion in the statement is of easy verification and is left to the reader. □

An important special case of great importance is when, besides having no sinks, every vertex in E^0 is regular, according to (35.13.iii) (so E cannot have any sources either). Then there are no paths in $E^\flat$ satisfying (37.4.ii) or (37.4.iii), so we get the following immediate consequence of (37.8):

Proposition 37.9. *If E is a graph with no sinks and such that every vertex $v \in E^0$ is regular, then $E^\flat = E^\infty$, and consequently there is a *-isomorphism*

$$\rho : C^*(E) \to C_0(E^\infty) \rtimes \mathbb{F},$$

such that

$$\rho(\tilde{s}_a) = 1_a \delta_a, \quad \forall\, a \in E^1,$$

where 1_a is the characteristic function of $X_a \cap E^\infty$.

Observe that if E^0 is finite then $E^\flat$ is compact, because the latter is closed in $E^\sharp$ by (37.4), and $E^\sharp$ is compact by (36.16). If, in addition, we are under the hypotheses of the last result, then clearly E^∞ is also compact.

Having described $C^*(E)$ as a partial crossed product, we now have a wide range of tools to study its structure. We begin with amenability.

Theorem 37.10. *Let E be a graph without sinks. Then:*

(a) *the semi-direct product bundle for the partial action describing $C^*(E)$ in (37.8) satisfies the approximation property and hence is amenable,*

(b) *there are natural isomorphisms*

$$C^*(E) \simeq C_0(E^\flat) \rtimes \mathbb{F} \simeq C_0(E^\flat) \rtimes_{red} \mathbb{F},$$

(c) *$C^*(E)$ is nuclear.*

Proof. The Fell bundle mentioned in (a) is clearly a quotient of the Fell bundle discussed in (36.20), so the conclusions in (a) follow from (21.33.ii) and (20.6). Point (b) follows from (37.8) and (a), while (c) is a direct consequence of (b) and (25.10). □

We will next study fixed points for the partial action τ of $\mathbb{F}$ on $\widetilde{E}^\sharp$. For this we should recall that by (36.10.i), unless a given element g in $\mathbb{F}$ has a standard form, τ_g is the empty map and hence g cannot possibly have any fixed points. For this reason elements not admitting a standard form are left out of the next result.

Proposition 37.11. *Let $g \in \mathbb{F}$, with $g \neq 1$, and suppose that g admits a standard form $g = \mu\nu^{-1}$.*

(i) *Then g has at most one fixed point in $\widetilde{E}^\sharp$.*

(ii) *If g has a fixed point, then $|\mu| \neq |\nu|$.*

(iii) *If g has a fixed point and $|\mu| > |\nu|$, then there exists a cycle[49] γ such that $\mu = \nu\gamma$ (whence $g = \nu\gamma\nu^{-1}$). In addition, the infinite path*

$$\alpha = \nu\gamma\gamma\gamma\ldots$$

is the unique fixed point for g in $\widetilde{E}^\sharp$.

Proof. Assume that α is a fixed point for g. Then

$$\alpha \in \widetilde{E}^\sharp_{g^{-1}} \cap \widetilde{E}^\sharp_g = X_\nu \cap X_\mu,$$

so we may write

$$\alpha = \nu\varepsilon = \mu\zeta,$$

for suitable (finite or infinite) paths ε and ζ, and moreover

$$\alpha = \tau_g(\alpha) = \tau_g(\nu\varepsilon) = \mu\varepsilon.$$

We then conclude that $\nu\varepsilon = \mu\varepsilon$, so either ν is a prefix of μ, or vice versa. If $|\mu| = |\nu|$, one would then necessarily have $\mu = \nu$, whence $g = 1$, contradicting the hypothesis. This proves (ii).

Speaking of (iii), let us add to the above assumptions that $|\mu| > |\nu|$, so we may write $\mu = \nu\gamma$, for some finite path γ such that $|\gamma| > 0$. We then have

$$\nu\varepsilon = \mu\varepsilon = \nu\gamma\varepsilon,$$

so $\varepsilon = \gamma\varepsilon$, whence $|\varepsilon| = |\gamma| + |\varepsilon|$, from where we deduce that $|\varepsilon| = \infty$. Moreover notice that this implies that

$$\varepsilon = \gamma\gamma\ldots,$$

so γ is necessarily a cycle. Consequently $\alpha = \nu\varepsilon = \nu\gamma\gamma\ldots$, proving (iii), and hence also (i). □

[49] A finite path $\gamma \in E^*$ is said to be a *cycle* if $|\gamma| > 1$, and $r(\gamma) = d(\gamma)$. Notice that in this case we may concatenate γ with itself as many times as we wish, forming a finite of infinite periodic path $\gamma\gamma\ldots$.

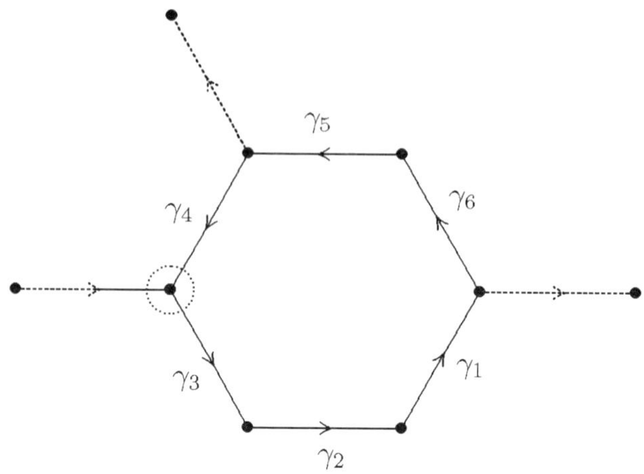

An entry to a cycle.

The reader must have noticed the omission of the case $|\mu| < |\nu|$, above. However, in this situation g^{-1} has precisely the same fixed points as g, and the condition expressed in (37.11.iii) evidently holds for g^{-1}.

Having understood fixed points, we will now study topological freeness. We must therefore analyze the interior of the fixed point sets F_g which, as seen above, have each at most one point. However, a singleton can only have a nonempty interior if its unique point is an isolated point. We must therefore discuss conditions under which a fixed point is isolated, and this hinges on the following important graph-theoretical concept:

Definition 37.12. Let $\gamma = \gamma_1 \dots \gamma_n$ be a cycle in E. We say that γ has an *entry*, if the range of some γ_i is the range of an edge other that γ_i.

Proposition 37.13. *Let α be an infinite path of the form*

$$\alpha = \nu\gamma\gamma\gamma\dots,$$

so that necessarily $d(\nu) = r(\gamma)$, and γ is a cycle. Then

(*i*) *α (or any other infinite path) is never an isolated point in $\widetilde{E}^\sharp$,*
(*ii*) *α is an isolated point in $E^\flat$ if and only if γ has no entry.*

PROOF. An infinite path $\alpha = \alpha_1\alpha_2\alpha_3\dots$ is always the limit of the sequence of finite paths obtained by truncating α, that is

$$\alpha = \lim_{n\to\infty} \alpha_1 \dots \alpha_n,$$

so α is an accumulation point, hence not isolated.

Focusing now on (ii), suppose that α is an isolated point of $E^\flat$. So, by (37.6), there is a large enough prefix μ of α such that

$$\{\alpha\} = X_\mu \cap E^\flat. \tag{37.13.1}$$

Therefore μ is a prefix of α, and by enlarging μ a bit, we may clearly assume that it is of the form

$$\mu = \nu \underbrace{\gamma\gamma\dots\gamma}_{n},$$

for some integer n. Assuming by contradiction that γ has an entry, we can find a path of the form

$$\gamma_1\gamma_2\ldots\gamma_k\varepsilon,$$

with $\varepsilon \neq \gamma_{k+1}$ (here $k+1$ stands for 1, in case $k = |\gamma|$), which we may extend as far as possible, obtaining a maximal path

$$\beta = \gamma_1\gamma_2\ldots\gamma_k\varepsilon a_1 a_2 a_3 \ldots$$

Considering the path

$$\alpha' = \mu\beta = \nu\underbrace{\gamma\gamma\ldots\gamma}_{n}\gamma_1\gamma_2\ldots\gamma_k\varepsilon a_1 a_2 a_3 \ldots$$

notice that α' is clearly also maximal hence $\alpha' \in E^\flat$. Since μ is a prefix of α', we see that

$$\alpha' \in X_\mu \cap E^\flat,$$

contradicting (37.13.1). This proves that γ has no entry.

Conversely, supposing that γ has no entry, it is easy to see that the only path in $E^\flat$ extending

$$\mu := \nu\gamma$$

is α, itself. So

$$X_\mu \cap E^\flat = \{\alpha\},$$

showing that α is indeed isolated in $E^\flat$. □

Putting together our findings so far we have:

Proposition 37.14. *For every graph E with no sinks, one has:*

(*i*) *The partial action τ of $\mathbb{F}$ on $\widetilde{E}^\sharp$, described in* (36.13), *is topologically free.*

(*ii*) *The restriction of τ to $E^\flat$ is topologically free if and only if every cycle in E has an entry.*

Proof. The main ingredients of this proof have already been taken care of, so all we need is to piece them together.

Given g in $\mathbb{F}$, one has by (37.11) that the fixed point set F_g is either empty or consists exactly of one infinite path. In the second case this infinite path is never isolated in $\widetilde{E}^\sharp$ by (37.13.i), so the interior of F_g is always empty, and hence τ is topologically free on $\widetilde{E}^\sharp$.

We next prove the only if part of (ii) via the contra-positive. If there exists a cycle γ with no entry, then

$$F_\gamma = \{\gamma\gamma\gamma\ldots\},$$

by (37.11), and moreover $\gamma\gamma\gamma\ldots$ is isolated in $E^\flat$ by (37.13.ii), whence F_g is open and consequently τ is not topologically free on $E^\flat$.

Conversely, observe that any nonempty fixed point set necessarily looks like

$$F_g = \{\nu\gamma\gamma\gamma\ldots\},$$

by (37.11.iii) (recall that if the hypothesis "$|\mu| < |\nu|$" in (37.11.iii) fails, then it will hold for g^{-1}, while $F_g = F_{g^{-1}}$), where γ is a cycle. If every cycle has an entry, then $\nu\gamma\gamma\gamma\ldots$ is not isolated in $E^\flat$ by (37.13.ii), so the interior of F_g is empty. This proves that τ is topologically free on $E^\flat$. □

We may then use the above result in conjunction with (29.5) to obtain the following result, sometimes referred to as the *uniqueness Theorem for graph C*-algebras*:

Theorem 37.15. *Let E be a graph with no sinks such that every cycle in E has an entry. Also let B be a C^*-algebra which is generated by a set*

$$\{p'_v : v \in E^0\} \cup \{s'_a : a \in E^1\},$$

where the p'_v are mutually orthogonal projections and the s'_a are partial isometries satisfying the relations defining $C^*(E)$, namely (35.1.i)–(35.1.iii). Then B is naturally isomorphic to $C^*(E)$ provided $p'_v \neq 0$, for every v in E^0.

PROOF. Consider the *-homomorphism

$$\varphi : C^*(E) \to B$$

mapping the p_v to p'_v, and the s_a to s'_a, given by the universal property of $C^*(E)$. Since B is supposed to be generated by the p'_v and the s'_a it is clear that φ is surjective. So the proof will be concluded once we prove that φ is injective. In order to do this we argue by contradiction, supposing that

$$\mathrm{Ker}(\varphi) \neq \{0\}.$$

Identifying $C^*(E)$ and $C_0(E^\flat) \rtimes_{\mathrm{red}} \mathbb{F}$, by (37.10.b), we may see $\mathrm{Ker}(\varphi)$ as an ideal of the latter algebra. So we have by (29.5) and (37.14.ii), that

$$K := \mathrm{Ker}(\varphi) \cap C_0(E^\flat) \neq \{0\}.$$

Moreover K is an invariant ideal of $C_0(E^\flat)$ by (23.11). So, writing $K = C_0(U)$, where U is an open subset of $E^\flat$, we have that U is nonempty and invariant.

Using (37.7) we may find a maximal path α in U, and then by (37.6) we see that there exists a finite path μ such that

$$\alpha \in X_\mu \cap E^\flat \subseteq U.$$

Adopting the notation $X^\flat_\mu := X_\mu \cap E^\flat$, we rewrite the above as

$$\alpha \in X^\flat_\mu \subseteq U.$$

The standard form of μ^{-1} is $d(\mu)\mu^{-1}$, so $\tau_{\mu^{-1}}$ maps $X^\flat_\mu$ to $X^\flat_{d(\mu)}$. Therefore the invariance of U implies that

$$X^\flat_{d(\mu)} = \tau_{\mu^{-1}}(X^\flat_\mu) = \tau_{\mu^{-1}}(X^\flat_\mu \cap U) \subseteq U.$$

Consequently the characteristic function of $X^\flat_{d(\mu)}$ lies in $C_0(U)$, and hence also in the null space of φ. Noticing that said function identifies with $p_{d(\mu)}$ under (37.10.b), we then conclude that

$$p'_{d(\mu)} = \varphi(p_{d(\mu)}) = 0,$$

a contradiction. □

Let us now discuss the question of simplicity of $C^*(E)$.

Definition 37.16. We shall say that a graph E is:

(i) *transitive* if, for any two vertices v and w in E^0, there exists a finite path δ such that $d(\delta) = v$, and $r(\delta) = w$.

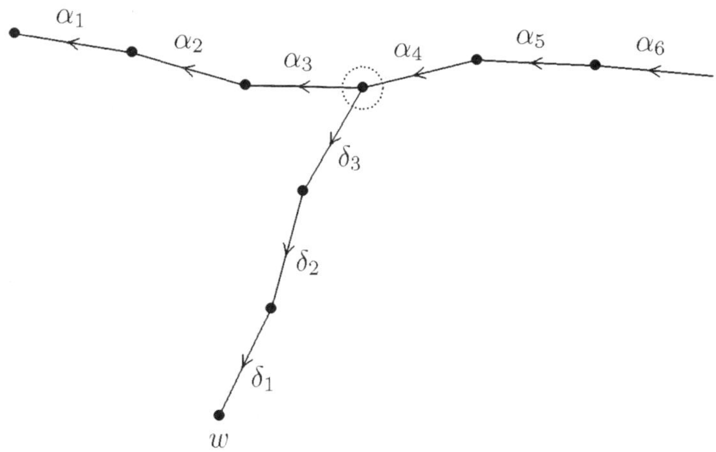

(ii) *weakly transitive* (also known as *co-final*) if for any path α in $E^\flat$ and any vertex w, there exists a finite path δ such that $d(\delta)$ is some vertex along α, and $r(\delta) = w$.

37.17. A pictorial way to understand weak transitivity is to imagine that you are standing at the end (range) of the path α, attempting to reach a given vertex w. Even if there is no path to get you there, in a weakly transitive graph you are able to back up a few edges along α before finding a path δ taking you to w.

Proposition 37.18. *The partial action τ of $\mathbb{F}$ on $E^\flat$ is minimal if and only if E is weakly transitive.*

PROOF. Supposing that E is weakly transitive, we will show that the orbit of every α in $E^\flat$ is dense. In view of (37.7), it is enough to prove that any maximal path lies in the closure of $\text{Orb}(\alpha)$.

So, let U be a neighborhood of a given maximal path β, and we must then prove that U has a nonempty intersection with the orbit of α. By (37.6) there exists a finite path μ such that

$$\beta \in X_\mu \cap E^\flat \subseteq U.$$

En passant we observe that μ is a prefix of β.

We next apply the hypothesis that E is weakly transitive to α and $d(\mu)$, obtaining a path δ whose source is some vertex v along α, and such that $r(\delta) = d(\mu)$.

Splitting α at v, let us write $\alpha = \alpha'\alpha''$, with $d(\alpha') = v = r(\alpha'')$. In particular $\mu\delta\alpha''$ is a well defined path. Assuming we have *backed up along* α as few steps as possible, as described in (37.17), we have that the rightmost edge of α' and the rightmost edge of δ differ, so that $\delta\alpha'^{-1}$ is in standard form. It then follows that $\mu\delta\alpha'^{-1}$ is also in standard form and the conclusion follows from the fact that

$$\tau_{\mu\delta\alpha'^{-1}}(\alpha) = \mu\delta\alpha'' \in X_\mu \cap E^\flat \cap \text{Orb}(\alpha) \subseteq U \cap \text{Orb}(\alpha).$$

Conversely, suppose that E is not weakly transitive. So we may pick a path α and a vertex w violating the condition in (37.16.ii). This means that it is impossible to replace a prefix of α by another, in order to obtain a path with range w. Observe that, by (36.11), the process of replacing prefixes is nothing but applying a partial

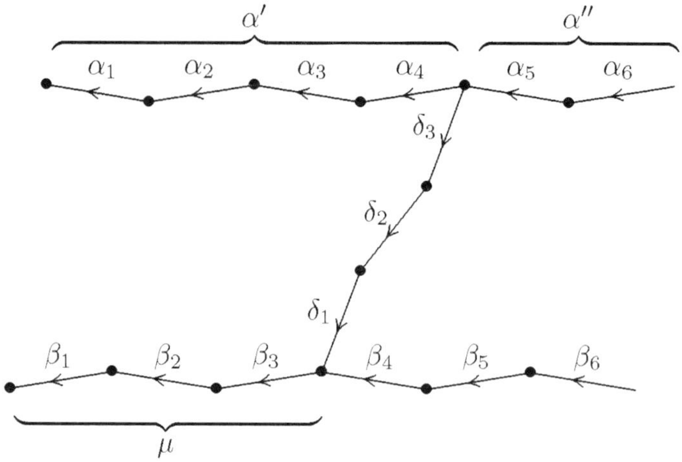

homeomorphism τ_g to a path, so we deduce from the above that

$$\mathrm{Orb}(\alpha) \cap X_w = \emptyset.$$

It follows that $\mathrm{Orb}(\alpha)$ is not a dense set, so its closure is an invariant subset, refuting the minimality of τ. □

As a consequence we obtain:

Theorem 37.19. *Let E be a weakly transitive graph with no sinks such that every cycle in E has an entry. Then $C^*(E)$ is a simple C^*-algebra.*

PROOF. Still referring to the partial action τ of $\mathbb{F}$ on $E^\flat$ giving rise to $C^*(E)$ in (37.8), we have that τ is topologically free by (37.14.ii), and minimal by (37.18). Thus $C_0(E^\flat) \rtimes_{\text{red}} \mathbb{F}$ is simple by (29.8). The conclusion then follows from (37.10.b). □

Our next main result will be an application of (29.9) to graph algebras. In order to succeed in this endeavor, we must therefore find sufficient conditions for τ to be topologically free on every closed invariant subset of $E^\flat$.

With this in mind we will now introduce another important graph theoretical concept.

Definition 37.20. Let γ be a cycle in E. We say that γ is:

(i) *recurrent* if there exists another cycle β, with the same range and source as γ, such

$$\gamma\beta\gamma\gamma\gamma\ldots \neq \gamma\gamma\gamma\ldots.$$

(ii) *transitory* if it is not recurrent.

In a recurrent cycle, the existence of the above path β means that it is possible to exit γ and return to it at a later time. In a transitory cycle this is impossible.

The cycle γ in the diagram above is recurrent because the path

$$\gamma\nu_1\nu_2\nu_3\gamma_9\gamma\gamma\gamma$$

temporarily exits γ before returning to it.

The comparison between infinite paths in (37.20.i) cannot be replaced by the comparison of $\gamma\beta\gamma$ with a finite path of the form $\gamma\gamma\ldots\gamma$. In fact, given a recurrent

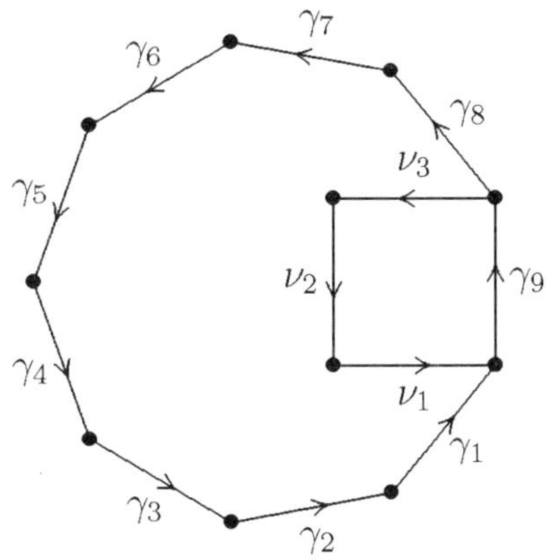

A recurrent cycle γ.

cycle σ, consider the cycle $\gamma = \sigma\sigma$. Then, taking $\beta = \sigma$, we have that the infinite paths $\gamma\beta\gamma\gamma\gamma\ldots$ and $\gamma\gamma\gamma\gamma\ldots$ coincide, whereas $\gamma\beta\gamma$ does not coincide with any path of the form $\gamma\gamma\ldots\gamma$. This is because

$$|\gamma\beta\gamma| = 5|\sigma|, \quad \text{and} \quad |\gamma\gamma\ldots\gamma| = 2n|\sigma|,$$

where n is the number of repetitions. Incidentally, notice that $\sigma\sigma$ is recurrent, provided σ is.

Observe that a recurrent cycle must necessarily have an entry.

Proposition 37.21. *Let E be a graph with no sinks. Then τ is topologically free on every closed τ-invariant subset of $E^\flat$ if and only every cycle is recurrent.*

PROOF. Suppose that every cycle in E is recurrent and let C be a closed invariant subset of $E^\flat$. To show that τ is topologically free on C we need to prove that, for every g in $\mathbb{F} \setminus \{1\}$, the interior of $F_g \cap C$, relative to C, is empty. This is obviously true if $F_g \cap C$ itself is empty, so let us assume otherwise.

Using (37.11) we may assume that g has a standard form $g = \nu\gamma\nu^{-1}$, where γ is a cycle, and that $F_g \cap C$ consists of the single point

$$\alpha = \nu\gamma\gamma\gamma\ldots$$

(for this we might need to replace g by g^{-1}, if necessary, observing that $F_g = F_{g^{-1}}$). In order to show that the interior of $F_g \cap C$ relative to C is empty, it is clearly enough to show that α is not isolated in C.

By hypothesis γ is recurrent, so we may choose a finite path β as in (37.20.i). For each n, consider the path

$$\alpha_n = \tau_{\nu\gamma^n\beta\nu^{-1}}(\alpha) = \nu\gamma^n\beta\gamma\gamma\gamma\ldots = \nu\underbrace{\gamma\gamma\ldots\gamma}_{n}\beta\gamma\gamma\gamma\ldots$$

Since C is invariant, it is clear that α_n lies in C. It is also easy to see that $\alpha_n \neq \alpha$, and

$$\alpha_n \overset{n\to\infty}{\longrightarrow} \alpha,$$

so α is an accumulation point of C, hence not isolated.

Conversely, suppose that τ is topologically free on every closed invariant subset and let γ be a cycle in E. Let

$$\alpha = \gamma\gamma\gamma\ldots \in E^\flat,$$

and let C be the closure of the orbit of α. Then it is evident that α is fixed by γ, so we have by (37.11) that

$$F_\gamma \cap C = \{\alpha\}.$$

Since τ is topologically free on C, by assumption, we have that the interior of $F_\gamma \cap C$ relative to C is empty. In particular $X_\gamma \cap C$, which is an open neighborhood of α, cannot be contained in $F_\gamma \cap C$. Consequently

$$\emptyset \neq (X_\gamma \cap C) \setminus (F_\gamma \cap C) = (X_\gamma \cap C) \setminus \{\alpha\} = \big(X_\gamma \setminus \{\alpha\}\big) \cap C.$$

So, $X_\gamma \setminus \{\alpha\}$ is an open set intersecting C, from where we see that

$$\big(X_\gamma \setminus \{\alpha\}\big) \cap \mathrm{Orb}(\alpha) \neq \emptyset.$$

Picking some $\alpha' \neq \alpha$ in $X_\gamma \cap \mathrm{Orb}(\alpha)$, notice that, since the action of $\mathbb{F}$ consists of prefix replacements (see (36.11)), we have that α', just like α, is an eventually periodic path ending in $\gamma\gamma\gamma\ldots$. Since α' is in X_γ, we see that γ is a prefix of α'. So α' must be a path of the form

$$\alpha' = \gamma\beta\gamma\gamma\gamma\ldots,$$

proving that γ is a recurrent cycle. □

With this we obtain the following important classification of ideals in graph C*-algebras.

Theorem 37.22. *Let E be a graph with no sinks, such that every cycle is recurrent. Then there is a one-to-one correspondence between open τ-invariant subsets of $E^\flat$ and closed two-sided ideals in $C^*(E)$.*

PROOF. Observe that the partial action in point satisfies both (29.9.i) & (29.9.ii), by (37.10.a) and because the free group is known to be exact.

So the statement follows from (37.21) and (29.9), once we realize $C^*(E)$ as $C_0(E^\flat) \rtimes_{\mathrm{red}} \mathbb{F}$, by (37.10). □

There is a lot more that can be said about the ideals of $C^*(E)$ under the above hypothesis. In [**12**] these are characterized in terms of *hereditary directed subsets* of E, plus some extra ingredients. Also, the quotient of $C^*(E)$ by any ideal is shown to be again a graph C*-algebra.

With some extra effort, these results may also be obtained via our techniques. We leave them as exercises for the interested reader.

Notes and remarks. Most of the results in this chapter may be generalized to Exel-Laca algebras, by employing very similar techniques [**57**].

The study of the ideal structure of graph C*-algebras was initiated in [**71**] for finite graphs, followed by [**12**], where the general infinite case is considered.

Graphs in which every cycle has an entry are said to satisfy condition (I). It is a generalization of a similar condition introduced in [**29**]. On the other hand, graphs in which every cycle is recurrent are said to satisfy condition (K) [**75**]. This in turn generalizes condition (II) of [**30**].

Most of the above results describing the various algebras associated to a graph E as partial crossed products, beginning with (36.14), could also be proved directly by first *guessing* the appropriate partial dynamical system and then proving that the corresponding crossed product algebra has the correct universal properties. This strategy would evidently be logically correct and a lot shorter than the one adopted here, but we have opted to follow the above more constructive, and hopefully more pedagogical approach, not least because it may be used in other situations where guesswork might not be an option.

Bibliography

[1] F. Abadie; "Sobre ações Parciais, Fibrados de Fell e Grupóides"; PhD Thesis, University of São Paulo, 1999

[2] F. Abadie, *Enveloping actions and Takai duality for partial actions*, J. Funct. Anal. **197** (2003), no. 1, 14–67, DOI 10.1016/S0022-1236(02)00032-0. MR1957674

[3] ———, *On partial actions and groupoids*, Proc. Amer. Math. Soc. **132** (2004), no. 4, 1037–1047, DOI 10.1090/S0002-9939-03-07300-3. MR2045419

[4] F. Abadie, M. Dokuchaev, R. Exel, and J. J. Simón, *Morita equivalence of partial group actions and globalization*, Trans. Amer. Math. Soc. **368** (2016), no. 7, 4957–4992, DOI 10.1090/tran/6525. MR3456167

[5] F. Abadie and L. Martí Pérez, *On the amenability of partial and enveloping actions*, Proc. Amer. Math. Soc. **137** (2009), no. 11, 3689–3693, DOI 10.1090/S0002-9939-09-09998-5. MR2529875

[6] C. Anantharaman-Delaroche, *Systèmes dynamiques non commutatifs et moyennabilité* (French), Math. Ann. **279** (1987), no. 2, 297–315, DOI 10.1007/BF01461725. MR919508

[7] P. Ara, R. Exel, and T. Katsura, *Dynamical systems of type (m, n) and their C^*-algebras*, Ergodic Theory Dynam. Systems **33** (2013), no. 5, 1291–1325, DOI 10.1017/S0143385712000405. MR3103084

[8] R. J. Archbold and J. S. Spielberg, *Topologically free actions and ideals in discrete C^*-dynamical systems*, Proc. Edinburgh Math. Soc. (2) **37** (1994), no. 1, 119–124, DOI 10.1017/S0013091500018733. MR1258035

[9] W. Arveson, *An invitation to C^*-algebras*, Springer-Verlag, New York-Heidelberg, 1976. Graduate Texts in Mathematics, No. 39. MR0512360

[10] ———, *C^*-algebras associated with sets of semigroups of isometries*, Internat. J. Math. **2** (1991), no. 3, 235–255, DOI 10.1142/S0129167X91000156. MR1104118

[11] D. Bagio, J. Lazzarin, and A. Paques, *Crossed products by twisted partial actions: separability, semisimplicity, and Frobenius properties*, Comm. Algebra **38** (2010), no. 2, 496–508, DOI 10.1080/00927870802614764. MR2598894

[12] T. Bates, J. H. Hong, I. Raeburn, and W. Szymański, *The ideal structure of the C^*-algebras of infinite graphs*, Illinois J. Math. **46** (2002), no. 4, 1159–1176. MR1988256

[13] B. Blackadar, *Shape theory for C^*-algebras*, Math. Scand. **56** (1985), no. 2, 249–275, DOI 10.7146/math.scand.a-12100. MR813640

[14] G. Boava and R. Exel; Partial crossed product description of the C*-algebras associated with integral domains; Proc. Amer. Math. Soc., (2010), arXiv: 1010.0967v2

[15] J.-B. Bost and A. Connes, *Hecke algebras, type III factors and phase transitions with spontaneous symmetry breaking in number theory*, Selecta Math. (N.S.) **1** (1995), no. 3, 411–457, DOI 10.1007/BF01589495. MR1366621

[16] L. G. Brown, *Stable isomorphism of hereditary subalgebras of C^*-algebras*, Pacific J. Math. **71** (1977), no. 2, 335–348. MR0454645

[17] L. G. Brown, P. Green, and M. A. Rieffel, *Stable isomorphism and strong Morita equivalence of C^*-algebras*, Pacific J. Math. **71** (1977), no. 2, 349–363. MR0463928

[18] L. G. Brown, J. A. Mingo, and N.-T. Shen, *Quasi-multipliers and embeddings of Hilbert C^*-bimodules*, Canad. J. Math. **46** (1994), no. 6, 1150–1174, DOI 10.4153/CJM-1994-065-5. MR1304338

[19] N. P. Brown and N. Ozawa, *C^*-algebras and finite-dimensional approximations*, Graduate Studies in Mathematics, vol. 88, American Mathematical Society, Providence, RI, 2008. MR2391387

[20] Z. Burdak, *On decomposition of pairs of commuting isometries*, Ann. Polon. Math. **84** (2004), no. 2, 121–135, DOI 10.4064/ap84-2-3. MR2112253

[21] A. Buss, S. Echterhoff and R. Willett; Exotic Crossed Products; preprint 2015, arXiv: 1010.0967v2

[22] A. Buss and R. Exel, *Twisted actions and regular Fell bundles over inverse semigroups*, Proc. Lond. Math. Soc. (3) **103** (2011), no. 2, 235–270, DOI 10.1112/plms/pdr006. MR2821242

[23] ———, *Fell bundles over inverse semigroups and twisted étale groupoids*, J. Operator Theory **67** (2012), no. 1, 153–205. MR2881538

[24] A. Buss, R. Exel and R. Meyer; Inverse semigroup actions as groupoid actions; Semigroup Forum, 85 (2012), 227-243, arXiv: 1104.0811v2

[25] L. A. Coburn, *The C^*-algebra generated by an isometry*, Bull. Amer. Math. Soc. **73** (1967), 722–726, DOI 10.1090/S0002-9904-1967-11845-7. MR0213906

[26] F. Combes, *Crossed products and Morita equivalence*, Proc. London Math. Soc. (3) **49** (1984), no. 2, 289–306, DOI 10.1112/plms/s3-49.2.289. MR748991

[27] J. Crisp and M. Laca, *On the Toeplitz algebras of right-angled and finite-type Artin groups*, J. Aust. Math. Soc. **72** (2002), no. 2, 223–245, DOI 10.1017/S1446788700003876. MR1887134

[28] ———, *Boundary quotients and ideals of Toeplitz C^*-algebras of Artin groups*, J. Funct. Anal. **242** (2007), no. 1, 127–156, DOI 10.1016/j.jfa.2006.08.001. MR2274018

[29] J. Cuntz and W. Krieger, *A class of C^*-algebras and topological Markov chains*, Invent. Math. **56** (1980), no. 3, 251–268, DOI 10.1007/BF01390048. MR561974

[30] J. Cuntz, *A class of C^*-algebras and topological Markov chains. II. Reducible chains and the Ext-functor for C^*-algebras*, Invent. Math. **63** (1981), no. 1, 25–40, DOI 10.1007/BF01389192. MR608527

[31] R. E. Curto, P. S. Muhly, and D. P. Williams, *Cross products of strongly Morita equivalent C^*-algebras*, Proc. Amer. Math. Soc. **90** (1984), no. 4, 528–530, DOI 10.2307/2045024. MR733400

[32] K. R. Davidson, *C^*-algebras by example*, Fields Institute Monographs, vol. 6, American Mathematical Society, Providence, RI, 1996. MR1402012

[33] M. Dokuchaev and R. Exel, *Associativity of crossed products by partial actions, enveloping actions and partial representations*, Trans. Amer. Math. Soc. **357** (2005), no. 5, 1931–1952, DOI 10.1090/S0002-9947-04-03519-6. MR2115083

[34] M. Dokuchaev, R. Exel, and P. Piccione, *Partial representations and partial group algebras*, J. Algebra **226** (2000), no. 1, 505–532, DOI 10.1006/jabr.1999.8204. MR1749902

[35] M. Dokuchaev, R. Exel, and J. J. Simón, *Crossed products by twisted partial actions and graded algebras*, J. Algebra **320** (2008), no. 8, 3278–3310, DOI 10.1016/j.jalgebra.2008.06.023. MR2450727

[36] ———, *Globalization of twisted partial actions*, Trans. Amer. Math. Soc. **362** (2010), no. 8, 4137–4160, DOI 10.1090/S0002-9947-10-04957-3. MR2608399

[37] M. Dokuchaev and B. Novikov, *Partial projective representations and partial actions*, J. Pure Appl. Algebra **214** (2010), no. 3, 251–268, DOI 10.1016/j.jpaa.2009.05.001. MR2559695

[38] ———, *Partial projective representations and partial actions II*, J. Pure Appl. Algebra **216** (2012), no. 2, 438–455, DOI 10.1016/j.jpaa.2011.07.007. MR2835207

[39] R. G. Douglas, *On extending commutative semigroups of isometries*, Bull. London Math. Soc. **1** (1969), 157–159, DOI 10.1112/blms/1.2.157. MR0246153

[40] ———, *On the C^*-algebra of a one-parameter semigroup of isometries*, Acta Math. **128** (1972), no. 3-4, 143–151, DOI 10.1007/BF02392163. MR0394296

[41] M. Enomoto and Y. Watatani, *A graph theory for C^*-algebras*, Math. Japon. **25** (1980), no. 4, 435–442. MR594544

[42] E. G. Effros and F. Hahn, *Locally compact transformation groups and C^*- algebras*, Memoirs of the American Mathematical Society, No. 75, American Mathematical Society, Providence, R.I., 1967. MR0227310

[43] G. A. Elliott, *Some simple C^*-algebras constructed as crossed products with discrete outer automorphism groups*, Publ. Res. Inst. Math. Sci. **16** (1980), no. 1, 299–311, DOI 10.2977/prims/1195187509. MR574038

[44] I. Erdélyi, *Partial isometries closed under multiplication on Hilbert spaces*, J. Math. Anal. Appl. **22** (1968), 546–551, DOI 10.1016/0022-247X(68)90193-5. MR0228998

[45] R. Exel, *Circle actions on C^*-algebras, partial automorphisms, and a generalized Pimsner-Voiculescu exact sequence*, J. Funct. Anal. **122** (1994), no. 2, 361–401, DOI 10.1006/jfan.1994.1073. MR1276163

[46] ______, *The Bunce-Deddens algebras as crossed products by partial automorphisms*, Bol. Soc. Brasil. Mat. (N.S.) **25** (1994), no. 2, 173–179, DOI 10.1007/BF01321306. MR1306559

[47] ______, *Approximately finite C^*-algebras and partial automorphisms*, Math. Scand. **77** (1995), no. 2, 281–288, DOI 10.7146/math.scand.a-12566. MR1379271

[48] ______, *Twisted partial actions: a classification of regular C^*-algebraic bundles*, Proc. London Math. Soc. (3) **74** (1997), no. 2, 417–443, DOI 10.1112/S0024611597000154. MR1425329

[49] ______, *Amenability for Fell bundles*, J. Reine Angew. Math. **492** (1997), 41–73, DOI 10.1515/crll.1997.492.41. MR1488064

[50] ______, *Partial actions of groups and actions of inverse semigroups*, Proc. Amer. Math. Soc. **126** (1998), no. 12, 3481–3494, DOI 10.1090/S0002-9939-98-04575-4. MR1469405

[51] ______, *Partial representations and amenable Fell bundles over free groups*, Pacific J. Math. **192** (2000), no. 1, 39–63, DOI 10.2140/pjm.2000.192.39. MR1741030

[52] R. Exel, Exact Groups, Induced Ideals, and Fell Bundles; preprint, arXiv: math.OA/0012091.

[53] ______, *Exact groups and Fell bundles*, Math. Ann. **323** (2002), no. 2, 259–266, DOI 10.1007/s002080200295. MR1913042

[54] ______, *Hecke algebras for protonormal subgroups*, J. Algebra **320** (2008), no. 5, 1771–1813, DOI 10.1016/j.jalgebra.2008.05.001. MR2437631

[55] ______, *Inverse semigroups and combinatorial C^*-algebras*, Bull. Braz. Math. Soc. (N.S.) **39** (2008), no. 2, 191–313, DOI 10.1007/s00574-008-0080-7. MR2419901

[56] R. Exel, *Non-Hausdorff étale groupoids*, Proc. Amer. Math. Soc. **139** (2011), no. 3, 897–907, DOI 10.1090/S0002-9939-2010-10477-X. MR2745642

[57] R. Exel and M. Laca, *Cuntz-Krieger algebras for infinite matrices*, J. Reine Angew. Math. **512** (1999), 119–172, DOI 10.1515/crll.1999.051. MR1703078

[58] ______, *Partial dynamical systems and the KMS condition*, Comm. Math. Phys. **232** (2003), no. 2, 223–277, DOI 10.1007/s00220-002-0713-4. MR1953065

[59] R. Exel, M. Laca, and J. Quigg, *Partial dynamical systems and C^*-algebras generated by partial isometries*, J. Operator Theory **47** (2002), no. 1, 169–186. MR1905819

[60] ______, *Partial dynamical systems and C^*-algebras generated by partial isometries*, J. Operator Theory **47** (2002), no. 1, 169–186. MR1905819

[61] R. Exel and F. Vieira, *Actions of inverse semigroups arising from partial actions of groups*, J. Math. Anal. Appl. **363** (2010), no. 1, 86–96, DOI 10.1016/j.jmaa.2009.08.001. MR2559043

[62] J. M. G. Fell, *An extension of Mackey's method to Banach *-algebraic bundles*, Mem. Am. Math. Soc. **90** (1969).

[63] J. M. G. Fell and R. S. Doran, *Representations of *-algebras, locally compact groups, and Banach *-algebraic bundles. Vol. 1*, Pure and Applied Mathematics, vol. 125, Academic Press, Inc., Boston, MA, 1988. Basic representation theory of groups and algebras. MR936628

[64] N. J. Fowler, M. Laca, and I. Raeburn, *The C^*-algebras of infinite graphs*, Proc. Amer. Math. Soc. **128** (2000), no. 8, 2319–2327, DOI 10.1090/S0002-9939-99-05378-2. MR1670363

[65] F. P. Greenleaf, Invariant means on topological groups, Mathematical Studies, vol. 16, van Nostrand-Reinhold, 1969

[66] M. Gromov, *Random walk in random groups*, Geom. Funct. Anal. **13** (2003), no. 1, 73–146, DOI 10.1007/s000390300002. MR1978492

[67] P. R. Halmos and J. E. McLaughlin, *Partial isometries*, Pacific J. Math. **13** (1963), 585–596. MR0157241

[68] E. Hewitt and K. A. Ross, *Abstract harmonic analysis. Vol. II: Structure and analysis for compact groups. Analysis on locally compact Abelian groups*, Die Grundlehren der mathematischen Wissenschaften, Band 152, Springer-Verlag, New York-Berlin, 1970. MR0262773

[69] E. Hopf and N. Wiener, Über eine Klasse singulärer Integralgleichungen; *S.-B. Preuß. Akad. Wiss., Phys.-Math. Kl.* **30/32** (1931), 696–706

[70] K. Horák and V. Müller, *Functional model for commuting isometries*, Czechoslovak Math. J. **39(114)** (1989), no. 2, 370–379. MR992140

[71] A. an Huef and I. Raeburn, *The ideal structure of Cuntz-Krieger algebras*, Ergodic Theory Dynam. Systems **17** (1997), no. 3, 611–624, DOI 10.1017/S0143385797079200. MR1452183

[72] K. K. Jensen and K. Thomsen, *Elements of KK-theory*, Mathematics: Theory & Applications, Birkhäuser Boston, Inc., Boston, MA, 1991. MR1124848

[73] T. Katsura, P. S. Muhly, A. Sims, and M. Tomforde, *Graph algebras, Exel-Laca algebras, and ultragraph algebras coincide up to Morita equivalence*, J. Reine Angew. Math. **640** (2010), 135–165, DOI 10.1515/CRELLE.2010.023. MR2629692

[74] S. o. Kawamura and J. Tomiyama, *Properties of topological dynamical systems and corresponding C^*-algebras*, Tokyo J. Math. **13** (1990), no. 2, 251–257, DOI 10.3836/tjm/1270132260. MR1088230

[75] A. Kumjian, D. Pask, I. Raeburn, and J. Renault, *Graphs, groupoids, and Cuntz-Krieger algebras*, J. Funct. Anal. **144** (1997), no. 2, 505–541, DOI 10.1006/jfan.1996.3001. MR1432596

[76] M. Laca, *From endomorphisms to automorphisms and back: dilations and full corners*, J. London Math. Soc. (2) **61** (2000), no. 3, 893–904, DOI 10.1112/S0024610799008492. MR1766113

[77] M. Laca and I. Raeburn, *Semigroup crossed products and the Toeplitz algebras of nonabelian groups*, J. Funct. Anal. **139** (1996), no. 2, 415–440, DOI 10.1006/jfan.1996.0091. MR1402771

[78] E. C. Lance, *Hilbert C^*-modules*, London Mathematical Society Lecture Note Series, vol. 210, Cambridge University Press, Cambridge, 1995. A toolkit for operator algebraists. MR1325694

[79] M. V. Lawson, *Inverse semigroups*, World Scientific Publishing Co., Inc., River Edge, NJ, 1998. The theory of partial symmetries. MR1694900

[80] M. Matsumura, *A characterization of amenability of group actions on C^*-algebras*, J. Operator Theory **72** (2014), no. 1, 41–47, DOI 10.7900/jot.2012sep07.1958. MR3246980

[81] K. McClanahan, *K-theory for partial crossed products by discrete groups*, J. Funct. Anal. **130** (1995), no. 1, 77–117, DOI 10.1006/jfan.1995.1064. MR1331978

[82] S. Montgomery, *Hopf algebras and their actions on rings*, CBMS Regional Conference Series in Mathematics, vol. 82, Published for the Conference Board of the Mathematical Sciences, Washington, DC; by the American Mathematical Society, Providence, RI, 1993. MR1243637

[83] K. Morita, *Duality for modules and its applications to the theory of rings with minimum condition*, Sci. Rep. Tokyo Kyoiku Daigaku Sect. A **6** (1958), 83–142. MR0096700

[84] G. J. Murphy, *C^*-algebras and operator theory*, Academic Press, Inc., Boston, MA, 1990. MR1074574

[85] A. Nica, C*-algebras generated by isometries and Wiener-Hopf operators; J. Operator Theory, 27 (1992), 17-52

[86] J. A. Packer and I. Raeburn, *Twisted crossed products of C^*-algebras*, Math. Proc. Cambridge Philos. Soc. **106** (1989), no. 2, 293–311, DOI 10.1017/S0305004100078129. MR1002543

[87] A. Paques and A. Sant'Ana, *When is a crossed product by a twisted partial action Azumaya?*, Comm. Algebra **38** (2010), no. 3, 1093–1103, DOI 10.1080/00927870902888250. MR2650393

[88] W. L. Paschke, *Inner product modules over B^*-algebras*, Trans. Amer. Math. Soc. **182** (1973), 443–468, DOI 10.2307/1996542. MR0355613

[89] G. K. Pedersen, *C^*-algebras and their automorphism groups*, London Mathematical Society Monographs, vol. 14, Academic Press, Inc. [Harcourt Brace Jovanovich, Publishers], London-New York, 1979. MR548006

[90] J. Phillips and I. Raeburn, *Semigroups of isometries, Toeplitz algebras and twisted crossed products*, Integral Equations Operator Theory **17** (1993), no. 4, 579–602, DOI 10.1007/BF01200396. MR1243998

[91] D. Popovici, *A Wold-type decomposition for commuting isometric pairs*, Proc. Amer. Math. Soc. **132** (2004), no. 8, 2303–2314, DOI 10.1090/S0002-9939-04-07331-9. MR2052406

[92] G. B. Preston, *Inverse semi-groups*, J. London Math. Soc. **29** (1954), 396–403, DOI 10.1112/jlms/s1-29.4.396. MR0064036

[93] J. C. Quigg, *Discrete C^*-coactions and C^*-algebraic bundles*, J. Austral. Math. Soc. Ser. A **60** (1996), no. 2, 204–221. MR1375586

[94] J. Quigg and I. Raeburn, *Characterisations of crossed products by partial actions*, J. Operator Theory **37** (1997), no. 2, 311–340. MR1452280

[95] I. Raeburn, *Graph algebras*, CBMS Regional Conference Series in Mathematics, vol. 103, Published for the Conference Board of the Mathematical Sciences, Washington, DC; by the American Mathematical Society, Providence, RI, 2005. MR2135030

[96] J. Renault, *A groupoid approach to C^*-algebras*, Lecture Notes in Mathematics, vol. 793, Springer, Berlin, 1980. MR584266

[97] ———, *The ideal structure of groupoid crossed product C^*-algebras*, J. Operator Theory **25** (1991), no. 1, 3–36. With an appendix by Georges Skandalis. MR1191252

[98] M. A. Rieffel, *Morita equivalence for C^*-algebras and W^*-algebras*, J. Pure Appl. Algebra **5** (1974), 51–96, DOI 10.1016/0022-4049(74)90003-6. MR0367670

[99] E. P. Scarparo; Álgebras de Toeplitz generalizadas, master's thesis, Universidade Federal de Santa Catarina, 2014

[100] C. F. Sehnem; Uma classificação de fibrados de Fell estáveis, master's thesis, Universidade Federal de Santa Catarina, 2014

[101] N. Sieben, *C^*-crossed products by partial actions and actions of inverse semigroups*, J. Austral. Math. Soc. Ser. A **63** (1997), no. 1, 32–46. MR1456588

[102] ———, *C^*-crossed products by twisted inverse semigroup actions*, J. Operator Theory **39** (1998), no. 2, 361–393. MR1620499

[103] M. Słociński, *On the Wold-type decomposition of a pair of commuting isometries*, Ann. Polon. Math. **37** (1980), no. 3, 255–262. MR587496

[104] Ş. Strătilă and D. Voiculescu, *Representations of AF-algebras and of the group $U(\infty)$*, Lecture Notes in Mathematics, Vol. 486, Springer-Verlag, Berlin-New York, 1975. MR0458188

[105] V. V. Wagner, *Generalized groups* (Russian), Doklady Akad. Nauk SSSR (N.S.) **84** (1952), 1119–1122. MR0048425

[106] Y. Watatani, *Index for C^*-subalgebras*, Mem. Amer. Math. Soc. **83** (1990), no. 424, vi+117, DOI 10.1090/memo/0424. MR996807

[107] H. Zettl, *A characterization of ternary rings of operators*, Adv. in Math. **48** (1983), no. 2, 117–143, DOI 10.1016/0001-8708(83)90083-X. MR700979

[108] R. J. Zimmer, *Hyperfinite factors and amenable ergodic actions*, Invent. Math. **41** (1977), no. 1, 23–31, DOI 10.1007/BF01390162. MR0470692

[109] ———, *Amenable ergodic group actions and an application to Poisson boundaries of random walks*, J. Functional Analysis **27** (1978), no. 3, 350–372. MR0473096

Index

SELECTED PUBLISHED TITLES IN THIS SERIES

224 **Ruy Exel,** Partial Dynamical Systems, Fell Bundles and Applications, 2017

222 **Alexandru Buium,** Foundations of Arithmetic Differential Geometry, 2017

221 **Dennis Gaitsgory and Nick Rozenblyum,** A Study in Derived Algebraic Geometry, 2017

219 **Richard Evan Schwartz,** The Projective Heat Map, 2017

218 **Tushar Das, David Simmons, and Mariusz Urbański,** Geometry and Dynamics in Gromov Hyperbolic Metric Spaces, 2017

217 **Benoit Fresse,** Homotopy of Operads and Grothendieck–Teichmüller Groups, 2017

216 **Frederick W. Gehring, Gaven J. Martin, and Bruce P. Palka,** An Introduction to the Theory of Higher-Dimensional Quasiconformal Mappings, 2017

215 **Robert Bieri and Ralph Strebel,** On Groups of PL-homeomorphisms of the Real Line, 2016

214 **Jared Speck,** Shock Formation in Small-Data Solutions to 3D Quasilinear Wave Equations, 2016

213 **Harold G. Diamond and Wen-Bin Zhang (Cheung Man Ping),** Beurling Generalized Numbers, 2016

212 **Pandelis Dodos and Vassilis Kanellopoulos,** Ramsey Theory for Product Spaces, 2016

211 **Charlotte Hardouin, Jacques Sauloy, and Michael F. Singer,** Galois Theories of Linear Difference Equations: An Introduction, 2016

210 **Jason P. Bell, Dragos Ghioca, and Thomas J. Tucker,** The Dynamical Mordell–Lang Conjecture, 2016

209 **Steve Y. Oudot,** Persistence Theory: From Quiver Representations to Data Analysis, 2015

208 **Peter S. Ozsváth, András I. Stipsicz, and Zoltán Szabó,** Grid Homology for Knots and Links, 2015

207 **Vladimir I. Bogachev, Nicolai V. Krylov, Michael Röckner, and Stanislav V. Shaposhnikov,** Fokker–Planck–Kolmogorov Equations, 2015

206 **Bennett Chow, Sun-Chin Chu, David Glickenstein, Christine Guenther, James Isenberg, Tom Ivey, Dan Knopf, Peng Lu, Feng Luo, and Lei Ni,** The Ricci Flow: Techniques and Applications: Part IV: Long-Time Solutions and Related Topics, 2015

205 **Pavel Etingof, Shlomo Gelaki, Dmitri Nikshych, and Victor Ostrik,** Tensor Categories, 2015

204 **Victor M. Buchstaber and Taras E. Panov,** Toric Topology, 2015

203 **Donald Yau and Mark W. Johnson,** A Foundation for PROPs, Algebras, and Modules, 2015

202 **Shiri Artstein-Avidan, Apostolos Giannopoulos, and Vitali D. Milman,** Asymptotic Geometric Analysis, Part I, 2015

201 **Christopher L. Douglas, John Francis, André G. Henriques, and Michael A. Hill, Editors,** Topological Modular Forms, 2014

200 **Nikolai Nadirashvili, Vladimir Tkachev, and Serge Vlăduţ,** Nonlinear Elliptic Equations and Nonassociative Algebras, 2014

199 **Dmitry S. Kaliuzhnyi-Verbovetskyi and Victor Vinnikov,** Foundations of Free Noncommutative Function Theory, 2014

198 **Jörg Jahnel,** Brauer Groups, Tamagawa Measures, and Rational Points on Algebraic Varieties, 2014

197 **Richard Evan Schwartz,** The Octagonal PETs, 2014

For a complete list of titles in this series, visit the AMS Bookstore at **www.ams.org/bookstore/survseries/**.